W9-CSB-814

APPROACHES
TO SUSTAINABLE
DEVELOPMENT

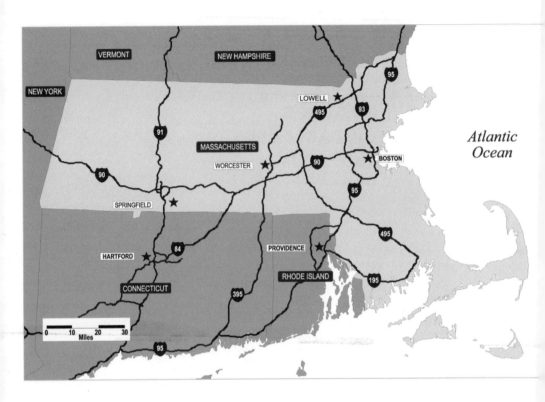

APPROACHES TO SUSTAINABLE DEVELOPMENT

The Public University
in the Regional Economy

Edited by Robert Forrant,
Jean L. Pyle,
William Lazonick, and
Charles Levenstein

University of Massachusetts Press   Amherst

Copyright © 2001 by
University of Massachusetts Press
All rights reserved

Printed in the United States of America
LC 2001027694
ISBN 1-55849-305-0 (cloth)
ISBN 1-55849-311-5 (paper)

Set in Sabon by Graphic Composition, Inc.
Designed by Dennis Anderson
Printed and bound by Thomson-Shore, Inc.

Library of Congress Cataloging-in-Publication Data

Approaches to sustainable development : the public university in the regional economy /
edited by Robert Forrant . . . [et al.].
    p. cm.
  Includes bibliographical references and index.
    ISBN 1-55849-305-0 (cloth : alk. paper) — ISBN 1-55849-311-5 (pbk. : alk. paper)
    1. Community and college—Massachusetts—Lowell—Case studies—Congresses. 2.
Community development—Massachusetts—Lowell—Case studies—Congresses. 3. Universities and colleges—Massachusetts—Lowell—Public services—Case studies—Congresses. I. Forrant, Robert, 1947–

LC238.3.L69 A66 2001
378.1′03—dc21
                                                                        2001027694

British Library Cataloguing in Publication data are available.

This book is published with the support and cooperation of
the University of Massachusetts Lowell.

Contents

Preface

FOR more than one hundred years, the University of Massachusetts Lowell has been providing for the common good in its neighboring communities and across the state. Through teaching, research, and service, UMass Lowell has contributed mightily to the development of its region. Today, we are building on this legacy by directing significant resources and our energy toward understanding how social and economic development works in defined geographic regions.

Some of our recent work concerning regional social and economic development is represented in this volume. The essays were presented as papers at two conferences organized by the university's Committee on Industrial Theory and Assessment (CITA). They provide a broad picture of the types and range of issues with which we at UMass Lowell are concerned. The volume is a direct result of a restructuring process we began at the university in the late 1980s. Our region's minicomputer industry was in precipitous decline, and a reduction in federal defense spending, on which the region had heavily relied, contributed to a bleak employment outlook. Many of us recognized that it was time to transform our activities so that UMass Lowell would remain a force in support of the region's development.

Herein we have expressed a vision of what constitutes a vital region: a defined geographic area with a sustainable economy, a place in which people can live and work in a healthy, productive, equitable, innovative society with a high quality of life that can be maintained in the long run. And we have given a name to our ongoing effort to combine theory and practice as a means of realizing the vision—The Lowell Model. Our ambition is to be a leading model for a technology-oriented public university in the land grant tradition.

Historically, certain regions have fared better than others have, and our aim is to identify the factors that lead to sustainable development within regions. Our own region in northeastern Massachusetts offers a ready-made laboratory for such study, given the historic role played by the Merrimack Valley in the nation's industrial development and the subsequent economic and social challenges the region has faced in the past 175 years.

UMass Lowell has oriented its work toward four factors that we believe are crucial to regional vitality: (1) the development of intellectual capital through wide-ranging support of K–12 education, community and corporate education, and lifelong learning; (2) the improvement of personal health and health care delivery among residents of the region; (3) the enhancement and protection of the physical environment; and (4) the strengthening of the region's social fabric and enrichment of people's daily lives. From this river valley, our work moves outward. We expect our action-oriented research will have applications nationally and globally. In turn, UMass Lowell researchers are drawing lessons from strategies in play in vibrant regions elsewhere.

Any university seeking to make a positive impact on the economy and society of its region must be guided by a reflective and dynamic region-wide discourse. Through our teaching, we strive to provide students in all our colleges (arts and sciences, education, engineering, health professions, and management) at the undergraduate and graduate levels with a fundamental knowledge of the changes in the economy, workplace organization and diversity, neighborhoods, political structure, and the environment that are transforming many other regions across the country and around the world. We encourage outreach activities through our research centers and institutes that have led to partnerships with local communities, neighborhood organizations, business enterprises and groups, and government agencies. These relationships enable us both to learn and to advise, and to improve our research and teaching.

To carry out our work, we sought to integrate research, teaching, and outreach across colleges, departments, institutes, and centers. In 1993, we created a body now called the Council for Regional Development (CRD) as a means of bringing together faculty and administrators who are deeply involved in development matters. Two faculty-run committees, the Committee for Federated Centers and Institutes (CFCI) and CITA have helped transform the ways in which UMass Lowell serves its various constituents.

CITA promotes new teaching, research, and outreach activities that engage faculty members in a multidisciplinary approach to sustainable regional development. The committee funds cross-disciplinary research, presents campus seminars, and sponsors annual conferences to gain feedback and develop wider collaborations. One major accomplishment of CITA has been the formation of a new graduate department of Regional Economic and Social Development (RESD) at UMass Lowell. This department works closely with the successful Department of Work Environment in the College of Engineering. In October 2000, CITA hosted an international conference on approaches to sustainable development and the role of the university in the globalizing economy which attracted scholars and development practitioners from Cambodia, India, Malaysia, Mexico, Norway, Slovenia, South Africa, Sweden, and the United States.

UMass Lowell's primary role is the advancement of learning across all sectors of society. We believe that the confluence of approaches and perspectives that our process generates can be a catalyst for sustainable development. We seek to go beyond a random approach to development, and also to demonstrate how public universities can "roll up their sleeves" to grapple with problems affecting individuals, families, and communities. We believe that there is a singular role to be played by the public university in the cultivation of a healthy, equitable, productive, and thus sustainable economy.

I hope that you will find this volume to be of value in your own work.

William T. Hogan
Chancellor

Acknowledgments

THE publication of this unique volume is due to the support, input and suggestions, and efforts of many people. First, a volume such as this would not have come into existence without the backing of Chancellor William T. Hogan of University of Massachusetts Lowell. He had the vision to set sustainable regional development as a major component of the mission of UMass Lowell. It was because of his support for an innovative multidisciplinary approach to sustainable regional development—both its theory and its applications—that we were afforded the time and space to develop interdisciplinary connections, research and write papers, develop linkages with the community, and hold workshops and conferences.

The papers in this volume have been widely read and commented on by others—and to all these people we owe thanks. First, we want to acknowledge the contributions of the many people on the Committee for Industrial Theory and Assessment (CITA) who read and commented on papers developed for each of the annual workshops. Having an internal review before the workshop presentations provided authors with additional feedback to further hone papers before presentation.

In addition, we invited prominent individuals from outside UMass Lowell—academics, community organizers, practitioners, activists, and those in government—to comment on papers presented at each of the previous conferences. Their careful and insightful comments were of great help in sharpening the linkages and strengthening the analyses. In this vein, we owe thanks to David Barkin, Sue Beaton, Frank Carvalho, Pierre Clavel, Barry Commoner, Thomas Hubbard, Marie Kennedy, Mel King, David Magnani, Mary O'Sullivan, and Charles Weiner.

We are indebted to Mary Lee Dunn of UMass Lowell for organizational and editorial support. We are especially grateful to Mary Capouya for her detailed and careful copyediting, to Carol Betsch, managing editor of the University of Massachusetts Press, for her patience and care in shaping this volume into its final polished form, and to Bruce Wilcox, director of the press, for his enthusiastic support of the volume.

The Editors: Robert Forrant, Jean L. Pyle,
William Lazonick, and Charles Levenstein

Acronyms

ADA   Americans with Disabilities Act
CERES   Coalition for Environmentally Responsible Economies
CRA   Community Reinvestment Act
DARPA   Defense Advanced Research Projects Agency
EEOC   Equal Employment Opportunity Commission
EPA   Environmental Protection Agency
FMS   Flexible Manufacturing Systems
JUSE   Japanese Union of Scientists and Engineers
LEJC   Lawrence Environmental Justice Council
MSDs   Musculoskeletal Disorders
MVEC   Merrimack Valley Environmental Coalition
MVC   Merrimack Valley Commission
NESWC   Northeast Solid Waste Committee
NIST   National Institute of Standards and Technology
NMCG   Northern Middlesex Council of Governments
NSF   National Science Foundation
OECD   Organization for Economic Cooperation and Development
OFCCP   Office of Federal Contract Compliance Programs
OIDA   Optoelectronics Industry Development Association
OSHA   Occupational Safety and Health Administration
QCRG   Quality Control Research Group
UNCSD   United Nations Commission on Sustainable Development
WCED   World Commission on Environment and Development

University of Massachusetts Lowell

BPRC   Biodegradable Polymers Research Center
CEMOS   Center for Electromagnetic Materials and Optical Systems
CFWC   Center for Family, Work and Community
CIC   Center for Industrial Competitiveness
CITA   Committee on Industrial Theory and Assessment
COPC   Community Outreach Partnership Center
IMPACT   Institute for Massachusetts Partnering and Commercialization
           of Technology
LCSP   Lowell Center for Sustainable Production
RESD   Department of Regional Economic and Social Development
TURI   Toxics Use Reduction Institute

APPROACHES
TO SUSTAINABLE
DEVELOPMENT

General Introduction

Sustainable Development for a Regional Economy

The Editors

THIS volume is the result of six years of research and discussion by scholars at the University of Massachusetts Lowell on the role of a public university as an active participant in the process of regional economic and social development. This project was carried out under the auspices of the Committee on Industrial Theory and Assessment (CITA), the cross-campus body that was created in 1993 to find new ways for UMass Lowell to engage in and foster sustainable regional development. Following four years of work devoted to developing cross-disciplinary discussions and encouraging collaborative research projects at UMass Lowell, CITA began to hold annual conferences in which faculty members presented their research. The essays included in this volume were crafted for CITA annual conferences held in the late 1990s that focused on "Approaches to Sustainable Development"—including the critical elements of sustainability, the role of the university in sustainable development, and the importance of innovation.

The chapters mark the convergence of several hitherto distinct fields, including planning and regional development, economic geography, ergonomics and work design, research on the innovation process in firms, manufacturing engineering, community, and neighborhood studies, gender and race/ethnic studies, and health-related and environmental studies. Given the industrial and immigrant history of Lowell, the background and expertise of UMass Lowell as a science and engineering institution, and the work being done to forge linkages across the sciences, engineering, and the social sciences, Lowell is indeed an appropriate place in which to ask several difficult questions about how academic institutions can engage in the regional development process. There is no consensus on the concept of development. Does it imply an expanded economic pie—for instance, a larger tax base, more jobs, new firms—without regard for how the additional slices are distributed? Does it foster increased stability in the economy, an end to the boom-and-bust cycles that communities such as Lowell have witnessed for much of the twentieth century? What of worker health and safety, energy efficiency, and efforts to eliminate toxics usage in production? Is a public role in the nurturing of start-up enterprises consistent with sustainable development? How can a more encompassing definition of development, one that moves beyond the focus on growth of gross domestic product (GDP), take shape? What roles ought a public university play in the regional development process; asked another way, what is the value of higher education to regional development (Thanki 1999)?

The University of Massachusetts Lowell

The university is a product of the region, its people and its political economy, although it has also helped shape the kind of development that has occurred. Its predecessor institutions, the Massachusetts State Normal School at Lowell and the Lowell Textile School, were created in 1894 and 1895, respectively, to serve the economic and social needs of the region at the time. The two went through several reorganizations in the twentieth century before merging in 1975 to become the University of Lowell. The creation of this new institution was thus concurrent with the resurgence of the Lowell economy of the 1970s and 1980s, primed by the rapid growth of the computer industry and defense spending. But this new prosperity, so reliant on one industry (as it had been in the nineteenth century with textiles), was not sustainable. In 1991, in the midst of a deep downturn, the university was once again reorganized and became the University of Massachusetts Lowell, part of a five-campus statewide system (for a more detailed history, see Blewett 1995). The crash of the minicomputer industry, sharp cuts in Pentagon spending, rising unemployment, and a series of devastating plant shutdowns compelled UMass Lowell to reconsider its role in the region.

During this last period of restructuring, UMass Lowell articulated a focus on sustainable regional development as one of its central priorities. In the downturn of the late 1980s and early 1990s, the university's leaders recognized that the socioeconomic problems of old industrial regions are complex and require creative, multidisciplinary solution strategies if they are to be resolved. Seeing that there was no panacea for the boom-and-bust cycles that shaped Lowell and the region's history, these leaders concluded that it was imperative to foster the purposeful interaction of social scientists and historians, engineers and scientists, and colleagues in the education and health professions, to approach creatively the difficult economic and social challenges confronting the region. Not satisfied to simply educate and purvey high-tech labor to the region's businesses, faculty and administrators became engaged in a lively discussion of alternative strategies for attaining sustainable development. Issues of workplace health, environmental sustainability, incorporation of diverse populations into the workforce, and community empowerment were being linked to the formulation of innovative social and economic policies.

To proceed conceptually and practically—for thinking *and* doing are essential—the administration formed and provided resources for two committees working closely together under an umbrella council and supported the creation of a new interdisciplinary academic department in the College of Arts and Sciences. The Committee of Federated Centers and Institutes (CFCI) was formed to facilitate interdisciplinary activities among faculty-led centers and foster their relations with industry and community-based organizations. CITA was established to develop new theoretical and pragmatic ways of thinking about sustainable regional development and to assess the success of these initiatives. Designed to explore the issues of sustainable regional development from interdisciplinary perspectives, CITA has drawn its membership from the Colleges of Engineering, Arts and Sciences, Health Professions, and Management. The new graduate-level department, the Department of Regional Economic and Social Development (RESD), is composed of faculty with academic backgrounds and careers in disciplines such as economics, history, political science, psychology, sociology, and urban planning. RESD faculty have a collective history of well over a hundred years of working with the community and industry.

This process results in a "virtuous circle" of improvement. Practically speaking, a healthy economy redounds to the benefit of the university in terms of an increased budget, greater enrollments, and larger alumni donations. By extension, a thriving university can increase its activities and thus play a more dynamic role in the wider world it inhabits. Although the chapters in this volume are a product of their environment, we firmly believe that the discussions engendered herein are relevant far beyond UMass Lowell and the Merrimack River Valley. There are myriad programs at colleges and universities throughout the world confronting the same questions that motivate our work. This volume provides a perspective on how to integrate the disparate approaches of academics to begin to trace the circumference of that "virtuous circle."

Lowell and the Regional Economy

Lowell is an old industrial city thirty miles northwest of Boston, poised on the edge of the Merrimack River. The river, with its powerful Pawtucket Falls, inspired merchant capitalists of Boston to establish a city with the first textile mills in the United States that integrated all aspects of production, launching the American industrial revolution in the 1820s and 1830s. Lowell was constructed with utopian fervor; the city was to be a well-planned, well-supervised, Puritan American alternative to the Dickensian "dark Satanic mills" in England. It was not possible, however, to sustain prosperity in a one-industry town, particularly when it was based on an industry looking for new markets, cheap labor, and tax breaks. By the 1920s, much of the Lowell and New England textile industry had headed south (for a more detailed history, see Gross 1993). Lowell, a place that had offered economic hope to waves of new immigrants, became a decaying city with high unemployment and rising poverty, a scene all too familiar in older industrial cities throughout the Northeast United States. Renascence occurred for a time in the 1970s and 1980s, as the computer giant Wang established its global headquarters in Lowell and the federal government turned some of the old mills into our first urban National Historical Park (for a discussion of boom-and-bust cycles, see Best and Forrant in part III herein).

Today Lowell is the fourth-largest city in Massachusetts, with a population of approximately 105,000. The daily and long-term problems that the city faces are far too common in older industrial cities and thus make the activities of UMass Lowell relevant to other urban university campuses. Lowell remains an immigrant city; although its total population has increased approximately 10 percent since the early 1980s, the immigrant and minority share of the population has risen from 5 percent to 31 percent. The city has sizable Laotian, Vietnamese, Dominican, and Puerto Rican populations, and its Cambodian population is the second-highest concentration of Cambodian immigrants in the United States. Indeed several neighborhood census tracts contiguous to the university have immigrant and minority concentrations greater than 50 percent. Economic prosperity in the early 1980s, especially in computer manufacturing, drew many of these newcomers to the city, but the sharp downturn in the late 1980s and early 1990s resulted in high unemployment and extreme economic hardship. And in spite of the city's overall unemployment rate of approximately 4 percent in recent years, joblessness in neighborhoods with minority and immigrant concentrations is at least three times higher than the city average. Numerous abandoned buildings and vacant lots are hazardous waste sites; thus, land available for economic development or housing construction is often contaminated. Jobs were available again in the latter 1990s, but largely in the surrounding suburban corridors not easily reached by Lowell residents without private transportation. Many in Lowell have been left behind in the so-called new economy.

What is the larger geographic area that we are working in? Route 128 and Route 495, the circumferential highways around Boston mark out the region (for a discussion of notions of region, see Gerson and Wooding, in part I herein). The Northeast region of Massachusetts is made up of older mill cities, such as Lowell, Lawrence, and Haverhill, and cities and towns that have undergone explosive growth fueled by the recent high-tech boom.

Since the end of the Second World War, three broad trends have influenced wild swings in the regional economy: the sharp decline of the textile industry; the meteoric emergence and precipitous fall of the minicomputer industries and the recent high-tech rebirth along Routes 128, 93, and 495, north and west of Boston; and the buildup of and drastic cut in defense spending. Between 1980 and 1994, the defense stimulus to companies, colleges, universities, and federal and private research laboratories in Massachusetts totaled approximately $90 billion, with much of this coming into its Northeast region. But the reduced defense budget—in terms of actual dollars, Massachusetts received $8.7 billion in 1989, $5.1 billion in 1994, and about $4 billion in 1998—eroded once-dependable markets, and the failure of Wang, the problems of Digital Equipment, and the demise of other computer firms had hurt the region in the late 1980s and early 1990s. The economic health of the region improved during the national economic expansion of the mid- to late 1990s, with high concentrations of employment in computer and communications equipment, software, electronic and electrical components, and aircraft production. Yet even before the downturn in the high-tech sector in 2000, Lowell and other older industrial communities grappled with jobless rates higher than the state average. The state's overall poverty rate in 1999 was higher than at any time since 1980, the year the Census Bureau began tracking this figure.

Defining Sustainability

Sustainability is not an abstract issue for Lowell. Twice in the twentieth century Lowell has been ravaged when industries that had dominated the region migrated or went into decline. The grinding poverty of regional recession is familiar to Lowell, Lawrence, and the Merrimack Valley. The university's origins are in the old textile industry. And in the heyday of the computer industry, the university was the technological handmaiden of this new world. Our history is one of technical and cultural servicing of industry: We have provided technical ideas, the engineers to implement them, and we have provided the cultural–community support by educating everyone from the music teachers in the high schools of the Merrimack Valley to the community organizers in social agencies in Lowell. These experiences are at the root of our discussions of the role of the university in *sustainable* regional development. There is no quarrel over the importance of the university's engagement in and service to the region: We have adopted the public university land-grant model, but within our ranks there are very different conceptions of sustainability.

Sustainability is not a vacuous concept: It developed as a critique of the "tunnel division" of economists and advocates of economic growth. "The momentum for sustainable development grew from the work of the World

Commission on Environment and Development that was headed by Gro Brundtland, who later became prime minister of Norway. From 1983 to 1987, the World Commission conducted public hearings throughout the world, from Brazil to Mexico, Zimbabwe to China, to review the concept of sustainability. The commissioners concluded in their unanimous report that our common future depends on sustainable development" (Lowrie, 1996, as quoted in IISD).

The World Commission defined sustainable development as meeting present needs without compromising the capability of future generations to meet their requirements. The United Nations Conference on Environment and Development in 1992, the "Earth Summit," arose out of efforts to popularize and build support throughout the world for this conception. "The assembled leaders (from 100 countries) signed the Framework Convention on Climate Change and the Convention on Biological Diversity; endorsed the Rio Declaration and the Forest Principles; and adopted Agenda 21, a 300-page plan for achieving sustainable development in the 21st century" (United Nations Commission on Sustainable Development 1999). There are numerous research efforts taking place around the world to develop indicators of sustainable production. Government programs, including the Environmental Protection Agency's Industry Partnership Program encourages companies to develop environmentally sound strategies.

Although the impetus for the Earth Summit was rooted in the environmental critique of Western economic development theories and practices, "sustainability" began to take on a broader meaning. According to Noel J. Brown, of the United Nations Environment Program, "Agenda 21 is both a foundation for global governments, a framework for global cooperation, as well as a catalyst for citizen action. It comprises 115 concrete program proposals that would promote improved air quality, protect the quality of the environment and land-based resources, and address the problems of waste, poverty, and lifestyles and disseminate environmentally sound technology" (Brown 1996, as cited in IISD). One organization for corporate responsibility has adopted the following dimensions of sustainability: protection of the biosphere; sustainable use of natural resources; reduction and disposal of waste; wise use of energy; risk reduction; marketing of safe products and services; disclosure of and compensation for damage; and use of environmental directors and managers.

The chapters in this volume discuss many of these dimensions, augment them, and provide some indication of their relevance and applicability for regional development. For example, the concept of sustainability described by Veleva and Crumbley in part II integrates the economic, social, and environmental realms of human activities. Social justice and popular participation, and the practical application of the principles of cleaner production are of particular concern to Lowell researchers, as is the process of innovation.

Innovation, Economic Development, and Sustainability

As used by economists, "innovation" is a process that generates *higher quality, lower cost* products than had previously been available, whereas the term "economic development" is a process that allocates resources to the production of goods and services that augment the standards of living of the population over time. Thus defined, innovation is a necessary, but not sufficient, condition for economic development to occur. It is a necessary condition because economic development requires the production of higher quality, lower cost goods and services. It is not a sufficient condition because the distribution of the costs and benefits of the innovation process may actually lower the standards of living of some members of the population. Innovation may render existing skills obsolete. It may be accomplished by demanding more onerous work or foisting more hazardous work conditions on certain participants in the economy. Innovation may also entail destruction of the physical environment. Future generations may end up paying the costs of innovation without reaping even a small share of the benefits relative to current generations. By the ways it transforms existing human and physical resources, and by the ways it distributes the resultant goods and services, innovation may undermine rather than promote sustainable development.

Of course, if we were to define "higher quality" and "lower cost"—the key manifestations of an innovation process—broadly enough, we could count as "innovation" only those processes that are consistent with a broad definition of sustainable development. But such a definition would only enable us to avoid the difficult issues of what we mean by "higher quality" and "lower cost," and how different groups within the population, now and in the future, can experience the innovation process very differently. The point of economic and social analysis is to confront, not ignore, such issues. How, then, can business and government policy promote innovation processes that enhance the standards of living of the population in an equitable and stable way?

We cannot answer this question abstractly apart from the particular social contexts in which innovation might occur. But we do need a theoretical framework to analyze how, in particular social contexts, business and government policy can encourage the kinds of innovation that contribute to sustainable development. We also must construct assessment tools to determine whether in fact the innovation processes taking place are indeed making the contributions to sustainable development that were intended.

It is with these analytical objectives in mind that CITA was created to encourage the type of research contained in this volume. Although some of the contributors to this volume are more explicit than others in specifying their analytical framework, all of them reflect a growing understanding among the researchers at UMass Lowell—itself in part the result of the collective learning process that CITA helped to set in motion—concerning the intellectual approach that such analytical work must take.

What are the key elements of this approach to the study of innovation and sustainable development? Perhaps the most important element is that innovation that supports sustainable development results from the interaction of industrial, organizational, and institutional transformations (Lazonick and O'Sullivan 2000). Industrial transformations revamp the technologies that we have available to us and the markets (or uses) that these technologies serve. Thus, industrial transformations are the very substance of the innovation process.

But how are such industrial transformations set in motion? The essence of the innovation process that supports economic development is a cumulative learning process that takes place in certain locations over time and a collective learning process that develops and integrates the capabilities of large numbers of people with different specialties and responsibilities (O'Sullivan 2000). To change the ways in which people engage in such cumulative and collective learning requires transformations in the ways in which they relate to one another at a point in time and over time. Such is the substance of the organizational transformations that generate industrial transformations.

This relation between organizational and industrial conditions occurs, moreover, in broader institutional contexts, characterized especially by particular norms and laws concerning employment relations (including educational institutions) and financial relations (including the increasingly important institutions that regulate the distribution of income between the retired and working populations). For both analytical and practical purposes, we generally take these institutional conditions as given when we look at how we can affect organizational transformations that permit the industrial transformations needed to generate the innovations that support sustainable development. But when such innovations are not occurring, or when they are not having the desired results, we must begin thinking about whether institutional transformations are required, and about what political processes must be set in motion to bring them about. Hence the need for our work to be rooted in, but at the same time critical of, the particular social environment that must be transformed.

Such action-oriented research requires not only a deep understanding of the industrial, organizational, and institutional conditions that face a particular region such as Massachusetts, but also the conditions that exist, and the transformations that are taking place in regions around the world. The region within which we are located both collaborates and competes with regions in other parts of the United States and abroad. In a world of industrial innovation, what was successful yesterday in our region may be a failure today because of the emergence of new, high quality, lower cost products from other regions. Hence the need for industrial renewal.

At UMass Lowell, we are committed to studying regional development and international competition on a global scale. In doing so, our prime charge is to find ways to encourage innovation that supports sustainable development in the region in which we are located. But we can only do so by learning from, and helping to train, people from around the world who are intent on pursuing similar goals. The global economy has become more interdependent, and we believe that, given all the forces that may be working to inhibit or even undermine sustainable development, the future lies much more in cooperation rather than competition among regions and nations. Yet such an ideal of global cooperation will remain only that if we cannot understand and promote innovation that supports sustainable development where we live and work. Thus, in participating in the process of regional development, our goals at UMass Lowell are to promote the next round of innovation before, not after, yesterday's industries can no longer compete, and to search for the kinds of innovation that can productively engage the skills and efforts of a broad base of people in the region.

The Shape of This Book

It is within these contexts—historical, regional, institutional, and intellectual—that we at UMass Lowell have developed the concepts and approaches in this volume. The process by which we have come together to formulate the new approaches to sustainability presented in the following four parts is perhaps as important as the substance, particularly in terms of the increasingly important need to generate knowledge and new solutions to social and technical issues via multidisciplinary approaches.

We have had a unique situation in which to work. By its designation of sustainable regional development as one of its main goals and by its establishment of CITA, the university administration provided an environment that enabled faculty and staff of centers to begin and to continue interdisciplinary, cross-college discussions about the theory and practice of sustainable regional development. This multiyear period of gestation was necessary, for scholars from diverse disciplines needed the time to cross disciplinary lines to consider the perspectives of others and the opportunity collaboratively to construct new approaches based on insights from the disciplines involved. Many faculty members and centers were already engaged in scholarly and practical endeavors related to particular aspects of sustainable development. This was the catalyst that fostered a major effort to rethink what "sustainable regional development" means and how the university can assist the region in achieving it.

As a result, the authors of the essays in this collection have spent a significant amount of time talking to one another in informal and formal venues, widening and deepening their views of appropriate conceptual models, innovative processes (whether technical or social), and practical tools for putting ideas into action in the region. Many of the jointly authored essays are the result of CITA funding and encouragement, which facilitated the coming together of scholars from diverse disciplines—people who otherwise would never have had the time or opportunity to explore interdisciplinary boundaries. CITA has sponsored events in which this work was presented, ranging from a seminar series at which we discussed working papers to our annual workshops, where more formal presentations are made, and outside commentators—academics and representatives from industry, the community, and government—are invited to share their views on what we are trying to do. There has evolved a continual process of discussion, constructive criticism, reflection, and reworking of ideas internally. In addition, most of the contributors regularly present their research at regional, national, and international conferences, garnering still more feedback.

Last, in putting together the major groupings of this volume, we held discussions among the authors contributing to each part to ensure that, in final revisions, they would be working together closely. We would like to emphasize however, that the authors speak in their own voices. Although we are developing a common understanding of what sustainable regional development means and what will foster it, we have deliberately avoided attempts to homogenize the discussion and thereby destroy the freshness of the various perspectives that are being brought to bear on these issues.

These four parts, respectively, provide an overview of some issues and factors we feel should underlie a discussion of sustainable regional development in today's political economy, offer a perspective on what we believe "sustainable regional development" must mean in order to benefit the most people in society, and illustrate how the university can move in the direction of achieving it, via relationships with firms that foster the types of innovation that contribute to sustainable regional development and by building strong ties and collaborations with a wide spectrum of organizations in the greater region to develop sustainable communities.

The first part, "The Political Economy of Sustainable Development," provides a broad framework for the discussion of sustainable regional development in subsequent parts. Laurence Gross argues that any understanding of the problems and prospects of the regional economy requires

historical perspective; the industrial, organizational, and institutional lega-
cies of the past help to explain both the constraints on and opportunities
for sustainable development in the present and future. William Lazonick
contends that analysis of regional development must include an understanding of the historical trajectories of those other regions and nations
that have, through their industrial development, confronted the established
industries in the region. At the same time, as Louis Ferleger and Lazonick
illustrate, it is necessary to view the employment opportunities of the region from the perspective of changes in industrial structure and the composition of employment opportunities taking place in the economy as a
whole. Indeed, as Meg A. Bond and Jean L. Pyle argue, at the level of
the business enterprise, even new trends in management practice and their
impacts on increasingly diverse workforces can be better understood in
terms of the economic and social changes that the region is going through.
Finally, the last chapter in this part, by Jeffrey Gerson and John Wooding
is devoted to the proposition that any discussion of regional development
must confront the question: What is the region? The answer will depend
on the particular sector of the economy on which one focuses, as well as
on the political processes that mobilize people who live in geographical
proximity to one another to attempt to transform the industrial, organizational, and institutional conditions under which they both work and live.

The essays in the second part, "Rethinking Sustainable Development:
Health, Work, and the Environment," make a strong argument that "sustainability" is not just about avoiding environmental degradation or destruction while fostering economic growth. They argue that sustainable
development must also involve a healthy work environment (jobs and a
workplace that are not damaging to one's health) and healthy communities
(places where people living together can create and enjoy a just and improving quality of life). The chapters by Laura Punnett and John Mac-
Dougall each reveal the substantial costs of not being attentive to these
broader aspects of sustainability—both in the work environment and in
the community. The following chapters show clearly there *are* means to
achieve this broader vision of sustainability—it is doable. These authors—
Margaret Quinn et al., Ken Geiser and Tim Greiner, Vesela Veleva and
Cathy Crumbley—present illustrations of concrete ways in which production can be designed to be more sustainable in this broader sense and how
sustainability can be measured.

In part III, "Rethinking Sustainable Development: Technology, Business, and the University," contributors discuss how the university can develop relationships with firms to generate innovations that promote this
broader definition of sustainable regional development. Michael Best and
Robert Forrant consider what forms of institutional collaboration between

the university and firms designed to generate innovation would be most effective in meeting the goals of competitive firms and in promoting sustainability. The second essay, by David Kriebel, Geiser, and Crumbley, suggests that firms and organizations can meet what might appear to be contradictory goals (such as economic viability and sustainability) by redesigning the production process. It provides examples of how a university center can assist in this process. The last chapter, written by Michael A. Fiddy, Dikshitulu K. Kalluri, and Julian P. Sanchez, is concerned with how to maintain the relationship between a university center working with leading-edge technology and relevant firms (a relationship important for the innovation process and for promoting sustainable regional development) when firms typically have short-term, immediate needs, and centers must have longer term financial support for research and graduate students.

In Part IV, "University–Community Collaboration," the authors examine the challenges for the university in trying to establish ways for the faculty and centers to develop innovative collaborations with the community that promote sustainable regional development. Forrant posits that, to be effective in the region, the university must establish ongoing discussions that are theoretically and methodologically grounded, involve key constituencies in the region, and advance learning across these organizations. Linda Silka emphasizes the importance of developing an applications component to learning—creating projects that involve faculty, students, and some community groups and simultaneously advance teaching and learning, research, and sustainable regional development. In the next chapter, administrator Nancy Kleniewski discusses ways the university can support applied community research by defining what it is and how it will be assessed, surmounting traditional institutional barriers to such endeavors. Finally, Krishna Vedula et al. address the importance of a well-educated workforce (in this case, engineering) that is versed in issues of sustainability and has had "hands-on" experience as part of the education process.

Universities, Learning, and Development: The Applicability of Our Work

UMass Lowell is not unique in its endeavor to shape a development infrastructure, and this collection of studies adds to the growing stock of knowledge on the role of higher education institutions in the regional development process. These institutions around the world are attempting to deliver education and training consonant with new modes of knowledge production in the fast-paced global economy (Tierney 1998). In a recent article, Kevin Morgan of the Department of City and Regional Planning, University of Wales Cardiff, states that "in recent years there has been a growing convergence between students of economic geography and students of innovation; the former are becoming more interested in innovation capacity as a way of explaining uneven regional development, while the latter are no longer impervious to spatial considerations in their work on technological change" (Morgan 1997, 494).

Urban and regional development centers can be found at such universities as the Georgia Institute of Technology, the University of Cambridge, the University of Molise, Italy, Department of Economic and Social Geography at the University of Cologne, Germany, the Chalmers University of Technology in Göteborg, Sweden, the Center for Urban and Regional Development Studies at the University of Newcastle upon Tyne, England, and the Urban Research Center Utrecht, Faculty of Geographical Sciences, University of Utrecht, Netherlands. International conferences on sustainable urban development now occur regularly; for example, in 1998 the Eighth Conference on Urban and Regional Research was held in Madrid, Spain, and it addressed issues regarding urban policy and sustainable economic development. It was cosponsored by the United Nations Commission for Europe (Tsenkova 1999).

There is a growing body of academic research on the role of universities in the regional development process finding its way into leading journals (see, for example, Beck et al. 1995; Dineen 1995; Huggins and Cooke 1996; Labrianidis 1995). Some journals ran special issues. *American Behavioral Scientist,* for example, produced a special issue in 1999 on universities in troubled times. *Regional Studies* focused on collective learning and innovation in a special issue the same year that covered research funded by the European Commission. European Planning Studies focused on universities and regional development in July 2001. Also in 1999, the United Nations Development Program issued its *Human Development Report* which described indicators of sustainable development. The national and local policy implications of these endeavors are also under review (Asheim 1996; Felsenstein 1996; Florida 1995). What is distinctive about the work at UMass Lowell is the inclination to collaborate across disciplinary lines and the importance placed on concrete practical activities; in other words, engagement *and* research *and* a reflective back-and-forth occur.

Felsenstein (1996), in his review of the typical development impacts associated with metropolitan universities, offers three approaches to analyzing the role of the university. The first attempts to correlate the "concentration of high-technology activity" with location factors "perceived as inducing this spatial clustering." The second approach reviews specific growth processes, such as skill and knowledge development, and examines the university's role in producing them. The third category is for traditional, straightforward impact studies that in the main seek to place a dollar value on the wages spent and the goods and services purchased by the university community. Yet the three fail to consider the university as a dynamic and consistent participant in the formulation, implementation, and analysis of long-term strategic initiatives to improve the quality of life for the inhabitants not only of the campus, but of the region in which the university is situated.

The array of activities at UMass Lowell is extremely important for the academic community and development practitioners to consider, for we are making a concerted effort to cross numerous academic and community–university boundaries that have heretofore proven difficult to bridge. We approach our work not simply to help firms acquire the latest technology, but to make it possible for underemployed and unemployed workers to receive the education and training required to work with that technology. It is imperative that our social scientists do not "study" the ethnic and immigrant groups that reside in Lowell as an anthropological outing, but that we work with residents and among businesses to study and learn together. Economic input–output models fail to capture the vital, cumulative impact of such complex human efforts, while traditional research paradigms that focus on business growth and technology diffusion to the exclusion of social, environmental, and cultural development also fall short of the mark when it comes to formulating innovative, long-term, sustainable development initiatives.

REFERENCES

American Behavioral Scientist 42 (Winter 1999). Special issue on universities in troubled times.

Asheim, B. 1996. Industrial districts as "learning regions": A condition for prosperity. *European Planning Studies* 4:379–400.

Beck, R., D. Elliott, J. Meisel, and M. Wagner. 1995. Economic impact studies of regional public colleges and universities. *Growth and Change* 26:245–60.

Blewett, M. 1995. *To enrich and to serve: The centennial history of the University of Massachusetts Lowell.* Virginia Beach, Va.: Donning Company.

Dineen, D. 1995. The role of a university in regional economic development: A case study of the University of Limerick. *Industry & Higher Education* (June): 140–48.

Felsenstein, D. 1996. The university in the metropolitan arena: Impacts and public policy implications. *Urban Studies* 33:1565–80.

Florida, R. 1995. Toward the learning region. *Futures* 27:527–36.

Gross, L. 1993. *The Course of Industrial Decline: The Boott Cotton Mills of Lowell, Massachusetts, 1835–1955.* Baltimore: Johns Hopkins University Press.

Huggins, R., and P. Cooke. 1996. The economic impact of Cardiff University: Innovation, learning and job generation. *GeoJournal* 41:325–37.

International Institute for Sustainable Development (IISD). [http://www.iisd.ca].

Labrianidis, L. 1995. Establishing universities as a policy for local economic development: An assessment of the direct impact of three provincial Greek universities. *Higher Education Policy* 8:55–62.

Lazonick, W., and M. O'Sullivan. 2000. Perspectives on corporate governance, innovation, and economic performance. Photocopy. France: European Institute of Business Administration (INSEAD), February. [http://www.insead.edu/cgep].

Lowrie, D. n.d. Excerpt from Global Tomorrow Coalition Tool Kit, as quoted on International Institute for Sustainable Development (IISD) Web site. Originally at [http://iisd.ca/educate/learn/].

Morgan, K. 1997. The learning region: Institutions, innovation and regional development. *Regional Studies* 31:491–503.

O'Sullivan, M. 2000. *Contests for corporate control: Corporate governance and economic performance in the United States and Germany.* New York: Oxford University Press.

Regional Studies. 1999. Special issue on collective learning and innovation.

Thanki, R. 1999. Do we know the value of higher education to regional development? *Regional Studies* 33:84–92.

Tierney, W., ed. 1998. *The responsive university: Restructuring for high performance.* Baltimore: Johns Hopkins University Press.

Tsenkova, S. 1999. Sustainable urban development in Europe: Myth or reality? *International Journal of Urban and Regional Research* 23:361–66.

United Nations Development Programme. 1999. *Human development report.* New York: Oxford University Press.

United Nations Commission on Sustainable Development. March 19, 1999. "About the Commission on Sustainable Development." [http://www.un.org/esa/sustdev/csd.htm].

I THE POLITICAL ECONOMY OF SUSTAINABLE DEVELOPMENT

Introduction

William Lazonick

SUSTAINABLE development is a process that links economics and politics. As such, it should be studied in terms of the industrial, organizational, and institutional conditions that can promote economic growth that is stable, equitable, and environmentally sound. The economic base for sustainable development is the transformation of technological (supply) and market (demand) conditions to generate higher quality, lower cost products. Such transformation requires the mobilization of labor and capital through organizations that can develop and utilize the society's productive resources. Conventional economics has it that this mobilization occurs through markets that harness the individual pursuit of self-interest. But the collective and cumulative character of the learning processes that are fundamental to the development and utilization of productive resources means that markets cannot do the job adequately. Hence the importance of organizations—both business and governmental—in the economy. In the United States, the government has controlled, and continues to control, the allocation of substantial resources to such developmental investments. But the central institution on which we rely for economic development is the business enterprise, and particularly the corporate enterprise characterized by the separation between the legal ownership of assets by shareholders and control over the allocation of corporate resources by managers.

But managers, whether in business or government, do not allocate resources just as they please. They make and implement decisions concerning investments in productive resources and the distribution of the returns from these investments, subject to a set of institutional conditions—particularly those related to employment and finance—that are

17

deeply embedded in the national social context. It is around the operation and persistence of these institutions that the main political contests concerning the goals and means of economic development are waged.

What, then, is the importance of the "region" to this process? From the perspective of sustainable development, we might define the region as the geographical area within which the interaction of transformative industrial, organizational, and institutional conditions takes place. The problem is that, in a nation such as the United States, industrial, organizational, and institutional conditions all tend to span geographical regions. Hence the problem arises that those engaged in the promotion of regional economic and social development face in implementing changes to industries, organizations, and institutions on a local basis. And hence the need exists to understand the political economy of sustainable development, if we are to find ways to refashion these conditions on a regional basis even in the presence of powerful national, and even global, changes that often operate to counter our efforts.

The essays in this part reflect some of our attempts at UMass Lowell to understand this problem. The first and last chapters here—one historical and the other contemporary—are specifically about the geographical region of which the university is a part. The three chapters in between are not about the geographical region in particular, but rather about the conditions that determine how members of the U.S. labor force are employed and the challenges that a region, such as the Merrimack Valley, Massachusetts, or New England, faces in promoting economic development that is equitable and sustainable.

In the first essay, "Revenge of the Federalists," Laurence F. Gross traces the emergence of the original "Lowell model" of economic development in the city that launched the American industrial revolution of the nineteenth century. It was a model of development that was highly corporatist from the outset, bringing together powerful Boston-based business elites who shaped the entire Massachusetts economy, and who did so by putting in place organizations that featured the building of managerial structures and the social segmentation between management and labor that is characteristic of business organization in the United States more generally. The historical process has had enduring effects.

In "Organizational Integration and Sustainable Prosperity," William Lazonick shows how this model of business organization, which had been such a powerful force for economic development, met its limits in the 1970s and 1980s, with the coming of Japanese industrial competition that confronted many of America's strongest industries by putting in place broader and deeper skill bases to develop and utilize productive resources. Lazonick draws the parallel between the hierarchical, functional, and strategic integration that has become increasingly important for equitable and stable economic development in the late-twentieth century and the organizational challenges facing a university that seeks to promote such development.

At UMass Lowell, these efforts focus especially on sustaining the viability of the manufacturing base. Yet within the economy, the inexorable shift to services employment continues apace. Next, in "Can the Shift to Services Support Sustainable Prosperity?" Louis Ferleger and Lazonick raise questions about the quality of employment opportunities that are being created in services, both in terms of the productive capabilities required in many fast-growing, low-paid job categories and the tendency for many of the higher paying jobs in services such as finance to contribute to instability and inequity in the development of the economy. This essay highlights the need for those concerned with sustainable development to engage with a theoretical perspective that can comprehend those forces that support and those that undermine their economic and social objectives.

The problems of economic inequality in the United States are bound up with marked differences in opportunities and rewards by gender, ethnicity, and race. In "Diversity Dilemmas at Work," Meg A. Bond and Jean L. Pyle show how the 1980s and 1990s saw a shift from government policy promoting equal employment opportunity and affirmative action to reliance on organizations in both business and government to "manage diversity." They contend that, as a means of equalizing employment opportunities across gender, ethnicity, and race in the United States, the workplace diversity movement faces constraints that reflect assumptions regarding equality and difference, backlash dynamics, and the resistance of organizational cultures to incorporate diversity. To overcome these constraints, they argue, the problem of managing diversity must be confronted in particular contexts, where those involved can develop a commitment to confronting these dilemmas.

In the final chapter, "Understanding the Political Geography of the Region," Jeffrey Gerson and John Wooding argue that the "region" with which the university must interact is for practical purposes ill-defined until one can identify the cultural and political ties that bind people in close geographical proximity together in the pursuit of common goals. They locate a major problem that afflicts the UMass Lowell region in the fragmentation of regional politics, based on a tradition of self-governance that is highly local and that is designed to exclude rather than include. They call for studies to identify the social forces that can engender a more encompassing regional identity and they raise questions—pursued at greater length elsewhere in this work—as to how the university can serve as a catalyst for regional movements that unite more than they divide.

Revenge of the Federalists

The Lowell Model and Its Persistence

Laurence F. Gross

HISTORIANS have long recognized and generally admired the ten original Lowell Corporations as an early form of industrial development in the United States. I argue that this Lowell Model represents a political and economic position on the part of its founders that established relationships among capital, management, technology, and labor appropriate to these attitudes, and that this style of corporate behavior persists.

The Lowell Model

In the decades preceding and following 1800, mercantilist capitalists had amassed major fortunes in New England through maritime trade. During this period, the Cabot, Lowell, Jackson, Lawrence, and other families moved into Boston to create or expand these fortunes as well as their social position. From the start, the men in these families saw their enterprises as in part a group effort, and they pursued them together both in trade and through an ever-widening kinship network produced by intermarriage (disparaged as inbreeding by them only in lower-class circles or rural New England).

Certain elements in the new Republic tied independence to renouncing the mercantile economy, and the War of 1812 further obstructed that path. Powerful forces resisted fundamental change, however, and wealthy traders, already organized for political and economic advantage in the Federalist Party, continued their efforts to develop economic and political prominence for this privileged elite. Yet at the same time they initiated efforts to shift the economic foundations of their position while maintaining their party.

Francis Cabot Lowell, a member of the group, visited Great Britain in 1810 and returned believing he had seen the future embodied in the powered spinning and weaving of cotton cloth. He brought together a few of his friends in a small and loosely organized group, later dubbed the Boston Associates, to embark on a prototype textile mill on the Charles River in

Waltham. The group hired an exceptional machinist–machine-builder–inventor, Paul Moody (also of their class), and began the work of reinventing the power loom, not yet available in this country and unavailable from the protectionist English. The colleagues acquired and built the full complement of machinery needed to run the first "integrated" textile mill in the country (Samuel Slater and others had only *spun* cotton at their Rhode Island mills); that is, raw cotton entered at one end, and cloth came out the other. The organization of the Boston Manufacturing Company (BMC) as a joint-stock company was as novel as its technology. Its combination of capital, powered equipment, factory organization, and financial acumen proved very successful.

The approach of the BMC showed careful planning and thorough rationalism. The headings in its ledger book listed the elements to be coordinated: Land and Water Power, Buildings, Machinery, Labor, and Administrative Supervision. As George Sweet Gibb put it in *The Saco-Lowell Shops,* "How successfully these factors could be combined by unpracticed men to turn out an unfamiliar product on equipment not yet invented was far from clear" (1950, 23). "In a factory to be peopled with workers neither present nor trained," Gibb might have continued. In any case, the ledger headings indicated that labor simply represented a commodity to be managed.

Initial success at Waltham led to the replication and expansion of this type of industrial development, first at Lowell and later elsewhere. The textile corporations the Associates created greatly resembled one another. They were closely held by a small overlapping group of investors. They produced a narrow range of cloth of a comparatively coarse type that required as little skill as possible in production. They intended, by keeping their products discrete, to minimize competition with one another. They expected to attract a native-born female workforce from the surrounding countryside to tend the machines under male supervision. Long seen as the forward-looking development of the economically adventuresome, Lowell has recently been revealed as the cautious development of the socially conservative. Robert F. Dalzell, Jr.'s *Enterprising Elite: The Boston Associates and the World They Made* does nothing to diminish the significance of their industrial activity, but it recasts their motives and makes visible the goals of their management of the new industry. Basically, Dalzell shows that the Boston Associates used textiles as part of a broad economic, political, and social strategy to perpetuate their families and class in ways closely attached to the past. The system represented conservative planning dedicated to the security of the original investors, not to continuing growth, and, once established, not to profit that entailed risk. Furthermore, they designed the system to operate without the direct interference of the principals, who would thus be permitted the time to play the formative or controlling roles they coveted in society: "Far from the production of wealth in the usual sense, the goal was the preservation [and enlargement] of fortunes already made, positions already won." The intention, in fact, was to restrict and contain the economy in the interest of the preservation, the security, of the Boston Associates (Dalzell 1987, 67). Their brilliance lay in their ability to combine novel technology, business organization, and labor policies to secure those ends.

Business Organization

The corporations formed were marked by "business policies that had been the foundation of Appleton's mercantile trade and were brought into use in manufacturing as well: High capital, careful accounting techniques, and a considerable authority invested in the board of directors in overseeing and directing company policy" (Farrell 1993, 42). The companies also had the benefit that "only three or four of those owners had to bother with the business at all" (Dalzell 1987, 43).

While at Waltham the Treasurer acted as resident manager; thereafter the investors operated the corporations from offices in Boston. Just as sea captains had taken their ships around the world, now Agents ran their mills. A hierarchical management structure composed of skilled males who had risen from the ranks here or, especially at the higher levels, in Great Britain, presided over the routinized labor on the factory floor. Their role permitted the Directors to address only major issues of product lines and shutdowns. In fact, separate companies known as Selling Houses, operated by a subset of the directors, handled marketing, arranged short-term financing, and advised on production (for a percentage of sales), further freeing those at the top from direct involvement in running the mills. Only the Treasurer, among them, had a consistent and direct involvement with the factories, but even he basically ordered raw materials and equipment to meet needs specified by the agent or by outside experts.

Just as they had as merchants, the Boston Associates expected the state and federal governments to support their endeavors through tariff protection and sympathetic legislation. Their steady efforts generally ensured the fulfillment of their expectations: "'There is no man whose beneficial influence in establishing satisfactory relations in relation to this manufacture exceeded that of Mr. John A. Lowell,'" Nathan Appleton wrote in 1858 (cited in Greenslet 1946, 223). Thus the original impetus of the industrialization of the Boston Brahmins and would-be elite continued down the generations. So thorough was the involvement of this class that, "by 1834, seven-eighths of the Boston merchants were identified with New England cotton mills as stock-holders, selling agents, or directors" (Jaher 1968, 195). They had both indicated and extended their financial position by establishing the first public commercial bank in Boston in 1774, the First National Bank, which remained the largest financial institution in the country until 1827 and was still third-largest in 1893. The Massachusetts Hospital Life Insurance Company, "designed to be an investment vehicle for the textile manufacturers," followed and displaced it. The largest banks in the city reflected the thinking of the Associates, who could exercise financial domination when necessary (Jaher 1968, 193). The Suffolk Bank, founded by the group in 1818, offered them the "extended institutional basis for economic control its founders envisioned and established in its structure and organization. . . . Coordinated banks could more easily meet high capital requirements of turnpike, canal, and railroad expansion and other large-scale interests" (Farrell 1993, 47). The need for transportation combined with desires for profitable investments placed the Associates in leading roles in the flow of capital to Western railroads in the 1850s and 1860s. By 1850 the Boston Associates controlled 20 percent of the cotton spindles in the country, 30 percent of Massachusetts railroad mileage, 39 percent of the state's insurance capital, and 40 percent of its banking resources (Jaher 1968, 193).

Thomas Jefferson Coolidge, one of the descendants and perpetuators of the Associates' position, later referred to these days as a time when "'everybody was at work trying to make money and money was becoming the only real avenue to power and success both socially and in the regard of your fellow men'" (cited in Jaher 1968, 196). Yet such a profound change took place in a conservative context, aimed at, first, preserving the positions of the families most directly involved, and then leading the way into the new industrial world. Pressure for dividends to serve their needs was consistent: maintaining family fortunes through trusts (avoiding division among heirs) de rigueur. As Dalzell notes, "The use of the corporate form, the high rate of payout of earnings, the conservative financial policies, and the increasing resistance to innovation all remained highly unusual until after the Civil War" (1987, 70).[1]

Definition of Labor

The new manner of production and economic organization also produced revolutionary change in the nature of work itself. For the first time people worked for wages at a pace set relentlessly by machines. As spools whirled and lays beat, someone had to service the spinning frames and looms continuously. An element of freedom previously available was lost to these workers. Even at the time, observers, including sympathizers with the new system, marveled at its novelty, at the degree of change it represented: In 1835, Britain's Andrew Ure described the monumental difficulty facing industrialists. He denigrated the accomplishment of inventing the machinery for cotton manufacture in contrast to that of overcoming "desultory habits of work" and creating through human labor the "unvarying regularity of the complex automaton," the factory: "To devise and administer a successful code of factory discipline, suited to the necessities of factory diligence, was the Herculean enterprise, the noble achievement of Arkwright. Even at the present day, when the system is perfectly organized, and its labor lightened to the utmost, it is found nearly impossible to convert persons past the age of puberty, whether drawn from rural or from handicraft occupations, into useful factory hands. After struggling for a while to conquer their listless or restive habits, they either renounce the employment spontaneously, or are dismissed by the overlookers on account of inattention" (Ure 1835, 15–16). The owners' needs required new attitudes to govern human relations, and only the establishment of a new work discipline could provide them. The task, then, was twofold: to attract workers and to teach them the Boston Associates' new way of work, their new way of life. Responses to the two needs dovetailed, with many measures serving both ends.

Not only did no labor force stand trained and ready for the work, but the bad press of the English textile factories, the "dark Satanic mills" of the romantic poets, had prejudiced U.S. residents against the very idea of factory work. The public relations efforts of the BMC were intended, in part, to overcome this feeling.

As part of their plan of operation, the new corporations looked to obtain production workers not by paying sufficient wages to draw males away from their occupations (from which they might also bring ideas about workers' perquisites and rights), but to attract a cheaper and less privileged workforce of rural female labor. Domestic service, the only widely available alternative employment paying cash wages, repulsed many women because of its subservient nature. Farm life had taught them how to work hard and not to fear hard work. If they could be attracted, they could do the work.

Every aspect of labor's treatment in Lowell reflected the owners' attitudes and goals regarding their employees. Lowell's isolated and previously undeveloped location provided an opportunity to create and then control the social matrix within which the factories operated. Through a strict system of supervision of the women's every hour, working and free, waking and sleeping, the Associates could overcome the reservations they and their parents might hold about factory life. They provided boardinghouses and matrons, their cost deducted from the workers' pay. In the words of Nathan Appleton, company housing "secures an excellent class of work[ing] people which tells materially in the prosperity of the company" (cited in Gregory 1975, 191). He saw the entire effort as an entity, a construct of the owners, not a joint endeavor with contributions from various participants: "We are building a large machine I hope at Chelmsford" (cited in

Bender 1975, 99, and Gross 1993, 10–11). Those ledger headings again, as housing, labor, power, buildings, and the rest, stood together as a construct of the BMC. Every aspect was to make its contribution, with the result seen as mechanical in toto as the looms themselves.

The long-term political–social outlook of the Boston Associates predisposed them to assume their separateness from and superiority to labor. They began as Federalists, people later characterized by John Quincy Adams as not only opposed to the French Revolution and all things French, but also holding "'a strong aversion to republics and republican government, with a profound impression that our experiment of a confederated republic had failed for want of virtue in the people.'" As their admirer Ferris Greenslet restated the case, "*Egalité* and *fraternité* in practice had never meant much to a race of shipmasters and seagoing owners accustomed to exact instant obedience from their crews, damn their eyes" (Greenslet 1946, 88–89). As aristocrats (always in their own eyes, increasingly in the terms of the new age), they easily extended the precapitalist distinction between family and strangers into new terms where "outsiders . . . were not to be trusted and who were therefore prey for exploitation and piracy." Capitalism turned all but family and joint actors (fellow corporate directors) into "others." These distinctions offered positive benefits: "Men used their families to help coordinate economic interests. The kin network provided individual businessmen with a structure that more effectively allowed the mobilization of capital and the coordination of economic activities" (Farrell 1993, 3, 9).[2] It also presented a negative corollary: The interests of those outside the families were not important and presented no impediment to their exploitation.

Henry Miles, prominent among the apologists for the investors, described the logic behind this point of view:

> The productiveness of these works depends upon one primary and indispensable condition—the existence of an industrious, sober, orderly, and moral class of operatives. Without this, the mills in Lowell would be worthless. Profits would be absorbed by cases of irregularity, carelessness, and neglect; while the existence of any great moral exposure in Lowell would cut off the supply of help from the virtuous homesteads of the country. *Public morals and private interests, identical in all places,* are here seen to be linked together in an indissoluble connection. Accordingly, the sagacity of self-interest, as well as more disinterested consideration, has led to the adoption of a *strict system of moral police.* (Miles 1845; emphasis added)

Miles's rationalization of this system of social control demonstrated the hubris and the pervasive self-justification of New England's financial powers; it left no room for "disinterested considerations." The arrogance of ownership appeared in its ability to define and regulate both "public morals and private interests" for its own advantage. A great distance separated it from those to whom it assigned habits of "irregularity, carelessness, and neglect," despite generations of hard work and success across New England. One wonders, in fact, how the sons and daughters of the "virtuous households" could tend so powerfully, upon their arrival in Lowell, toward such base habits.

Similarly, a Lowell mill owner spoke frankly about the reasons for and utility of providing schooling. Educated help produced more, exhibited "better morals," and was "more ready to comply with the wholesome and necessary regulation" of the mill, even at times of wage cuts, while "the ignorant and uneducated I have generally found the most turbulent and troublesome, acting under the impulse of excited passion and jealousy" (Special Commission 1866, 13). Once the rationale for the industry was based on profit rather than individual or social need, it required new techniques to motivate workers (as Ure had noted). Educating them to believe in a supposed identity of goals helped predispose them to accept a discipline serving capital's interests. Thus, as Philip Scranton points out, the capacity to reason is tied to letters: Those without schooling can't reason and are thus not quite human. It was, indeed, a new world being created.[3]

The relative positions of the two parties also appeared clearly in the terms of employment: "All persons entering into the employment of the Company are considered as engaged to work twelve months, if the Company require their services so long." All the demands were made on the employees by the company, with no obligation on its part to employ for a period, only for the workers to stay while they were useful.[4]

Owners saw the workers not as people like themselves, but as a commodity to be managed; workers who expected those in charge to deal with them as social equals, fellow Yankees, were naive. A letter from Boott Mill Agent B. F. French to William Boott in 1844 indicated a willingness to exploit workers: "The evils which constant employment and want of amusements are calculated to produce, if persisted in too long, are to a very great extent counteracted by periodical visits to their friends."[5] While admitting willingness to expose the women to "evils" consequent to their employment, the owners still recognized other reasons to encourage temporary employment. Miles noted the importance of homes elsewhere to which workers could return in slow times in order that the manufacturers could save their wages without fear they would "sink down here a helpless caste, clamouring for work, starving unless employed, and hence ready for a riot, for the destruction of property, and repeating here the scenes enacted in the manufacturing villages of England." Although schools taught compliance, experience was educating the workers in the reasons they might rebel. Miles admitted the absence of any identity of interests between owners and workers and hoped to avoid disruptions through turnover. Factory work, he expected, would teach the children of "virtuous homesteads" that they had no right to expect any return beyond an hourly wage when they were of service (Miles 1845, 75–76). Previously, corporations such as towns maintained responsibility not only for their own success, but also for the well-being of their members. In extreme cases, that meant poorhouses or public work. The new corporation devoted itself entirely to its financial success, denied reciprocal responsibility to employees, and left their care when no longer needed at the factory to their families or public agencies. As a citizen lost rights by crossing political boundaries, so an employee lost them by leaving work, regardless of the impetus.[6] Of course, once the less independent Irish arrived, any thought of avoiding their "sinking down here a helpless caste" was forgotten (Gross 1993, 11).

Management

Managers on site faced the task of regulating production of several mills (making up one corporation) operating largely independently of one another in the early years, and of increasingly varied and intermingled production later. Complex machinery and inexperienced labor remained constants, as, by midcentury, did limited reinvestment in new construction and equipment that could have facilitated management tasks.

From the start, the ownership group recognized difficulties inherent in this structure. Technical innovation came primarily from machineshop employees, leading quickly to contracts reserving the rights to their ideas to the owners. Authority had to be parceled out to enable managers to deal effectively with unexpected problems and with employees. Yet at the same time, managers were not family and required new inspiration to inspire effectiveness. Contracts and piecework incentives for managers aimed to overcome individual rights and a culture initially shared by workers and managers alike.

The Lowells, Appletons, Lawrences, and the rest saw that although they felt they had every right to complete control of their factories, the issue required some care. As Dalzell puts it: "Control would have to remain in the hands of management, but that control would have to stop short of becoming in fact, or in appearance, too oppressive. Workers would have to submit to the routine of factory organization without exploiting the uses of organization to further their own, as opposed to their employers', interests" (1987, 32). Just as they dedicated the fruits of the textile industry to their own families, they also hoped to retain the forms of the new system, as well as its innovations and skills, for their exclusive use or benefit. Efforts to do so involved a variety of balances and control mechanisms. In the first reference in the records of the Boston Manufacturing Company to the employees as a group, in October 1819 and March 1820, the Directors discuss building a meetinghouse for the workers, but also considered how "to draw such a contract as the company wish to make with the

Clergy man who may be engaged to the same."[7] Contracts with higher level employees routinely reserve to the company "all improvements made or suggested by him in the construction or management of these works," but they have to determine how much the company owes to Patrick Tracy Jackson and the estate of Francis Cabot Lowell for *their* improvements to the power loom.[8] Directors held control so closely that top employees had to be "authorized to order out and expel from the Buildings and Lands of the Company any persons whom they shall think fit."[9] Even Superintendent Jackson and Agent Moody signed contracts promising to devote their full attention to the company and, at least in Moody's case, promising that any improvements he made or suggested belonged to the BMC. Recognizing the problem created by the supervisors' constrained roles, the Directors voted Moody and one or two others small holdings in the corporation, "Whereas [they] . . . are in the service of the company and it is desirable to have them interested in the Stock." Lesser workers received bonuses rewarding such behavior as putting out a fire and saving the company.[10] The Directors recognized that the limits they placed on employees' roles, activities, and benefits left workers with little incentive to pursue company interests.

Laborers' wages, basically women's, were lower than those of the average English male worker, "which made possible the higher salaries that the American managers had to give the technician. Their [women's] use reduced the total costs of labor in the American industry, making the overall costs of production less than in England" (Gregory 1975, 190). Even in the early years of Waltham, the combination of low wages, restricted responsibility for and involvement of the workers, along with their increasing realization of the social and financial remove at which the owners held themselves, led to a level of exertion the owners found insufficient. By 1835 at least, the Directors had instituted a "Memo of Agreement with Overseers," which provided a premium for overseers and second hands in each mill according to the pounds of cloth produced by those they supervised. Later that year the formula was changed to a yardage-produced basis, "as cloth has grown heavier and cotton more expensive."[11] It appears the overseers had adjusted production to produce heavier cloth for their own advantage, underlining the difficulty of obtaining even from the highest rank of employees an unselfish dedication to the new order.

Technology

Owners used both factory organization and machinery to minimize the cost of labor and maximize the control of labor. They aimed to segregate tasks until they could be learned quickly, to diminish the capability of the employees to bargain over wages or conditions, or to affect production. The machine tenders were intended to be easily replaceable. The means was to develop "labor-cheapening" machinery, cost-saving, not labor-saving. The Lowell system sought and incorporated every technique by which it could reduce the workers' importance (skill, independence, responsibility, creativity) and cost. They stuck to the production of comparatively coarse cloth (leaving the finer grades to such centers as Rhode Island and Philadelphia), which fit their rather crude machinery and general approach. They escaped the need for the skilled help associated with fine goods or woolens (e.g., mulespinners, cloth finishers). The emphasis on stop-motions exemplified this intent. These mechanical devices stopped a machine in the case of malfunction or lack of raw material, just as an attentive operator would. They represented, from the Waltham days, the Lowell model's endless search for automaticity. Their idea that stop-motions permitted stretch-outs and speedups without cost in worker effort is belied by endless testimony from workers and consultants that these additions to machines did not work as advertised and/or were not found on machines supposedly so equipped (e.g., in 1901, of 500 looms operating in Waltham, "none have stop-motions").[12] Similarly, when the Draper automatic loom was in the process of development in the 1890s, all the stop-motions associated with its operation had to be reinvented on account of the insufficiency of those available.

Conduct of Labor

Skill was still required, and the common portrayal of the women employees of Lowell, the famous "factory girls," as unskilled operatives deserves more careful consideration than it has traditionally received. The young women of the region had abilities and training that made them well suited for work in textile factories. They were accustomed to hard work and familiar with the mechanical operations seen at the pervasive carding, saw, grist, and fulling mills, and numerous smaller textile mills that dotted the area. As part of a domestic production unit likely involved in preparing and spinning fiber, making and repairing diverse textile products, they had developed a full range of abilities related to handling fiber and thread and were prepared to adapt to and be adapted by their experience in Lowell's manufactories. Because these skills were common among them, they did not lead to high wages or create a strong bargaining position. Since owners wanted wages to remain low, they encouraged the definition of the women's jobs as unskilled. A look, however, at some of the tasks expected of a weaver, for example, offers another point of view.

A weaver, a woman at this time, operated two or three looms. Maintaining the filling supply represented the primary limiting factor. Each time she observed a shuttle carrying a nearly empty bobbin, she had to stop the loom in the proper position, move the lay, or beater, to free the shuttle, remove and replace it, and restart the loom. Careful procedure was needed to prevent unevenness in the cloth. She also had to adjust warp let-off and cloth take-up as the size of these two rolls varied (in inverse proportion), detect and repair broken warp threads before they led to escalating damage to the product, detect broken filling threads, and generally assure successful operation of the loom, calling a fixer when necessary. Each task required attentiveness; most required skills, both of judgment and execution. Adjustments were made by evaluation, not measurement. Warp repair required a particular knot, reinsertion of a thread into the correct heddle in the correct harness and the proper dent in the reed. Failure in any aspect of any chore produced defective cloth, marketable only at a sacrifice. Given piecework wages, speed was of the essence for the workers. Only those who did not (or do not) perform such work would be likely to describe it as unskilled. In fact, a now-anonymous writer in the Boott's Letterbook in 1859 recognized their skills. Addressing the importance of worker persistence, he wrote, "They will be worth more to us the last six months than they are the first twelve," if new workers could be persuaded to stay. Doubling their value after a year's experience implied a considerable period of learning and skill development (in Shlakeman 1969, 147).

To describe these jobs as unskilled is simply to devalue dexterity, quickness, keen eyesight, and substantial grace under pressure. Given the female origins of such jobs in this country, this devaluation comes as no surprise. Parallel skills of male handweavers previous to power weaving had not been similarly denigrated. Furthermore, the assessments that worked to devalue these skills were those of the men who paid for them—a self-serving judgment (Gross 1993, 13).

Jeanne Boydston has shown how the value of women's work at home had fallen in the eyes of society during the later colonial and early national periods. This devaluation related, among other things, to the separation of products from producers, a dependence on arbitrarily set cash wages, and facilitated the manufacturers' success in freeing themselves from the obligation of providing sufficient pay for the subsistence of the wage earner— let alone a family. Boydston describes how the devaluation of women's work at home led to the reduction in cost of all labor. In a time of comparatively scarce capital, it enabled owners to retain a larger, and growing, proportion of labor's value, increasing the amount it could treat as surplus to finance the growth of the industrial system (Boydston 1990).

Adding financial insult to moral injury, labor was also seen from the outset as the system's economic shock absorber: When business became less profitable as a result of national economic circumstances in the early 1840s, pay was cut in the mills of Lowell. Marcus Rediker points out the precedent in the background of the owners upon which they may have drawn, the practice of "deflecting many of capital's risks in transoceanic

commerce on the collective back of the maritime working class" by dock-
ing their wages in tough times (Rediker 1987, 145). In 1843, a stockholders' committee set up to investigate the Boott's affairs recommended cutting
the salaries of the treasurer (John Amory Lowell) and agent from $5,000
per year to $3,000. The treasurer, it noted, held the same position "in sev-
eral others [corporations], in one of which he receives the same salary as
in this." Although such cuts at top levels did not have a serious impact on
costs, they might mollify stockholders temporarily deprived of dividends.
More significantly, the committee hoped that the cuts "would tend to reconcile the operatives at the mills to the reduction which has already taken
place in their wages and even induce them to submit cheerfully to a further
reduction if required." This concept of "cheerful" impoverishment on the
part of the workers makes visible the model of expected deference held by
the corporate directors.[13]

Dalzell shows how the Boston Associates devised the Lowell-style operation in order to ensure themselves and their descendants places of prosperity, safety, and power. They intended to protect their class and to "check
the dangerous potential of industry to undermine the peace and tranquil
order of society." Investors made every effort to make government a silent
partner in their efforts. They "cultivated" politicians, initiated favorable
legislation, while steadily augmenting the economic power that they used
to foster special treatment of their interests. He makes it clear that their
campaign of self-aggrandizement proceeded without regard to its cost to
the workers who, because of the emphasis on profit and limited investment
(particularly reinvestment), bore steadily increasing pressure to increase
production at lower cost to maintain dividends (Dalzell 1987, 67).

As the system became more firmly established and the workers more
acculturated, the machine operation and flow smoother, the managers
raised the number of machines per employee, the "stretch-out," as well as
the operating speeds of the machines, the "speedup." At the same time,
piece rates, and total pay, went down. By the 1840s, the young Yankee
females were leaving Lowell in droves as, indeed, they always had (generally
staying a few days, months, or years, depending on the individual's tolerance and need). But this time their New England sisters refused to replace
them. They judged work in Lowell's mills not acceptable.

What had begun as an experiment had developed into a political and
social revolution in management, control, and economy. Gibb compares
the new system to an older one: "This was a pattern of feudalism in which
the mill was the castle and the agent was the bailiff. Some would have said
that, on the one day in the week when he visited the mill, the treasurer was
the Lord High Executioner. . . . The position of the unskilled or semiskilled laborer in Lowell was one of considerable economic and social subservience" (Gibb 1950, 70, 89). The trade from family control on the farm

to this subservience did not fit the expectations or plans of many of these women. They did not tolerate it for long.

Responses to the New System

The attitudes forming Lowell's new system and their consequences in it did not go unnoticed at the time. One worker noted, "Everyone here is wrapped up in self and looks out for No. 1." It seems a very modern place. His sister Mary soon came, despite his advice. Her difficult adjustment to employment there appeared when, after an absence, she returned to find that her overseer, Mr. Gaye, had taken away "her" winding frame and given it to another worker. Although her new frame was almost as good, she was furious and believed Gaye was too ashamed of his act to speak to her. On her part, she wouldn't speak until he did, "and then I shall give him a piece of my mind for he knows he had no business to give it away." Clearly, she admitted to no hierarchical social separation between her and the overseer, despite his authority. She did not appear to have learned the lessons intended for her. Instead, she exemplified the continuing difficulties in harnessing the traditionally and potentially independent Yankees to the industrial yoke.[14]

This system faced continuing opposition from all levels of society. Ralph Waldo Emerson stood for the widespread opposition to the nature of Lowell's development expressed in "intellectual" (read "publishing") circles: "They are an ardent race and are as fully possessed with that hatred of labor, which is the principle of progress in the human race, as any other people. They must & will have the enjoyment without the sweat. So they buy slaves where the women will permit it; where they will not, they make the wind, the tide, the waterfall, the steam, the cloud, the lightning, do the work, by every art & device their cunningest brain can achieve" (Gilman 1973, 102–3). Emerson described this "hatred of labor" as if innocently, but what concept could seem more slanderous, more heretical, in puritan New England, where the devil provided work for idle hands? To women he assigned the moral sense (particularly ironic given their subservient position as "mill girls"), and to these men cunning, the sly, underhanded desire for pleasure without effort, the producer of slavery and factories. He offered a biting commentary on the relationship between the two slaveries cited by North and South alike, that of the black, or chattel, slave, and that of the wage-slave selling his or her time to the industrialist without a share in the product. His resentment of the idea that some need not labor, that their triumph was defined as progress, expresses a strong opposition to the beliefs and propaganda of the capitalists.

Workers protested in parallel terms. They renounced a system in which they could only participate by selling their time, in which they gained no stake in the product of their exertion, as they had previously. They, too, complained that they were made wage-slaves by a system that reduced their role to that of automatons, undivided in treatment or role from the machines they tended.

Strikes protesting such conditions hit textile mills as early as 1820 and continued despite the social disapprobation of such public conduct by women, who recognized the inapplicability of old standards to new situations. Sarah Bagley, one of the workers' leaders and spokespersons, noted that inmates in Massachusetts prisons worked two hours per day less than she and had more time to use their library than did Lowell workers. She observed to an incarcerated forger: "You might have selected some game equally dishonest, that would . . . have made you looked up to as a man of wealth. . . . You might have performed some 'hocus-pocus' means of robbery, without forgery, and passed as an Appleton, a Lawrence, or an Astor in society" (Josephson 1949, 252). Strong condemnation of the nature of the system in which she worked by this "inarticulate" laborer—but recognizing and protesting a situation did not empower her to alter it, only to escape.

Early workers voted with their feet, and they voted early and often, making their way from mill to mill, from town to town, in their search for employment where they would be better treated. They pursued new situations in which they might profit from their Lowell experience and training.[15]

In the words of Emerson again, "The ways of trade are grown selfish to the borders of theft, and supple to the borders (if not beyond the borders) of fraud" (Josephson 1949, 116). The new order brought a new morality, and in Emerson's view only a new morality would tolerate this order. The depth of labor's opposition appeared in the words of Amelia Sargent, another Lowell worker: "We will soon show these drivelling cotton lords, this mushroom aristocracy of New England, who so arrogantly aspire to lord it over God's heritage, that our rights cannot be trampled upon with impunity" (Wright 1979, 409 and n.). The close parallels between her complaints and Emerson's indicated the widespread nature of the objection to those who made themselves a ruling class to dominate both nature and people in new ways. If Sargent's statement referred to Yankee women, her statement rings true, for they soon left Lowell's mills, only to have Ireland's famines send their replacements. Sargent, Emerson, and many others recognized the extent, the fundamental nature, of the changes accomplished at Lowell. Massachusetts had moved far from its namesake "commonwealth," one of mutual responsibility and effort in which all shared in both the product of their labor and in the deprivation of hard times.

Lowell's first decades had witnessed the institution of great changes, the initiation of a modern factory system. The nature of work and many of the bases of human relationships were drastically altered. Although work took on new guises, workers remained able to decorate their spaces with plants, confront their overseers without social subservience (if Mary Metcalf is a fair example), and exercise considerable mobility between jobs and across the landscape. The social status and industrial skills the women brought to the mills gave them grounds for pride and self-worth in their jobs, for resistance if necessary, for the sorts of protest expressed by Bagley, Sargent, and others.

As Gibb points out, "The managerial principles adopted at Waltham . . . not only established basic precedents for the guidance of the machine-building trades, but also set the pattern for the development of big business in America" (1950, 62). Factory labor, however, rejected the pattern and demanded reconsideration of its terms and conditions. If not for the fortuitous coming of the Irish, driven off their land by the potato blight and imperial policy (the Federalists' aristocratic role models serving once again), the Boston Associates would have had to bargain with these workers, too. Without the immigrants, the textile industry would have had to reach new agreements on relative responsibilities, benefits, control, and conditions of work. After the arrival of the Irish, an industrial standard rejected by native labor, even the disenfranchised and economically disrespected women, prevailed. One cannot ignore the irony of the desperate and abhorred Irish rescuing the New England elite from the catastrophe its policies had created.

The Lowell Model: After Yankee Labor's Flight

Criticism of the new industrial model did not readily abate, even after the Yankee women left. Corporate management came under heavy fire from stockholders outside the inner circle of Boston Associates as well as workers during the 1850s and 1860s. One of the most vociferous critics was stockholder James Cook Ayer.[16] He had made his fortune in the patent medicine business in Lowell and then invested substantially in a number of textile mills. He stood outside the original circle of investors on all counts: geography, family, training, and the other ties they shared.

In 1863, Ayer, in *Some of the Uses and Abuses in the Management of Our Manufacturing Corporations,* assailed corporation management. He complained of stockholders' meetings held in tiny Boston offices, called on short notice, and/or simultaneously for different companies, papered with false proxies, and scheduled to begin before the arrival of the Lowell train. He alleged that corruption was a way of life (as Emerson had feared), as instanced by an $89,000 "bribe for Congress" noted, without comment, at the board meeting of one of the Lowell corporations, Middlesex Manufacturing.

Ayer found the benefits of the industry distributed among none but sell-ing agents and managers controlled by and beholden to a few of the inves-tors, and that "while these grow rich, the [other] owners, skilled workmen, and operatives are poor." With no check on these officers, he charged that their power expanded while their efforts for the companies diminished. When the legislature did pass a law in 1859 requiring mild reform, the companies ignored it until they were able to secure its repeal, partly through petitions coerced from the workers by threats. Ayer characterized the situation as one of "wanton robberies" of the stockholders. He cited as an example two selling houses, A. & A. Lawrence and Company and J. W. Paige, who received $200,000 apiece annually in commissions and as much again in incidentals, without capital, risk, or expense. Boott stock-holders' efforts to reduce payments to Lawrence & Co. had been foiled by the presence of a principal of that firm on the Boott's board of directors, he charged. Abbot Lawrence, a director, had positioned himself to receive large profits from his roles in both the selling houses and the corporation. Large commissions for purchasers of cotton, coal, wood, and oil appeared to Ayer as similar exemplars of the nepotistic robberies taking place. Ex-tremely high salaries for the officers went to "men who have failed to be very valuable in any other pursuit—the son, son-in-law, nephew, or relative of some Director, who, in turn, allows the other Directors to put their de-pendents in good positions, also." The result for the workers in the mill was evident: "The labor employed in them has been pressed from low, even lower, and down to the least per diem that will support life." Ayer con-demned not the individual managers-Directors, but found "that the *system* is thoroughly bad." Reform could come only through state regulation: "So vicious had this system become in its influences, that it might almost seem to make swindlers of virtuous men, and it should be changed, not alone for the preservation of the great properties it concerns, but also to save from contamination those who administer it" (18–24). While echoing Em-erson's sentiments, Ayer attacked only those at the very top, primarily in the Boston offices. He portrayed the industry as bloated by success, mo-nopoly, and a willingness to operate for the benefit of its smallest com-ponent.

Reserving benefits for the few, the Associates continued to mistrust management and employed family to run all they could handle. A letter of recommendation for a candidate for the post of agent concentrated on moral issues rather than knowledge. The writer vouched for the candidate's trustworthiness and honesty, his energy and willingness to "devote himself soul and body to the interests of your corporation." Regarding his ability, the writer could only say, "As far as I can learn, [he has] sufficient skill and knowledge of manufacturing."[17] Character traits that produced loyalty to the corporation rather than to self marked the attractive candidate, implying a scarcity of such people—a lack of success in the reworking of human nature to fulfill the demands of the new order at even the highest employee level. The continuing difficulty of creating a situation in which workers were led to identify with, adopt, or even serve the owners' interests rather than those of their cohorts created another level of operational difficulty, heightening the contrast with the common image of a smoothly running giant operation associated with the Lowell system.

Expansion of Lowell Model: Pre–Civil War

Immense profits led to reproduction of the pattern, generally with heavy involvement by the same investors, all over New England and beyond. Lawrence, Holyoke, and Chicopee Falls, Massachusetts, Manchester and Nashua, New Hampshire, Saco–Biddeford and Lewiston, Maine, spread the new pattern across the countryside. The system worked, and corporate planners came from all over the country to study the original.

The money made in Lowell and its companion cities altered the course not only of local industrial history but also of national development. The Boston Associates' profits, political connections, and diversification led to control of textiles, railroads, insurance companies, banks, and more, giving them a dominant position in the New England economy and significant influence in Pennsylvania, New York, and the Midwest. This closed club of large investors developed substantial political power, serving on the state and national level and supporting chosen candidates, such as Daniel Webster, who adopted their views after a gift of stock in the Merrimack Company. Given legislative monopolies in railroads, creating them in textiles, approaching them in other areas, they became multimillionaires. Nathan Appleton alone was involved in thirty-one textile companies, and defender Edward Everett's hyperbole in 1863 described the kind of power he and his cohorts wielded: "Without the aid of William and Nathan Appleton and Abbott Lawrence, no new venture could be launched in New England" (cited in Gregory 1975, 196). Such dominance previously had been unheard of, and it represented a level of individual economic and political power that violated the precepts of the new nation's ideology of political and social equality, while heralding these men's success in the system they were creating.

Persistence of the Lowell Design

Dalzell offers a strong claim for the power of Lowell's example. He notes that

> the institutional arrangements that pool and preserve Boston capital, and structure its transmission from generation to generation, are not [now] at all unique. Neither is the control of certain key charitable, cultural, and educational institutions by a self-conscious elite—the sturdy foundation

that still sustains so much of the city's tradition of service. Over the years the two together, operating in tandem, have become the norm in American society. Written into law and accepted by the public at large, they are in fact the principal means by which the existence of wealth is justified and finally sanctioned in the United States.

And it was the Associates who, more than any other single group of individuals, created that system. To be sure, someone like Amos Lawrence was far less wealthy than a Morgan or a Rockefeller, but Rockefeller himself understood full well what he owed Bishop Lawrence's forbearers and gratefully acknowledged his debt. (Dalzell 1987, 230)

Yet lacking self-conscious emulation by modern business leaders citing the Lowell model, how does one measure the nature and extent of their impact on American corporate (and domestic) life?

Many types of institutions carried forward the ideas and values of the Boston Associates and their descendants. They founded the Lowell Institute to promote speakers advocating their idea of moral Christian doctrines, as well as addressing science and its application to the arts and utility (Greenslet 1946, 210–11). Abbot Lawrence donated money to Harvard for scientific studies, $50,000 to the Lawrence Scientific School, specifying support for "'applied engineering, mining, metallurgy, and the invention and manufacture of machinery,'" as well as advocating the teaching of management. Harvard's failure sufficiently to embrace his ideas led to the formation of the Massachusetts Institute of Technology (Noble 1979, 22). The Boston Associates dominated the Harvard University Corporation by the 1830s, funded and controlled its board, outlook, and values for decades, as they supported and participated in the direction of MIT, as well (Farrell 1993, 30). No wonder Emerson wrote in 1859 that State Street dominated Harvard: "'Everything will be permitted there which goes to adorn Boston Whiggism [successor to Federalism]; . . . every generosity of thought is *suspected,* and gets a bad name'" (Greenslet 1946, 235, 322–23, 236). Abbott Lawrence Lowell, as president of Harvard, introduced the case study method, which promoted the training of a managerial elite as administrators for American business to the Business School.[18] Although the descendants were not always confident of the sufficiency of these efforts, as Charles Eliot Norton expressed it, "The rise of democracy in America is the rise of the uncivilized, whom no scholarly education can suffice to provide with intelligence and reason" (Federalism redux), their impact was immeasurable, culturally and in business (cited in Jaher 1968, 199–200).

The Associates' descendants perpetuated their outlook just as the old trusts and institutions perpetuated them as a class. Federalist thought appeared in the activity of the respectable and influential Immigration Restriction League late in the century, where Boston intellectuals "left a lasting imprint on the immigration debate by drawing a line between the 'old' and 'new' foreigners, separating the Anglo-Saxons and, by necessity, the Irish, from the new and lesser peoples." Similarly, Josephine Shaw Lowell, improver of the unfortunate, opposed the "'socialistic teaching'" that the poor would be given charity; instead, unemployed men should be given only work that was "'continuous, hard, and underpaid,'" another tradition of the New England textile mills. The Associates similarly adopted the Spencerian view that "survival of the fittest" meant the fates of rich and poor equally represented their just desserts (Brier et al. 1992, 153–54, 138).

The industrial practices of the Lowell Model spread through industry via many paths. The textile mills and machine shops the members of the elite had founded and ran persisted as significant employers for well over a hundred years and as examples of their approach and training grounds for its perpetuators. Although the many corporations they founded ran well into the twentieth century, the conduct of their machine shops offers a particularly telling example of their attitudes and their consequences. Whereas the several machine shops had originally served as rich sources for research and development, by the Civil War the locus of innovation in textiles had shifted to the shop floor and to independent machine shops. Interestingly enough, in power technology, which benefitted owners most directly, the employees of the closely held power company, the Proprietors of Locks and Canals, and the Lowell Machine Shop continued to play prominent roles until the turn of the century.

In these bastions of highly skilled male workers, originally the sources of innovation and the builders of the tools of production, labor had always been far more valued than on the factory floor. Gibb (1950) identifies the Saco Water Power Machine Shop's "greatest asset of all—the skills accumulated in half a century of machine-building" (406). Yet "labor was considered along with tools and metal as a factor of production, to be purchased for as low a price as possible in as small amounts as possible when required to fill orders on the books" (397). Thus, these highly skilled workers "were called in from the farms, fishing boats, and clam flats to work at utmost capacity on orders which had been received . . . [,] then released to get along as best they could until business revived" (389). They worked long hours at low wages, "but these conditions were universal" (394). Furthermore, the "atmosphere in the shop was a throwback to the early days at Lowell when the lives of the operatives were dominated by the religious, political, and economic views of the mill agents" (396). The company had been run into the ground by the turn of the century, "its assets shrunk and deteriorated, and the steady pouring-out of dividends had drained a once handsome surplus" (403), as was typical of the Lowell model. Similarly, the once-mighty Lowell Machine Shop sold out when "capitulation [was] profitable and therefore honorable" (305). The two men who bought it out immediately repaid themselves their purchase price with a special dividend (332–33). These corporations, as always, had been run by traders and merchants, not manufacturers (331). Financial talent in control at the start of the twentieth century ran them for the exclusive interest of their primary stockholders, drained, and closed them despite the great talent and flexibility for which machine shops are known. The contrast with the Toyoda Loom Company, which at a comparable period in its history determined to shift from a concentration on power looms to the manufacture of automobiles, makes their choice all the more telling.

Directly and indirectly, the influence of the Lowell model was spread through industry. The mills were the training grounds of thousands of overseers and machinists, and tens of thousands of textile workers, who carried their experience throughout industry for decades, as did companies controlled by the Associates and their banks and trusts. Textile practices were spread nationally by consultants such as Lockwood-Greene and Charles T. Main, both trained in the New England corporations.[19] New England machine builders and manufacturers held primary positions as stockholders in many Southern mills, as well as those closer to home.

Even as conservative a business historian as Gibb wonders about the "true cost" of the absolute resistance to organized labor and the dismissal of labor's interests in the activities of men who reorganized the machine shops and closed the textile mills in the early twentieth century. He cites the closing of Waltham Watch as comparable to the machine shops in the ferocity of its attitude toward labor. Frederic Dumaine, the orchestrator at Waltham Watch, received his training in the banks of the Associates before taking charge of the biggest textile mill in the world, Amoskeag, in Manchester, New Hampshire, which he then decimated in response to a strike in the 1920s (Gibb 1950, 501, 502; Landes 1983, 330–33). The power derived from decades of such practice and the success of the Associates' educational and cultural efforts are revealed in the attitude cited by David Brody: "American workers might engage in pressure tactics, but, as Sumner Slichter remarked, they knew they were breaking the rules. It would be hard to imagine a more insidious check on so fundamental a phenomenon as the self-activity of the work group" (Brody 1980, 206). The culture of the Lowell model, as spread by education, religion, work experience, and the media of the day established a perspective overcome only with the greatest difficulty.

The investing group known as the Boston Associates/Boston Brahmins grew to include names such as Higginson, Forbes, Cabot, Lee, Lowell, Fiske, Cushing, Jackson, Storrow, Fuller, Channing, Copley, Agissiz, Sargent, Sears, Lawrence, Perkins, and Wentworth, among others, who became as prominent in the Social Register as on the boards of industry. Given the insurance of the wealth generated by textiles and maintained in trusts, they remained active players and risk-takers in the late nineteenth- and early twentieth-century economy. The Associates and their descendants played influential roles in the major banks of Boston, which directed investment and often policy not only in the region but also in a broader sphere, including Western railroads (Gibb 1950, e.g., 418; Greenslet 1946, 318–19; Farrell 1993, 42, 47): "Throughout the nineteenth and into the early twentieth century, Boston capital was instrumental in funding large-scale investment projects, including railroads and western mining ventures" (Farrell 1993, 2). Charles Francis Adams, Jr., headed the Union Pacific, and John Murray Forbes and his son-in-law ran the Chicago, Burlington, and Quincy Railroad (Jaher 1968, 202). Men from the same circle created the hugely profitable Calumet and Hecla copper mines in Michigan, bringing in Lowell Machine Shop–trained talent to run the operation and drawing on their Lawrence Scientific School training to keep the operation in the nineteenth-century tradition: "The companies would deal their workers into appropriate occupations, houses, and social settings, and the men were supposed to settle for what they got. The company was the father, and father knew best" (Lankton 1991, 44–45, 18–21).

Through the investment houses of Lee, Higginson, & Co. and Kidder, Peabody, this economic and social elite kept Boston money at the highest level in manufacturing and finance: "As such they participated in many of the new industrial combinations, dominating some of the more technologically advanced and speculative," General Electric and AT&T among them. (The electrical industry was one of the most influential in setting the course for twentieth-century business.) Lee, Higginson, in the person of James Jackson Storrow, reorganized General Motors in 1910. Henry Lee Higginson himself took time from his investing to manage the Boston Symphony, exemplifying the group's cultural role as well as its consistency: "He insisted that musicians seeking to form a union be fired" (Kolko 1967, 343–63).[20] Godfrey Cabot Lowell founded Cabot Carbon Black Co. and became the richest man in Boston. Charles Francis Adams III sat on fifty-six boards, including Raytheon's. Boston banks played major roles in some of the biggest companies and conglomerates of the turn-of-the-century. In all these ways, these merchants turned financiers continued to shape industrial practice in this country (Jaher 1968, 230–32; Noble 1979, 7).

David Noble's discussion of the course of industry in the early twentieth century cites a variety of commonly accepted aspects of industrial behavior, which can be seen to reflect longtime Lowell-style practice. For example, he notes the "ever-widening perception of the exigencies of capitalist growth and stability prompted corporate leaders to band together, to pursue governmental regulation, or otherwise to counter the vagaries of the competitive market" in the period after 1900. The Boston Associates had been doing just such things since their start: overlapping investments in otherwise competing companies, acquiring tariff protection from the government, building their own transportation networks, controlling wages across cities and regions, starting and stopping their factories in unison, providing both mutual fire insurance and mutual strike insurance, and fixing prices.[21] They had addressed industrial relations as a means of seeking employee cooperation and loyalty, for example, through their boarding-houses, libraries, and churches. They had devised a system presaging most of the elements of Frederic Taylor's "scientific management": They introduced the narrowest possible division of work, systematic routing and scheduling of work, a "functional foremanship," which delivered workers' skills into management's hands and permitted central planning and control of work, as well as the pressure of incentive pay for both workers and foremen—in effect, most of Taylor except his time and motion studies (Noble 1979, 284, 288, 269). While labor continued to report to work daily with a full range of motor and cognitive abilities, every aspect of the owners' culture and industrial plan aimed to utilize less and less of them.

When Walter Licht, in his admirable summary of business' development in *Industrializing America,* describes the characteristics of industry into the early twentieth century, he offers a litany of the features initiated in the Lowell mills. Work rules, flow charts, arbitrary foremen in charge of hiring, firing, discipline, promotion, and the favoring of friends and relatives all stand out. Similarly, he cites rabidly antiunion stances, the drive to mechanize and specialize labor in order to replace skill, the corporate consolidation of resources (as in banks and trusts), the turn to the use of unskilled and semiskilled immigrants, and the substitution of technical personnel and instrumentation for decisions by labor (one thinks of the innumerable hank-clocks, pick counters, gram scales, tensile-strength testers of the textile mills). Furthermore, the use of gratuities and benefits in the commodification of allegiance and discipline, career ladders (within factory management) for loyalty, the perfection of machine tools and standardized parts (the former largely a function of the early textile machine shops of New England), increased output from increases in expectations from labor, as well as from organization and technology, mass producers with national and international markets, the separation of ownership and management—all these characteristics offer direct connections to the Lowell model, with or without the intermediaries of railroads, chemicals, and the electrical sectors in which Lowell money and personnel played a prominent role. Although the Lowell corporations did not display all the structures of a modern company, they established and promulgated the attitudes and tactics that continue to govern American business (Licht 1995, 129–33, 152, 154).

The Lowell model also exhibits many of the characteristics associated with the corporate raiders of the 1980s. Lowell always and increasingly relied on borrowed money, drained resources from healthy companies to serve immediate stockholder interests, ignored the interests of workers, even for continued employment, and saddled operations with the costs of loans, payouts, and costs incurred elsewhere. They moved the textile industry out of the region once their style of operation made continuance there more effortful, burdened, for example, by the costs of intermittent employment and hazardous work: Unemployment Insurance and Workers' Compensation, as well as outmoded buildings and worn-out machinery.

Many "developed" countries had the confrontation with workers that the flight of the Yankee women and the influx of the Irish made unnecessary in the United States. In Japan it led to the defeat of unionism and the establishment of lifelong employment. In Western Europe it variously produced systems of benefits, public and private, which serve the working class better than it is served here, where such a debate was not successfully joined until the Depression, by which time Big Capital was so entrenched, so powerful, and Federalist thinking so ingrained in the public psyche, that it needed to cede only enough to prevent meaningful change. In recent decades, even those slight accommodations of workers have continually dwindled.

NOTES

1. See, for example, Gibb 1950, 508; Farrell 1993, 52; Dalzell 1987, 70.

2. See also p. 13 where Farrell invokes Joseph Schumpeter as a proponent of the theory that family defines class interest.

3. Personal communication.

4. Regulations signed by Mary P. Johnson, presumably another employee of the day, but one of the many leaving no other record. Private Collection.

5. B. F. French, Agt., to William Boott, Esq., 3/22/1844, Massachusetts House of Representatives, Document no. 50, 1845, 24.

6. Both Paul Hudon and Philip Scranton have been helpful in developing this point.

7. Boston Manufacturing Company [BMC], Baker Library, Mss 442, 1:43, 59.

8. For example, BMC Unbound papers, Box 2-A; BMC Mss 442, vol. 2, Directors Reports [DR], 42–45.

9. BMC Mss 422, 1:61.

10. BMC Unbound papers, Box 2-A, Mss 44; BMC Mss 442, DR, 2: 20, 39; BMC Mss 442, 1:63.

11. BMC, 6:54, 61.

12. Parker Report, BMC Unbound papers, Box 1-A.

13. Report of Committee Appointed at Annual Meeting, Boott Mills, Mar. 6/43 to Investigate the Affairs of the Company, BMC, Mss 442.

14. Metcalf Letters, American Museum of Textile History.

15. See, for example, Preston Letters; Prude 1983.

16. James C. Ayer to Coolidge, 12/26/60, Box 29, Flather Collection, UMass Lowell Special Collections.

17. Unknown to Coolidge, 7/10/1861, Private Collection; Lubar 1983 finds similar attitudes (52–53).

18. Conversation with Al Bartovics.

19. Gibb, for one, notes Lockwood-Greene's role (1950, 381).

20. Brahmin descendants also controlled American Bell Telephone Company from 1879 to 1900 and took over Thomson-Houston Electric in 1887, which merged into General Electric.

21. For example, see BMC Mss, vol. 3 (n.p., 1/1898) regarding mutual strike insurance, and Box 2-A (1894), regarding fixing of prices.

REFERENCES

Ayer, J. C. 1863. *Some of the uses and abuses in the management of our manufacturing corporations.* Lowell.

Bender, Thomas. 1975. *Toward an urban vision: Ideas and institutions in nineteenth-century America.* Lexington: University Press of Kentucky.

Boydston, Jeanne. 1990. *Home and work: Housework, wages, and the ideology of labor in the early Republic.* New York: Oxford University Press.

Brier, Stephen, et al. 1992. *Who built America: Working people and the nation's economy, politics, and society.* New York: Pantheon Books.

Brody, David. 1980. *Workers in industrial America: Essays on the twentieth-century struggle.* New York: Oxford University Press.

Dalzell, Robert F., Jr. 1987. *Enterprising elite: The Boston associates and the world they made.* Cambridge: Harvard University Press.

Farrell, Betty G. 1993. *Elite families: Class and power in nineteenth-century Boston.* Albany: State University of New York Press.

Gibb, George Sweet. 1950. *The Saco-Lowell shops: Textile machinery in New England, 1813–1949.* Cambridge: Harvard University Press.

Gilman, William H., ed. 1973. *The journals and miscellaneous notebooks of Ralph Waldo Emerson.* Vol. 10. Cambridge: Belnap Press of Harvard University Press.

Greenslet, Ferris. 1946. *The Lowells and their seven worlds.* Cambridge, Mass.: Houghton Mifflin Company.

Gregory, Frances W. 1975. *Nathan Appleton, merchant and entrepreneur, 1779–1861.* Charlottesville: University of Virginia Press.

Gross, Laurence F. 1993. *The course of industrial decline: The Boott Mills of Lowell, Massachusetts, 1835–1955.* Baltimore: Johns Hopkins University Press.

Jaher, Frederic Cople. 1968. The Boston Brahmins in the age of capitalism. In *The age of industrialism capitalism,* ed. F. C. Jaher. New York: Free Press.

Kolko, Gabriel. 1967. Brahmins and business, 1879–1914: A hypothesis on the social basis of success in American history. In *The critical spirit: Essays in honor of Herbert Marcuse,* ed. Kurt H. Wolff and Barrington Moore Jr. Boston: Beacon Press.

Josephson, Hannah. 1949. *The golden threads: New England's mill girls and magnates.* New York: Duell, Sloan, and Pearce.

Landes, David. 1983. *Revolution in time: Clocks and the making of the modern world.* Cambridge: Harvard University Press.

Lankton, Larry. 1991. *Cradle to grave: Life, work, and death at the Lake Superior copper mines.* New York: Oxford University Press.

Licht, Walter. 1995. *Industrializing America: The nineteenth century.* Baltimore: Johns Hopkins University Press.

Lubar, Steven David. 1983. *Corporate and urban contexts of textile technology in nineteenth-century Lowell, Massachusetts: A study of the social nature of technological knowledge.* Ph.D. dissertation. University of Chicago.

Miles, Rev. Henry A. 1845. *Lowell, as it was, and as it is.* Lowell.

Noble, David. 1979. *America by design: Science, technology, and the rise of corporate capitalism.* Oxford: Oxford University Press.

Prude, Jonathan. 1983. *The coming of industrial order: Town and factory life in rural Massachusetts, 1810–1860.* Cambridge, U.K.: Cambridge University Press.

Rediker, Marcus. 1987. *Between the devil and the deep blue sea: Merchant seamen, pirates, and the Anglo-American maritime world, 1700–1750.* Cambridge: Cambridge University Press.

Shlakman, Vera. 1969. *Economic history of a factory town: A study of Chicopee, Massachusetts.* Northhampton, Mass.: Department of History of Smith College.

Ure, Andrew. 1835. *The philosophy of manufactures: Or, an exposition of the scientific, moral, and commercial economy of the factory system of Great Britain.* London.

Wright, Helena. 1979. Sarah G. Bagley: A biographical note. *Labor History* 20:409 and note.

Organizational Integration and Sustainable Prosperity

William Lazonick

A PRIME mission of the University of Massachusetts Lowell is to support regional economic development. In pursuing this mission, the university has certain unique technological capabilities to offer the regional economy, and the regional economy possesses an industrial base that can, and does, make use of these capabilities. Yet a deeper relationship between UMass Lowell and the regional economy could help to foster economic development on a scale and a scope that far exceed the impacts that are currently being achieved.

Corporate Investment and Sustainable Prosperity

The particular challenge that faces a university such as UMass Lowell is the promotion of regional economic development that generates "sustainable prosperity"—higher standards of living for more and more people over a prolonged period of time. During the 1950s and 1960s, the U.S. economy was capable of generating not only economic growth but also increasing income equality. Since the 1970s, however, a persistent feature of the U.S. economy, evident also in Massachusetts, has been an increasing income inequality, to the point where the United States now has the most unequal distribution of income among the advanced industrial economies (Atkinson, Rainwater, and Smeedling 1995; Levy and Murnane 1992; Moss 2001). Since the late 1990s, the U.S. economy has been booming, but it is by no means clear that a new trajectory of sustainable prosperity has been established. The challenge for those engaged in a regional university such as UMass Lowell is to take advantage of the boom to reorient the regional economy toward an enduring process of economic development that spreads its benefits more broadly and deeply among the regional population. It is important to recognize that, in the late twentieth century, our attempts to promote sustainable prosperity built on a highly innovative economy. In the early 2000s, after more than two decades of intense competitive challenges, the United States retains international leadership in a range of science-based industries such as computer electronics and pharmaceuticals as well as in service sectors such as finance and fast food. Yet, if the U.S. economy is capable of innovation, the growing income inequality of the past quarter-century suggests that it may be incapable of sustainable prosperity. Is it possible to structure regional institutions and organizations so that innovation and increased income equality can go hand in hand in the regional economy? The answer to this question is critical for a technological institution such as UMass Lowell that hopes to implement a regional economic development strategy that will result in sustainable prosperity.

In a project on the performance of the U.S. economy, the UMass Lowell Center for Industrial Competitiveness explored the hypothesis that the coexistence of innovation and growing income inequality in the U.S. economy in the 1980s and 1990s reflected a systematic bias of major U.S. corporations against making investments in *broad and deep skill bases* (Almeida 2001; Forrant 2001; Lazonick 2001; Lazonick and O'Sullivan 2001a; Lazonick and O'Sullivan 2001b; Moss 2001; O'Sullivan 2000; Tilly 2001). Rather, these corporations, which exercise significant control over the allocation of resources and returns in the economy, are choosing to invest, and are best able to innovate, in the production of goods and services that use narrow and concentrated skill bases to develop and utilize technology. Skill bases are broad when they include large numbers of people in collective and cumulative learning processes. Skill bases are deep when they include groups of workers engaged in shop-floor production as well as within subcontracting enterprises in collective and cumulative learning processes. Investment in broad and deep skill bases can generate the productive capabilities that make sustainable prosperity possible.

America's major industrial competitors—and especially the Japanese—have successfully challenged once-dominant U.S. producers across a number of industries by investing in skill bases that are broader and deeper than those that previously had been best practice in those industries. U.S. industrial corporations could, and some indeed have, responded by broadening and deepening the skill bases in which they invest. As a general rule, however, in the face of international competition, major U.S. corporations are choosing to produce those goods and services for which competitive advantage can still be gained by investing in narrow and concentrated skill bases; that is, skill bases of small numbers of highly educated and highly trained personnel.

Why are "skill bases" important to the economy? They form the foundations on which people engage in collective and cumulative—or organizational—learning, which is, in turn, central to the process of economic development. Case study evidence suggests that the manufacturing industries in which the U.S. economy was most severely challenged by high-wage foreign competition—industries such as automobiles, consumer electronics, machine tools, and commodity semiconductors—were those in which innovation and sustained competitive advantage demand investments in broader and deeper skill bases. If the "skill-base hypothesis" is valid, then it may well be that innovation and equality can go hand in hand. From a policy perspective, the relevant issue is how business enterprises can be induced to make innovative investments in broad and deep skill bases.

The skill-base hypothesis adds an important dimension to American debates on the relation between investments in "technology" and sustainable prosperity. On one side have been those who stress the weakened innovative capabilities of the U.S. economy in international competition (Cohen and Zysman 1987; Dertouzos, Lester, and Solow 1989; Tyson 1992). They have called for the U.S. government and businesses to allocate more resources to education, training, research, and cooperative investment projects that can support the United States in making a competitive response. These arguments assume, often more implicitly than explicitly, that these innovative responses will promote sustainable prosperity in the United States.

On the other side have been those who argue that income inequality cannot be blamed on international competition. The volume of world trade, they argue, is not large enough to have a significant impact on the distribution of income in the United States. Rather they attribute growing inequality to the employment impacts of "new technology" (Krugman 1997; Krugman and Lawrence 1994). If the United States has problems keeping people employed at high wages, it is because, for a given level of investment, technologies of the computer age do not create the same quantity and quality of employment opportunities for Americans as did the technologies of the past. Income inequality has grown, they argue, because new technologies displace employment opportunities that used to be well paid. Pay attention to raising the levels of both investment and relevant skill in the U.S. economy, and the income distribution will improve.

The skill-base hypothesis views both international competition and technological change as important determinants of the distribution of income, but it is embedded in a theory of innovation and economic development in which the impacts of international competition and technology on income distribution depend on corporate investment strategies. Across U.S. industrial corporations, these strategies, and the investment in skill bases that they entail, are in turn influenced by American institutions of corporate governance and corporate employment. The rise of powerful international competition based on investments in broader and deeper skill bases may lead U.S. corporations to seek to remain innovative by investing in technologies that only require investments in narrow and concentrated skill bases.

Powerful support for the skill-base hypothesis can be found in the experience of Japanese–U.S. industrial competition over the past few decades.

Japan took on and surpassed the United States in many industries in which it had been the world leader. The foundations of Japanese success in international competition, I argue, were investments in broad and deep skill bases to generate organizational learning. The problems of both innovation and equality in the United States in the 1980s and 1990s have not been inherent in technology. Rather they derive from corporate strategies to develop and utilize technology.

U.S. corporations, I contend, have been investing in narrow and concentrated skill bases in a world of international competition in which innovation has increasingly come from investing in broad and deep skill bases. If the skill-base hypothesis is correct, the solution to the problem of reversing the trend toward income inequality in the United States goes much deeper than growth policies or industrial policies. It requires transformation of the way industrial corporations are governed and the way people are employed.

Organizational Integration

Almost all the major industrial corporations in the U.S. economy in the post–World War II era made investments in managerial learning from the early decades of the twentieth century, if not before. Many of the productive and competitive advantages of these investments in managerial organization still accrued to these corporations decades after the particular individuals involved in those collective learning processes had left the corporate scene.

In a comparative international perspective, U.S. industrial corporations were not unique in building their managerial organizations into formidable sources of sustained competitive advantage. What made them unique among their counterparts in the advanced economies was their dedication to a strategy of taking skills, and hence the possibilities for craft learning— much less corporate learning—off the shop floor (Lazonick 1990; Lazonick and O'Sullivan 2001b; see also Jürgens, Malsch, and Dohse 1993). This process of transforming skilled craft work into "semiskilled" operative work was a prolonged one, constrained as it was by the development of new technology through managerial learning. But, as reflected in the distinction between "salaried" and "hourly" personnel, the strategy of relying exclusively on the managerial organization for the development of new productive capabilities has been, throughout the twentieth century, a distinctive characteristic of U.S. industrial development.

The American corporate strategy of confining organizational learning to those employed within the managerial structure enabled the United States to become the world's leading industrial power during the first half of the twentieth century (Lazonick and O'Sullivan 1997a, 1997b). On the basis of this leadership, U.S. industrial corporations were able to provide high pay and stable employment not only to managerial employees but also to shop-floor workers, whether skilled or semiskilled.

Over the past few decades, however, powerful international competitors have arisen that have developed productive capabilities by integrating managers and workers into their organizational learning processes. The *hierarchical segmentation* between managers and workers that the American "managerial revolution" entailed became a major institutional barrier to making investments in organizational learning required to sustain prosperity in the U.S. economy. In an era of intense international competition in which sustained competitive advantage went to those enterprises and nations that made investments in, and integrated, the organizational learning of both managerial and shop-floor personnel, the investment strategies of most U.S. industrial corporations focusing only on managerial learning fell short.

The competitive problem that has faced U.S. industrial corporations is that, over time for a particular product, the innovation process, of which the organizational learning process is its social substance, has become increasingly *collective* and *cumulative*. Organizational learning has become increasingly collective because innovation—the generation of higher quality, lower cost products—depends on the integration of an ever-increasing array of specific productive capabilities based on intimate knowledge of particular organizations, technologies, and markets. Organizational learning has become increasingly cumulative because the collective learning that an organization has built up in the past increasingly forms an indispensable foundation for the augmentation of organizational learning in the present and future.

The increasingly collective and cumulative character of organizational learning means that, for a particular product, an innovative investment strategy is one that entails investments in *broader and deeper skill bases*— divisions of labor that extend further down the organizational hierarchy and involve more functional specialties. The investments in skill bases are not simply investments in the learning of large numbers of individuals performing a wide variety of functions. For these investments in broader and deeper skill bases to generate higher quality, lower cost products requires *organizational integration,* a set of social relations that provides participants in a complex division of labor with the incentives to cooperate in contributing their skills and efforts toward the achievement of common goals.

At any point in time, the technological possibilities and organizational requirements of the innovation process vary markedly across industries in terms of the extent of the skill base in which the innovating enterprise must invest. In industries, such as pharmaceuticals, in which value-added comes mainly from research, design, and marketing, *narrow and concentrated skill bases* of scientists, engineers, and patent lawyers remain sufficient for generating higher quality, lower cost products. Thus, in such industries, U.S. industrial enterprises have been able to remain world leaders, but in industries, such as automobiles, where value-added derives mainly from manufacturing processes that combine a complex array of physically distinct components, international competitive challenges have been based on investments in broader and deeper skill bases. The investments in organizational learning occur not only within corporate management structures but also on the shop floor and in the vertical supply chain. In those industries in which international competition demands investments in such broad and deep skill bases, once-dominant U.S. industrial enterprises have lost substantial competitive advantage.

In the U.S. automobile industry, American-based companies have regained some of the markets they had lost—or at least have stemmed the loss of market share. The skill-base hypothesis posits that they have done so by investing in broader and deeper skill bases than before. In responding to these competitive challenges, moreover, the organizational problem that has faced U.S. industrial enterprises over the past few decades has gone beyond the hierarchical segmentation between managers and workers. Even within the managerial structure—the traditional locus of organizational learning in U.S. enterprises—organizational integration appears to have given way to *functional* and *strategic* segmentation.

Compared with both the integrated organizational structures of foreign competitors and the integrated managerial structures that characterized the most successful U.S. companies in the past, organizational learning within the managerial structures of U.S. enterprises has been limited by the *functional segmentation* of different groups of technical specialists. Specialists in marketing, development, production, and purchasing may be highly skilled in their particular functions, but in U.S. enterprises, relative to their counterparts abroad, they tend to respond to incentives that lead them to learn in isolation from one another. Functional segmentation makes it difficult, if not impossible, for such isolated specialists to solve complex manufacturing problems that require collective and cumulative learning.

In addition, in comparative and historical perspective, a distinctive characteristic of U.S. industrial enterprises since the 1960s has been the *strategic segmentation* of those top managers who control enterprise resources from those lower in the managerial hierarchy on whom the enterprise has relied for organizational learning. In allocating vast amounts of resources, top managers of major U.S. industrial corporations have increasingly lost the incentive to remain cognizant of the problems and possibilities for organizational learning within the enterprises over which they exercise control. Within a particular enterprise, tendencies toward hierarchical, functional, and strategic segmentation may be mutually reinforcing, thus making it all the more difficult for an enterprise, or group of enterprises, to invest in organizational learning once it has embarked on the organizational-segmentation path.

The skill-base hypothesis seeks to test these propositions concerning the growing importance of hierarchical, functional, and strategic integration for attaining and sustaining competitive advantage, and the increasing tendency toward organizational segmentation along these three dimensions in U.S. industrial corporations, in historical and comparative perspective. The skill-base hypothesis, and the theoretical perspective on innovation and economic development in which it is embedded, derives from our historical and comparative analyses of the role of organizational integration in shifts in international competitive advantage (Lazonick 1991, ch. 1; Lazonick and O'Sullivan 1996, 1997c; O'Sullivan 2000). The empirical evidence required to test the hypothesis must be derived from in-depth analyses of the investment strategies, organizational structures, and competitive performance of particular companies based in different nations that have engaged in head-to-head competition in particular industries.

Our purpose is to motivate such a research agenda by drawing on some of the findings of a now-vast range of literature on the interaction of organization and technology in U.S.–Japanese industrial competition. This evidence, much of it deriving from the experiences of management consultants and case studies by business academics, provides substance to the skill-base hypothesis. In this essay, I focus on differences in hierarchical integration and organizational learning in Japanese and American enterprises, positing that understanding hierarchical integration of technical specialists and production operatives forms an indispensable foundation for understanding the functional integration of technical specialists themselves—a subject that now dominates much of the management literature on technological competition. A discussion of strategic integration and segmentation—the analytical interface between issues of corporate governance and organizational learning—has been treated at length elsewhere (Lazonick and O'Sullivan 1997a, 1997b, 1997c, 2001; O'Sullivan 2000). I am concerned here with the social structures that generate organizational learning rather than with the social structures that allocate resources to building different types of skill bases.

Organizational Learning

If there is one nation that challenged the United States for international industrial leadership in the last half of the twentieth century, that nation is Japan. In 1950, Japan's GDP per capita was only 20 percent of that of the United States; in 1992, it was 90 percent (Maddison 1994, 22). The Japanese challenge came, moreover, not from those industries in which American companies were weak or had been neglectful of. On the contrary, the challenge has been in industries such as automobiles, electronics, and machine tools, in which the United States had attained a seemingly invincible position as the world's leading mass producer.

Since the 1980s much has been written about the institutional and organizational sources of the Japanese competitive advantage. Social institutions such as lifetime employment and cross-shareholding and organizational practices such as total quality management and consensus decision making have been critical elements of Japan's phenomenal rise from the ashes of defeat after World War II. But these institutions and organizations would not have generated the so-called economic miracle in the 1950s and 1960s had Japan not still possessed even in the immediate aftermath of the war, an accumulation of technological capabilities. Japan had been accumulating its capabilities in mechanical, electrical, and chemical technologies since the late nineteenth century when the Japanese "managerial revolution" had begun. At the time of the Meiji Restoration in 1868, Japan had little in the way of modern industrial capabilities (Morris-Suzuki 1994). Under the slogan "Rich Nation, Strong Army," the Restoration government implemented a strategy for industrial development that was heavily dependent on borrowing knowledge, technologies, and even institutions from abroad (Samuels 1994; Westney 1987). In the first half of the 1870s, private and public interests set up institutions of higher education—most notably, Keio University, the Institute of Technology (later part of Tokyo Imperial University), and the Commercial Law School (which became Hitotsubashi University)—to supply key personnel for an innovative industrial economy. (Abe 1996; Hirschmeier and Yui 1981, 166; Hunter 1984, 47). By the 1880s Japan had a steady supply of both indigenous graduates and teachers (Iwauchi 1989; Uchida 1989; Yonekawa 1984, 193–218).

Large numbers of university graduates were lured into industry, with the *zaibatsu* (large industrial conglomerates) taking the lead (Yonekawa 1984). From 1900 to 1920, for example, the employment of graduate engineers increased from 54 to 835 at Mitsui and from 52 to 818 at Mitsubishi (Uchida 1989, 108). These highly educated employees were not only eagerly recruited but also well paid by the companies that employed them. In addition, companies often incurred the considerable expense of sending them abroad for varying lengths of time to acquire more industrial experience (Hirschmeier and Yui 1981, 154; Iwauchi 1989, 99).

During the interwar period, the overall development strategy of the Japanese economy became increasingly dominated by the investment requirements of militarization and imperial expansion. Relying heavily on the *zaibatsu*, Japan devoted considerable resources to building capabilities in mechanical, electrical, and chemical engineering. In the immediate aftermath of World War II, as the Allied Occupation engaged in the dissolution of the once-powerful *zaibatsu* (Adams and Hoshii 1972; Bisson 1951; Hadley 1970), Japanese scientists and engineers organized to seek new ways to develop and utilize their capabilities.

In 1946 they formed the Japanese Union of Scientists and Engineers (JUSE), an association devoted to promoting the nation's technological development through education, standard setting, and the diffusion of information. Influenced by U.S. occupation officials versed in statistical quality control (SQC) techniques that the United States had used for military production during the war, JUSE focused on the application of quality control in an economy based on production for commercial markets. In 1949, JUSE established the Quality Control Research Group (QCRG), which included participants from academia, industry, and government.

The following year JUSE sponsored an eight-day seminar on SQC with Dr. W. Edwards Deming, a physicist who had been working for the U.S. government in developing the sampling methods for SQC (Ishikawa 1985, 16). These techniques were used to monitor mass-produced output for systematic deviations from "quality" standards as a prelude to controlling (identifying and correcting) quality problems. Deming's lectures were well received as was the edition of the lectures that JUSE promptly published. The author donated his royalties to JUSE, which in turn used the funds to establish the now-famous Deming Application Prize, awarded annually since 1951 to an industrial company for its achievements in the application of quality control (QC) methods (Nonaka 1995).

One of the key figures in applying QC methods to Japanese industry was Kaoru Ishikawa, an engineering professor at the University of Tokyo. Starting in 1949, under the auspices of QCRG, Ishikawa began teaching the QC Basic Course to industrial engineers, using translated British and American texts. "After conducting the first course," Ishikawa recalled, "it became clear to us that physics, chemistry, and mathematics are universal and are applicable anywhere in the world. However, in the case of quality control, or in anything that has the term 'control' attached to it, human and social factors are strongly at work. No matter how good the American and British methods may be, they cannot be imported to Japan as they stand. To succeed, we had to create a Japanese method" (Ishikawa 1985, 16–17). Ishikawa, with others, developed that Japanese method in the 1950s through their direct involvement with Japanese manufacturing companies, particularly in the fledgling automobile industry (Mizuno 1984; Nonaka 1995, 143).

What was different about Japanese conditions that made it necessary to "create a Japanese method"? And how by the 1970s and 1980s did that Japanese method become the world's most powerful manufacturing approach for setting new standards of high quality and low cost? In particular, how did Japanese manufacturing for mass markets differ from the system that Americans had previously developed in the first half of the twentieth century when U.S. industry established itself as the world's leading mass producer?

The fundamental difference between the Japanese and American organization of mass production was on the shop floor. The American system of mass production that dominated the world economy by the mid-twentieth century was based on the production of long runs of identical units by expensive special-purpose machines tended by "semiskilled" operatives (Hounshell 1984). The transformation of the high fixed costs of these mass-production technologies into low unit costs of final products required the cooperation of these shop-floor workers in the repetitive performance of narrow manual functions needed to maintain the flow of work-in-progress through the interlinked mechanical system.

The American machine operatives themselves were not involved in either monitoring the quality of work-in-progress or searching for solutions to quality problems in the manufacturing process. By design, they were excluded from the process of organizational learning that generated the American system of mass production (Lazonick 1990, chs. 7–9). Reflecting the American practice of confining organizational learning to the managerial structure and developing technologies that displaced the need for skill on the shop floor, quality control evolved in the United States as a strictly managerial function.

Leading American mass producers were willing and able to provide greater employment security and higher wages to shop-floor workers to ensure their cooperation in keeping pace with the expensive high-speed, special-purpose machinery. These companies, that is, established incentives to gain the cooperation of operatives in the *utilization* of technology. But the managers of these companies were unwilling to grant these operatives any role in the development of technology. Rather they confined such organizational learning to the managerial structure. Indeed, in the American companies considerable managerial learning was devoted to organizing work and developing mass-production technologies (Lazonick 1990, chs. 7–10; Lazonick 1992, pt. 2).

In the post–World War II Japanese automobile industry, companies like Toyota and Nissan did not have the luxury of long runs. In 1950, reflecting Japan's low level of GDP per capita, the entire Japanese automobile industry produced 31,597 vehicles, which was about the volume that U.S. companies produced in one and a half days (Cusumano 1985, 75, 266). In that year, Nissan accounted for 39 percent of production and Toyota 37 percent, while for the industry as a whole, 84 percent of the vehicles produced were trucks (Cusumano 1985, 75). As production increased over the course of the 1950s, with cars becoming a larger proportion of the total, Nissan or Toyota had to produce an increasing variety of vehicles to survive. In responding to these demand-side conditions, therefore, these companies could not possibly achieve low unit costs by simply adopting American mass-production methods.

On the supply side, over the course of the twentieth century, Japanese industry had developed capabilities that enabled companies like Toyota and Nissan to develop and utilize technology in a profoundly different way. These companies could draw on a sizable supply of highly educated and experienced engineers. Many Toyota employees, for example, had accumulated relevant technological experience over the previous decades in working for a predecessor company that was Japan's leading producer of textile machinery (Mass and Robertson 1996). In addition, the automobile industry was able to attract many engineers who had gained experience in Japan's aircraft industry before and during the war (Wada 1995).

Before the war, moreover, many Japanese companies had integrated foremen into the structure of managerial learning so that they could not only supervise but also train workers on the shop floor. Whereas, in the United States, the foreman, as "the man in the middle," served as a buffer between the managerial organization and the shop floor, in Japan, the foreman was an integrator of managerial and shop-floor learning. From the late nineteenth century on, a prime objective of U.S. managerial learning had been to develop machine technologies that could dispense with the skills of craft workers on the shop floor. In contrast, in Japan, without an accumulation of such craft skills the problem that had confronted technology-oriented managers from the late nineteenth century had been to develop skills on the shop floor as part of a coordinated strategy of organizational learning.

The rise of enterprise unions in the early 1950s both reflected and enhanced the social foundations for this hierarchical integration. During the late 1940s, dire economic conditions and democratization initiatives gave rise to a militant labor movement of white-collar (technical and administrative) and blue-collar (operative) employees. The goal of the new industrial unions was to implement "production control": the takeover of idle factories so that workers could put them into operation and earn a living (Gordon 1985, 343; Hiwatari 1996; Moore 1983). As an alternative to the "production control" strategy of militant unions, leading companies created enterprise unions of white-collar and blue-collar employees. In 1950, under economic conditions deliberately rendered more severe by the Allied Occupation's anti-inflationary "policy," companies such as Toyota, Toshiba, and Hitachi fired militant workers and offered enterprise unionism to the remaining employees. The post–Korean War recession of 1953 created another opportunity for more companies to expel the militants and introduce enterprise unionism. The continued and rapid expansion of the Japanese economy in the "high-growth era" ensured that enterprise unionism would become an entrenched Japanese institution (Cusumano 1985; Gordon 1985, ch. 10; Halberstam 1986, pt. 3; Hiwatari 1996).

The prime achievement of enterprise unionism was "lifetime employment," a system that gave white-collar and blue-collar workers employment security to the retirement age of fifty-five or sixty. Foremen and supervisors were members of the union, as were all university-educated personnel for at least the first ten years of employment before they made the official transition into "management." Union officials, who were company employees, held regularly scheduled conferences with management at different levels of the enterprise to resolve issues concerning remuneration, working conditions, work organization, transfers, and production (Nakamura 1997; Shimokawa 1994, ch. 3).

These institutional conditions supported the integration of shop-floor workers into a companywide process of organizational learning. Top managers had ultimate control over strategic investments, and technical specialists designed products and processes, typically on the basis of technology borrowed from abroad. But, given these managerial capabilities, the unique ability of Japanese companies to transform borrowed technology to generate new standards of quality and cost depended on the integration of shop-floor workers into the process of organizational learning.

Through their engagement in processes of cost reduction, Japanese shop-floor workers were continuously involved in a more general process of improvement of products and processes that, by the 1970s, enabled Japanese companies to emerge as world leaders in factory automation. This productive transformation became particularly important in international competition in the 1980s as Japanese wages approached the levels of the advanced industrial economies of North America and Western Europe. During the 1980s and 1990s, influenced as well by the impact of Japanese direct investment in North America and Western Europe, many Western companies have been trying, with varying degrees of success, to implement Japanese high-quality, low-cost mass-production methods.

Especially since the 1980s, a huge English-language literature has emerged on Japanese manufacturing methods, much of it written by industrial engineers with considerable experience as employees of, or consultants to, manufacturing companies in Japan and the West. In addition, there is a growing body of academic research on the subject, although it tends to focus more on functional integration than on hierarchical integration. By summarizing this body of evidence here, I intend to show that, in comparison with the once-dominant American mass producers, a fundamental source of Japanese manufacturing success has been the hierarchical integration of shop-floor workers in the process of organizational learning and also indicate how, within Japanese companies, hierarchical integration contributed to the generation of higher quality, lower cost products as part of a process of organizational learning that included integration across specialized functions.

In a comprehensive account of Japan's manufacturing challenge, Kiyoshi Suzaki (1987), a former engineer at Toshiba who then turned to consulting in the United States, contrasts the operational and organizational characteristics of a "conventional" (traditional American) company and a "progressive" (innovative Japanese) company in the use of men, materials, and machines in the production process (see table 1).

In the generation of higher quality, lower cost products, the integration of Japanese shop-floor workers into the process of organizational learning contributed to (1) the more complete *utilization of machines,* (2) superior *utilization of materials,* (3) improvements in *product quality,* and (4) *factory automation.* In summarizing the ways in which hierarchical integration contributed to these innovative outcomes in Japan, I indicate how and why Japanese practice differed from the hierarchical segmentation of shop-floor workers that was, and still largely remains, the norm in American manufacturing.

Table 1. Operational and organizational characteristics of American and Japanese manufacturing

	American company	Japanese company
Operational characteristic		
Lot size	Large	Small
Setup time	Long	Short
Machine trouble	High	Low
Inventory	Large	Small
Floor space	Large	Small
Transportation	Long	Short
Lead time	Long	Short
Defect rate	High	Low
Organizational characteristic		
Structure	Rigid	Flexible
Orientation	Local optimization	Total optimization
Communication	Long chain of command	Open
Agreement	Contract-based	Institution-based
Union focus	Job-based	Company-based
Skill base	Narrow	Broad
Education/training	Low quality	High quality
Training	Insignificant	Significant
Supplier relations	Short-term/many competitors	Long-term/selected few

Source: Adapted from Kiyoshi Suzaki, *The New Manufacturing Challenge* (New York: Free Press, 1987), 233.

Utilization of Machines

In the decade after the war, the Japanese pioneered in cellular manufacturing—the placement of a series of vertically related machines in a U-shape so that a worker, or team of workers, can operate different kinds of machines to produce a completed unit of output. Used particularly for the production of components, cellular manufacturing requires that workers perform a variety of tasks, and hence that they be multiskilled.

The Japanese system differed from the linear production system used in the United States in which shop-floor workers specialized in particular tasks, passing the semifinished unit from one specialized worker to the next. Historically, this fragmented division of labor resulted from the successful strategy of American managers in the late nineteenth century to develop and utilize mechanized technologies that could overcome their dependence on craft contractors who had previously controlled the organization of work (Montgomery 1987). To supervise better the "semiskilled" workers who operated the new mechanized technologies, American managers then sought to confine adversarial shop-floor workers to narrow tasks. After the rise of industrial unionism in the 1930s, shop-floor workers used these narrow job definitions as a foundation for wage setting, thus institutionalizing this form of job control in collective bargaining arrangements.

The prevalence of adversarial bargaining and job control only served to increase the resolve of most U.S. corporate managers to keep skill and initiative off the shop floor in the decades after World War II. Meanwhile, developing and utilizing the capabilities of the multiskilled shop-floor worker in myriad ways, Japanese companies created new standards of quality and cost. This continuous improvement, which the Japanese called *kaizen* (Imai 1986), enabled Japanese companies to outcompete the Americans, even in their own home markets, even as Japanese wages rose and the yen strengthened in the 1980s and 1990s.

With the need to use mass-production equipment to produce a variety of products in the 1950s, Japanese companies placed considerable emphasis on reducing setup times. Long setups meant excessive downtime, which meant lost output. Once set in motion, a search for improvements often continued over years and even decades. For example, in 1945, the setup time for a thousand-ton press at Toyota was four hours; by 1971, it was down to three minutes. A ring-gear cutter at Mazda that took more than six hours to set up in 1976 could be set up in ten minutes four years later (Suzaki 1987, 43).

By the 1980s, the extent of the market that Japanese manufacturers had captured meant that small-batch production was no longer the necessity it had been thirty years earlier. But the ability of these companies to do what the Japanese call "single-digit" (under ten minutes) setups enabled them to use the same production facilities to produce a wide variety of customized products. Single-digit setups had become a powerful source of international competitive advantage.

The reduction of setup times involved the redesign of fixtures, the standardization of components, and the reorganization of work. Shop-floor workers had to be willing and able to perform as much of the setup operations as possible for the next product batch while machines were still producing the current product batch. The reorganization of work needed to reduce setups represented another productive activity that could take advantage of the incentive and ability of Japanese shop-floor workers to engage in a variety of tasks. The broader knowledge of the production process that these workers possessed was in turn used to find new ways to reduce setup times.

In the United States, in contrast, the problem of reducing setup times was neglected in part because of long runs and in part because of the unwillingness of American management to invest in shop-floor skills. In Japan a dynamic learning process was set in motion in which the learning of shop-floor workers was critical. In the United States, hierarchical segmentation meant that, when the production of long runs of identical output was no longer a viable competitive strategy, corporations were left without the skill bases required for reducing setup times.

If shop-floor skills can prevent downtime through quick setups, they also can do so through machine maintenance. Keeping machines trouble free requires the involvement of shop-floor workers in continuous inspection and daily maintenance as well as engineers to solve chronic problems and to train the shop-floor operatives. As Suzaki (1987, 123) has put it, "zero machine troubles can be achieved more effectively by involving operators in maintaining normal machine-operating conditions, detecting abnormal machine conditions as early as possible, and developing countermeasures to regain normal machine conditions. This requires development of a close working relationship among operators, maintenance crews, and other support people as well as skill development and training to increase the abilities of those involved."

In American mass production, shop-floor workers have not only lacked the skills to maintain machines, they have also been denied the right to maintain machines by managers who feared that, far from reducing downtime by keeping machines trouble free, such shop-floor intervention would be used to slow the pace of work. Indeed one role of first-line supervisors employed on American mass-production lines has typically been to ensure that production workers do not interfere with machine operations on the assumption that such intervention will make the machines more trouble prone.

Cellular manufacturing, quick setups, and machine maintenance all contribute to higher levels of machine utilization and lower unit costs. But ultimately unit costs are dependent on how quickly products can be transformed from purchased inputs into salable outputs. That is, unit costs depend on cycle time.

As Jeffrey Funk (1992, 197) described it on the basis of his experience working at Mitsubishi Electric Corporation for a year: "The reductions in cycle time were achieved through numerous engineer and operator activities." The engineers were primarily responsible for making systemwide improvements concerned with identifying and resolving production bottlenecks, and with developing "product families" of different types of chips that undergo the same processes, thus reducing setup times and eliminating mistakes. The operators were primarily responsible for identifying possibilities for localized improvements on the wafer and assembly lines. Each operator was in a working group that met once or twice a month, through which they made numerous suggestions for improvements, a high proportion of which were acted upon by engineers. Operators responsible for wafer furnaces contributed, for example, to improvements in the delivery, queuing, and loading systems, all of which reduced cycle time. At Mitsubishi Electric between 1985 and 1989, cycle time for semiconductor chips was reduced from 72 days to 33 days, even as the number of chip styles more than doubled to 700 and the number of package types assembled increased from 20 to 70.

A comparison of the Mitsubishi wafer department with a U.S. factory using similar equipment found that the Japanese factory produced four times the number of wafers per direct worker, employed fewer support workers per direct worker, had a higher ratio of output to input in the wafer process, and had a cycle time that was one-fourth of that achieved by the U.S. factory. These improvements, according to Funk (1992, 198–204), "lead to shorter cycle time, higher yields, less wafer breakage, and higher production of wafers per direct worker. The multifunctional workers enable Mitsubishi to employ fewer support staff. Since the direct workers perform many of the activities typically performed by support staff in a U.S. factory, the direct workers can determine which activities are most important and how to improve the efficiency of these activities."

Utilization of Materials

Perhaps the most famous Japanese management practice to emerge out of the "high-growth era" was the just-in-time inventory system (JIT). By delivering components to be assembled as they are needed, the carrying costs and storage costs of work-in-progress can be dramatically reduced. But JIT only works if the parts that are delivered just in time are of consistently high quality. JIT only yields lower unit costs when component suppliers, be they in-house or external subcontractors, have the incentive and ability to deliver such high-quality parts. It was to ensure the timely delivery of such high-quality components, for example, that in 1949 and 1950 the first step taken by Taichi Ohno in developing JIT at Toyota was to reorganize the machine shop into manufacturing cells that required multiskilled operatives (Wada 1995, 22).

In the Japanese assembly process, JIT demands high levels of initiative and skill from production workers. Using the *kanban* system, it is up to assembly workers to send empty containers with the order cards—or *kanban*—to the upstream component supplier to generate a flow of parts. The assembly worker, therefore, exercises considerable minute-to-minute control over the flow of work—a delegation of authority that American factory managers deemed to be out of the question in the post–World War II decades on the assumption that shop-floor workers would use such control to slow the speed of the line. To prevent a purported shortage of components from "creating" a bottleneck in the production process, American managers kept large buffers of in-process inventory along the line.

The Japanese assembly worker also has the right to stop the line when, because of part defects, machine breakdowns, or human incapacity, the flow of work cannot be maintained without sacrificing product quality. When a problem is discovered and a worker stops the line, a light goes on to indicate its location, and others in the plant join the worker who stopped the line in finding a solution to the problem as quickly as possible. To participate in this process, therefore, shop-floor workers must develop the skills to identify problems that warrant a line stoppage, and they must contribute to fixing the problem. Without hierarchical integration, JIT and *kanban* cannot work (Urabe 1988).

Product Quality

The willingness of Japanese companies to develop the skills of shop-floor workers led to a very different mode of implementing quality control in Japan from that used in the United States. Statistical quality control (SQC), as already mentioned, originated in the United States. In American manufacturing, however, SQC remained solely a function of management, with quality-control specialists inspecting finished products after they came off the line. Defective products had to be scrapped or reworked, often at considerable expense. Defects that could not be detected because they were built into the product would ultimately reveal themselves to customers in the form of unreliable performance, again at considerable expense to the manufacturing company, especially when higher quality competitors came on the market.

For American companies, from the 1970s on, the higher quality competitors were typically the Japanese. In Japan, the integration of shop-floor workers into the process of organizational learning meant that product quality could be monitored while work was in the production process, and thus that defects could be detected and corrected before they were built into the finished product. The result was less scrap, less rework, and more revenues from satisfied customers. In the 1950s, American managers could justify the exclusion of shop-floor workers from participation in quality control on the grounds that the SQC methods in use were too complicated for the blue-collar worker. Only more highly educated employees were deemed capable of applying these tools. Given the quality of education received by young Americans destined to be "semiskilled" factory operatives, the managers of U.S. companies had a point. With mass education being controlled and funded by local school districts, most future blue-collar workers received schooling of a quality that was consistent with the minimal intellectual requirements of repetitive and monotonous factory jobs. This correspondence between schooling and prospective skill requirements in hierarchically segmented workplaces helps to explain why to this day the United States ranks among the lowest of the advanced economies in terms of the quality of mass education and among the highest in terms of the quality of higher education.

In Japan, even in the 1950s, blue-collar workers for manufacturing companies were high school graduates. But as part of a national system of education of uniformly high standards, they received much the same quality education as those who would go on to university. Even then, the involvement of Japanese shop-floor workers in SQC was accomplished by making the methods more easily accessible to, and usable by, blue-collar workers. As Kaoru Ishikawa (1985, 18), the pioneer in the implementation of SQC in Japan, put it: "We overeducated people by giving them sophisticated methods where, at that stage, simple methods would have sufficed."

The reliance of Japanese companies on the skill and initiative of shop-floor workers for superior machine utilization and reductions in materials costs made these employees ideal monitors of product quality. Relying on this skill base, SQC became integral to the Japanese practice of building quality into the product rather than, as in the United States, using SQC to inspect completed products that had had defects built in.

In the 1960s the involvement of shop-floor workers in improving machine utilization, materials costs, and product quality became institutionalized in quality control (QC) circles. In addition to initiatives undertaken by individual companies to apply QC methods in particular factories, a series of radio broadcasts by JUSE in the late 1950s had diffused an awareness of the potential of quality control. Then, in 1960, JUSE issued a publication, *A Text on Quality Control for the Foreman,* which was widely used by first-line supervisors in the workplace (Ishikawa 1985, 21). The success of this publication led to a monthly magazine, *Quality Control for the*

Foreman (F*Q*C). In the process of gathering information for the magazine, JUSE found that, in many factories, foremen and workers had formed themselves into small groups to discuss quality control and its application to specific problems. The editorial board of F*Q*C (of which Ishikawa was the chairman), in issuing the following statement, effectively launched the QC circle movement:

> 1. Make the content [of FQC] easy for everyone to understand. Our task is to educate, train, and promote QC among supervisors and workers in the forefront of our work force. We want to help them enhance their ability to manage and to improve.
> 2. Set the price low to ensure that the journal will be within the reach of everyone. We want as many foremen and line workers as possible to read it and benefit from it.
> 3. At shops and other workplaces, groups are to be organized with foremen as their leaders and include other workers as their members. These groups are to be named QC circles. QC circles are to use this journal as the text in their study and must endeavor to solve problems that they have at their place of work. QC circles are to become the core of quality control activities in their respective shops and workplaces. (Ishikawa 1985, 138)

QC circles could be registered with and announced in, F*Q*C. Beginning in 1963, a national QC circle organization was created, complete with central headquarters, nine regional chapters, conferences, seminars, and overseas study teams. Twenty years later there were almost 175,000 QC circles registered with nearly 1.5 million members (Ishikawa 1985, 138–39; Nonaka 1995).

QC circles became extremely effective in generating continuous improvements in the quality and cost of Japanese manufactured products. In participating in the continuous improvement of these production systems, shop-floor workers did not solve problems in isolation from the rest of the organization but rather as part of a broader and deeper process of organizational learning that integrated the work of engineers and operatives. The foreman as team leader served as the conduit of information up and down the hierarchical structure.

The QC circle movement, led by JUSE, helped to diffuse throughout Japanese industry the organizational and technological advances made at the leading companies. For example, in the mid-1960s there were frequent breakdowns of a newly installed automatic metal-plating machine in the assembly division of Toyota's Motomachi Plant. The relevant QC circle systematically considered possible causes and through testing came up with solutions. In reporting the work of this QC circle, F*Q*C stated: "The supervisor may understand the design of the machine and how to run it, but is probably unaware of its detailed tendencies or weaknesses. The people who know best about the condition of the machine are the workers, and quality circles provide an opportunity to get important information from them" (quoted in Nonaka 1995, 154).

In solving problems in machine utilization, QC circles found that the solutions invariably entailed improvements in product quality as well. As Izumi Nonaka (1995, 151) has put it in his account of the history of quality control at Toyota and Nissan: "Toyota production methods, such as just-in-time, kanban, and jidoka (automation) are well known, but it should be stressed that, in relation to quality control, if 100 per cent of the parts reaching a given process are not defect free, Toyota methods will not work smoothly. In other words, quality is the foundation of Toyota production methods. From about 1963, just-in-time and jidoka were adopted in all Toyota factories, and a close relationship between these methods and quality was immediately established."

The QC circle movement focused Japanese workers on the goal of achieving "zero defects"—detecting and eliminating defects as the product was being built rather than permitting defects to be built into the product. In recounting why an incipient zero-defect (ZD) movement (initiated by the U.S. Department of Defense for its contractors) failed in the United States in the mid-1960s, Ishikawa put the blame squarely on the failure of American companies to integrate shop-floor workers into the process, as was being done in Japan. "The ZD movement became a mere movement of will," Ishikawa (1985, 151–52) observed, "a movement without tools . . . It decreed that good products would follow if operation standards were closely followed." In the Japanese quality control movement, however, it was recognized that "operation standards are never perfect. . . . What operations standards lack, experience covers. In our QC circles we insist that the circle examine all operation standards, observe how they work, and amend them. The circle follows the new standards, examines them again, and repeats the process of amendment, observance, etc. As this process is repeated there will be an improvement in technology itself."

Not so, however, in the United States, where management practice "has been strongly influenced by the so-called Taylor method." In the United States, according to Ishikawa (1985, 151–52), "engineers create work standards and specifications. Workers merely follow. The trouble with this approach is that the workers are regarded as machines. Their humanity is ignored. [Yet] all responsibilities for mistakes and defects were borne by the workers . . . No wonder the [ZD] movement went astray."

In the late 1960s and early 1970s, on the eve of the Japanese challenge to U.S. manufacturing, many American industrial managers began to worry not so much about the quality of the products they were generating as about the quality of shop-floor work itself. The alienated worker was fingered as the source of lagging productivity (U.S. Department of Health,

Education and Welfare 1972; Walton 1979). During the first half of the 1960s, the annual average rate of increase in manufacturing productivity in the United States had been 5.1 percent while that of manufacturing wages had been 3.9 percent. But in the second half of the 1960s, when the annual rate of increase of manufacturing productivity averaged a mere 0.6 percent, manufacturing wages rose at a rate of 5.9 percent (Lazonick 1990, 280–84). Amid an escalation of absenteeism and unauthorized work stoppages, the productivity problem sparked a search among U.S. manufacturing companies for new structures of work organization that would secure the cooperation of shop-floor workers in realigning the relation between work and pay.

Within the automobile industry, the United Auto Workers joined corporate management in a National Joint Committee to Improve the Quality of Worklife. The problem was to convince workers that programs of "job enrichment" and "job enlargement" were not merely new ways to speed up production and reduce employment. Unfortunately, during the 1970s, even many promising experiments at work reorganization that had already yielded significant productivity gains were cut short when middle managers and first-line supervisors realized that the ultimate success of the programs would entail a loss of their power in the traditional hierarchically segmented organization (Marglin 1979; Walton 1975; Zimbalist 1975). Indeed, in general, the more pervasive response to the productivity problem in American manufacturing in the 1970s was an increase in shop-floor supervision rather than the transformation of work organization. From 1950 to 1970, the number of foremen per 100 workers in American manufacturing increased from 3.4 to 4.8; by 1980, this ratio had shot up to 8.0 (Lichtenstein 1989, 166).

During the 1980s, in the face of intense and growing competition from the Japanese, many companies throughout the United States sought to introduce Japanese-style "quality programs" into their workplaces. In their comprehensive survey of available case studies of these "experiments in workplace innovation," Eileen Appelbaum and Rose Batt found that "U.S. companies have largely implemented innovations on a piecemeal basis and that most experiments do not add up to a coherent alternative to [traditional U.S.] mass production." They contended that "quality circles and other parallel structures [of work reorganization] were a 'fad' in the early 1980s and have since been discredited in most U.S. applications as either not sustainable or providing limited results. . . . The overwhelming majority of cases show that firms have introduced modest changes in work organization, human resource practices, or industrial relations—parallel structures such as quality circles involving only a few employees, a training program, or a new compensation system. We consider these to be marginal changes because they do not change the work system or power structure in a fundamental way" (Appelbaum and Batt 1994, 10; see also Cole 1989; Kochan, Katz, and Mower 1984; Lawler, Ledford, and Mohrman 1989).

The fundamental problem, I would argue, was lack of resolve on the part of those who governed these corporations to effect the organizational integration of "hourly" shop-floor workers and "salaried" managerial employees. What is more, it appears that hierarchical segmentation in U.S. industrial enterprises fostered functional segmentation. Distant from the realities of problem solving in the actual production process, U.S. technical specialists sought to solve problems by using the tools of their own particular disciplines, putting up barriers to communicating even with other specialists within the managerial organization, and throwing partially solved problems "over the wall" into the domains of other functional specialists (Clark and Fujimoto 1991; Funk 1992; Okimoto and Nishi 1994; Westney 1994).

In Japan, by contrast, the hierarchical integration of technical specialists in a learning process with production workers created lines of communication and incentives to solve problems in concert with other specialists. The result of functional integration for Japanese manufacturers, in comparisons with their competitors in the United States, has been not only superior product quality but also more rapid new product development.

The different ways in which quality control systems were implemented in Japan and the United States is a demonstration of the disparity in national organizational learning. In Japan, QC was embedded in the whole structure of organizational learning. It was, as Nonaka (1995) has put it, "the responsibility of all employees, including top and middle management as well as lower-level workers, from planning and design, to production, marketing, and sales . . . [in] contrast with the American reliance on specialist quality control inspectors." Ishikawa (1985, 23) has emphasized the functional segmentation of American QC inspectors: "In the United States and Western Europe, great emphasis is placed on professionalism and specialization. Matters relating to QC therefore become the exclusive preserve of QC specialists. When questions are raised concerning QC, people belonging to other divisions will not answer, they will simply refer the questions to those who handle QC.

In Western countries, when a QC specialist enters a company, he is immediately put in the QC division. Eventually he becomes head of a subsection, a section, then of the QC division. This system is effective in nurturing a specialist, but from the point of view of the entire business organization, is more likely to produce a person of very limited vision.

For better or for worse, in Japan little emphasis is placed on professionalism. When an engineer enters a company, he is rotated among different divisions, such as design, manufacturing, and QC. At times, some engineers are even placed in the marketing division.

Factory Automation

In the late 1970s, American manufacturers continued to attribute the mounting Japanese challenge to low wages and the persistent productivity problem at home to worker alienation. By the 1980s and 1990s, however, the innovative reality of the Japanese challenge became difficult to ignore, as the Japanese increased their shares of U.S. markets across a range of key industries, even as Japanese wage rates rose rapidly and the yen steadily strengthened.

Even then, there appeared to be a way out for U.S. manufacturers that did not require imitation of the Japanese way by building broader and deeper skill bases. Since the 1950s American management had envisioned the "Factory of the Future"—a completely automated production facility that would do away with the need to employ production workers altogether (Noble 1984, ch. 4). Yet, notwithstanding massive investments by U.S. corporations and the U.S. government in factory automation, attempts by American companies to create the "factory of the future" failed (Noble 1984; Thomas 1994).

In sharp contrast, building on their investments in broad and deep skill bases, and decades of continuous improvement of production processes, Japanese companies succeeded in this effort. By the end of 1992, the Japanese had installed about 349,500 robots compared to 47,000 in the United States and 39,400 in Germany (Tsuneta Yano Memorial Society 1993, 191). The Japanese also developed and utilized flexible manufacturing systems (FMS): computer-controlled configurations of semi-independent work stations connected by automated material handling systems—in advance of, and on a scale that surpassed, other nations (Jaikumar 1989). Japan's success in machine tools and factory automation reflected their leadership in the integration of mechanical and electronic technologies, or what since the mid-1970s the Japanese have called "mechatronics" (Hunt 1988; Kodama 1995, 193).

For example, in his case study of the introduction of FMS at the machine tool company, Hitachi Seiki, Ramchandran Jaikumar (1989, 126) found that the first two attempts, undertaken between 1972 and 1980, had failed because of insufficient coordination across functions. In 1980, therefore, the company set up the Engineering Administration Department that "brought together a variety of different functions from machine design, software engineering, and tool design." The new structure of organizational learning, which built on the lessons of the previous failures, led to success. The development teams on the two failed attempts had, according to Jaikumar (1989, 126),

> integrated the different components of their systems through machinery design rather than through general systems engineering concepts. They had

viewed flexible manufacturing systems as technical problems to be solved with technical expertise. The difficulty of evaluating trade-offs whenever conflicts arose over design specifications or procedures convinced Hitachi Seiki that it was problems of coordination among people that was stymieing systems development. The company realized that what was needed was to view FMS as a manufacturing problem to be solved with both manufacturing and technical expertise. Consequently, the third phase of FMS development at Hitachi Seiki was a radical departure from the previous two.

In his comparisons of Japanese and U.S. FMS in the first half of the 1980s, Jaikumar found that, even though the FMS installations in both countries contained similar machines doing similar kinds of work, the Japanese developed the systems in half the time and produced over nine times as many parts per system. Average annual volumes were about one-seventh of American practice, with much greater automation and utilization rates. "Differences in results," said Jaikumar (1989, 129), "derive mainly from the extent of the installed base of machinery, the technical literacy of the work force, and the competence of management. In each of these areas, Japan is far ahead of the United States." More specifically, he described how the Japanese developed the reliability of FMS to achieve untended (automated) operations and system uptime levels of over 90 percent, in the process transforming not only shop-floor technology but also the job of a "shop-floor operator."

> The entire project team remains with the system long after installation, continually making changes. Learning occurs throughout and is translated into ongoing process mastery and productivity enhancement. . . . Operators on the shop floor, highly skilled engineers with multifunctional responsibilities, make continual programming changes and are responsible for writing new programs for both parts and systems as a whole. Like designers, they work best in small teams. Most important, Japanese managers see FMS technology for what it is—flexible—and create operating objectives and protocols that capitalize on this special capability. Not bound by outdated mass-production assumptions, they view the challenge of flexible manufacturing as automating a job shop, not simply making a mass production line flexible. The difference in results is enormous. (Jaikumar 1989, 130)

Central to factory automation have been teams of highly educated and highly trained engineers who had mastered their technical specialties but who were also able and willing to integrate across electronic, mechanical, and chemical specialties. As stated earlier, that the Japanese could even consider entry into complex manufacturing industries such as automobiles and consumer electronics after World War II was because of the learning that their scientists and engineers had accumulated in the decades before as well as during the war. But the Japanese experience in hierarchical integration of traditional blue-collar workers into the development and utilization of manufacturing technology laid the basis for functional integration as technology became more and more complex.

The accumulated learning of Japan's scientists and engineers after the war was in and of itself no match for the North Americans'. Yet, during the postwar decades, Japanese scientists and engineers developed and utilized their collective capabilities in manufacturing as part of an organizational learning process that integrated the capabilities of shop-floor workers in making continuous improvements to the manufacturing process. In the 1980s and 1990s, this tradition of hierarchical integration played a significant role in fostering the functional integration that has been key to Japan's success, relative to the United States, in factory automation.

The importance of taking organizational learning to the shop floor also applies in the semiconductor industry, the most complex and automated of manufacturing processes. As Daniel Okimoto and Yoshio Nishi (1994, 193) argue in their excellent comparative study of Japanese and U.S. semiconductor manufacturing:

> Perhaps the most striking feature of Japanese R&D in the semiconductor industry is the extraordinary degree of communication and "body contact" that takes place at the various juncture and intersection points in the R&D processes—from basic research to advanced development, from advanced development to new product design, from new product design to new process technology, from new process technology to factory-site manufacturing, from manufacturing to marketing, and from marketing to servicing. Owing to pragmatic organizational innovations, Japanese semiconductor manufacturers have excelled—where many American and European manufacturers have faltered—at the seemingly simple but extremely difficult task of making smooth "hand-offs" at each juncture along the long-interconnected R&D pipeline.

The key links in this pipeline in Japanese semiconductor R&D are between divisional labs and factory engineering labs. Engineers from these labs, according to Okimoto and Nishi (1994, 195), "continually meet and interact in seeking to iron out problems that inevitably arise in mass-manufacturing new products."

Okimoto and Nishi continue, stressing the importance of the integration of R&D with manufacturing:

> The largest concentration [of engineers] is usually found at the FELs [factory engineering laboratories], located at factory sites where the messy problems of mass production have to be worked out. The majority of Japanese engineers have at least some exposure to manufacturing engineering as part of their job rotation and career training. Not only is there no stigma attached to manufacturing assignments; the ladder of promotion leading up to higher reaches of executive management—and beyond (including

amakudari, or post-career executive entry into new companies)—pass through jobs that involve hands-on manufacturing experience. It is almost a requirement for upward career and post-career mobility.

In the United States, by contrast, manufacturing engineers carry the stigma of being second-class citizens. To the manufacturing engineers falls the "grubby" work of production—for which they receive lower pay and lower prestige compared with the "glamorous" design jobs. In how many U.S. semiconductor companies can it be said that the majority of engineers are engaged in manufacturing? Few, if any. And, looking at the large number of merchant semiconductor houses in Silicon Valley, we see that only a minority even possess manufacturing facilities, much less factory engineering laboratories. (Okimoto and Nishi 1994, 195)

It would appear more generally that, by focusing the skills and efforts of engineers on continuous improvements in quality and cost in the production process, hierarchical integration provided a foundation for functional integration in Japanese manufacturing. If, in the first half of the 1980s, most Western analyses of the sources of Japanese competitive advantage focused on the integration of the shop-floor worker into the organizational learning process, over the last decade or so the emphasis has shifted to the role of "cross-functional management," "companywide quality control," or "concurrent engineering" in generating higher quality, lower cost products. Much of the discussion of functional integration has been focused on its role in "new product development" from an international comparative perspective (Clark and Fujimoto 1991; Nonaka and Takeuchi 1995). But, I would argue, the key to understanding the influence of functional integration on innovation and international competitive advantage is the integration of product and process development, and the skill-base strategy that such integration entails (for an insightful comparative analysis of organization integration with suppliers in the Japanese automobile industry, see Sako 1998). Such an understanding of organizational integration requires an analysis of functional integration in relation to the legacy of hierarchical integration or segmentation.

International Competition

If valid, the skill-base hypothesis can reconcile the fact that many U.S. industrial enterprises still remain innovators in international competition even though income inequality has become worse in the United States. A systematic bias of U.S. industrial corporations toward competing for product markets by investing in narrow and concentrated skill bases could provide a significant explanation for the income inequality trends over the last two decades or so. Testing the skill-base hypothesis may help provide answers to a number of related questions concerning the ways in which, in particular industries and activities, U.S. industrial corporations have responded to international competitive challenges:

To what extent have U.S. companies exited from particular industries, and particular activities within a particular industry, in which they have been challenged by enterprises that have invested in broader and deeper skill bases as an alternative to transforming their strategies and structures to make the requisite investments in organizational learning?

To what extent have the attempts of U.S. companies to respond to these competitive challenges been hampered by their failure to confront and transform sufficiently the strategic, functional, and hierarchical segmentation that they have inherited from the past?

What can we learn about the incentive and ability of U.S. companies to make investments in broader and deeper skill bases by comparing the strategy, organization, and performance of different companies in the same industry—for example, Ford, GM, and Chrysler in automobiles—that have sought to respond to the same international competitive challenges?

What has been the importance of foreign direct investment—for example, Japanese "transplants" in the United States—as distinct from international trade in shaping the responses of U.S. companies to international competitive challenges?

What has been distinctive about the investment strategies and organizational structures of U.S. companies that have become or remained leaders in international competition in the 1980s and 1990s? Did an historical legacy of investments in broader and deeper skill bases, and a relative absence of organizational segmentation, enable an older company like Motorola or 3M to continue to make such investments in the 1980s and 1990s, thus representing the exceptions that prove the rule in U.S. industry? Have newer companies such as Intel and Microsoft become world leaders through the organizational integration of narrow and concentrated skill bases?

Such questions indicate that testing the skill-base hypothesis and its immediate implications requires in-depth research of particular companies that compete in particular industries in different national economies in different and, typically, over prolonged periods of time. The more limited objective of this study has been to elaborate the analytical framework for testing the skill-base hypothesis by synthesizing available evidence on differences in organizational learning in industries in which the United States and Japan compete head-to-head.

What are those industries, and how has competitive advantage been shifting between the United States and Japan? Tables 2a-c show the structure of bilateral Japanese–U.S. trade from 1979 to 1995. As useful as these

Table 2a. Japan–U.S. bilateral merchandise trade, 1979, 1987, and 1995 (in millions of current U.S. $)

	1979		1987		1995	
	Exports	Imports	Exports	Imports	Exports	Imports
Foodstuffs	189.0	4,422.9	404.1	6,778.9	303.4	15,951.4
Raw materials	136.5	6,927.3	167.3	7,039.8	380.9	9,329.1
Light goods	2,200.6	1,660.7	6,465.5	3,037.6	7,979.4	8,745.8
Chemical goods	653.1	2,053.3	2,080.8	4,035.3	4,826.2	7,072.7
Metal goods	3,939.6	481.1	4,101.8	901.0	4,045.1	2,190.4
Machinery	19,008.3	4,310.2	69,493.9	9,075.4	100,182.5	30,515.6
Office machines	679.9	530.1	7,373.7	1,589.9	14,183.7	4,862.5
Electrical machinery	4,393.3	1,349.9	17,050.1	3,008.9	29,384.8	12,746.4
Transportation equipment	10,106.4	985.5	32,050.3	1,854.7	32,023.9	5,987.7
Precision instruments	1,515.9	357.9	4,325.0	620.1	6,545.7	1,844.5
Reexports, unclassified	275.4	575.3	866.5	622.5	3,141.4	1,603.0
Total	26,402.5	20,430.8	83,579.9	31,490.5	120,858.9	75,408.1

Table 2b. Percentage of Japan–U.S. trade growth, 1979–1995 (1979 = 100%)

	Japanese Exports to United States			United States Exports to Japan		
	1979	1987	1995	1979	1987	1995
Foodstuffs	100	214	161	100	153	361
Raw materials	100	123	279	100	102	135
Light goods	100	294	363	100	183	527
Chemical goods	100	319	739	100	197	344
Metal goods	100	104	103	100	187	455
Machinery	100	366	527	100	211	708
Office machines	100	1,085	2,086	100	300	917
Electrical machinery	100	388	669	100	223	944
Transportation equipment	100	317	317	100	188	608
Precision instruments	100	285	432	100	173	515
Reexports, unclassified	100	315	1141	100	108	279
Total	100	317	458	100	154	369

Table 2c. Proportionate shares of Japan–U.S. bilateral merchandise trade, 1979, 1987, and 1995 (percentage of annual bilateral exports)

	Japanese Exports to United States			United States Exports to Japan		
	1979	1987	1995	1979	1987	1995
Foodstuffs	0.7	0.5	0.3	21.6	21.5	21.2
Raw materials	0.5	0.2	0.3	33.9	22.4	12.4
Light goods	8.3	7.7	6.6	8.1	9.6	11.6
Chemical goods	2.5	2.5	4.0	10.1	12.8	9.4
Metal goods	14.9	4.9	3.3	2.4	2.9	2.9
Machinery	72.0	83.1	82.9	21.1	28.8	40.5
Office machines	2.6	8.8	11.7	2.6	5.0	6.4
Electrical machinery	16.6	20.4	24.3	6.6	9.6	16.9
Transportation equipment	38.3	38.3	26.5	4.8	5.9	7.9
Precision instruments	5.7	5.2	5.4	1.8	2.0	2.4
Reexports, unclassified	1.1	1.0	2.6	2.8	2.0	2.1
Total	100.0	100.0	100.0	100.0	100.0	100.0

Source: Ministry of International Trade and Industry, *White paper on international trade* (Tokyo: Ministry of International Trade and Industry), 1980, 1988, 1996.

data are as points of departure, they have important limitations for defining the comparative case studies needed to test the skill-base hypothesis. The importance of foreign direct investment, cross-border outsourcing, and third-country exports means that trade data provide only a partial picture of shifts in head-to-head competitive advantage. Moreover, as we shall see, for example, in the category of "aircraft engines and parts," hidden within a narrowly defined industrial classification of traded goods, may be important international divisions of labor that reflect investments in different types of skill bases.

In 1995 Japan exported $120.9 billion of goods to the United States (27.3 percent of all Japanese exports) and imported $75.4 billion from the United States (22.4 percent of all Japanese imports) for a merchandise trade surplus of $45.5 billion. The United States is by far Japan's foremost trade partner for both exports and imports. Japan's next-largest trade partners in 1995 were, for exports, South Korea (7.1 percent of Japan's total) and, for imports, China (10.7 percent of the total) (*Nikkei Weekly* 1997, 107).

Of Japan's exports to the United States in 1995, 82.9 percent fell under the broad category of "machinery." This category included, among the major classifications, office machines (11.7 percent of all goods exports), electrical machinery (24.3 percent), transportation equipment (26.5 percent, of which automobiles were 18.3 percent and automobile parts 6.5

percent), and precision instruments (5.4 percent) (see table 2c). The remainder of Japanese exports to the United States consisted largely of chemical goods (4.0 percent), metal goods (3.3 percent), and light industrial products (6.6 percent).

What did the United States export to Japan? Machinery accounted for 40.5 percent of U.S. exports, consisting mainly of office machines (6.4 percent), electrical machinery (16.9 percent, of which semiconductors and integrated circuits were 7.1 percent), and transportation equipment (7.9 percent). The remainder of U.S. *manufactured* exports to Japan consisted mainly of an assortment of light products (11.6 percent, including textiles, paper products, records and tapes, and sporting goods) and chemical goods (9.4 percent). But all manufactured goods only accounted for less than two-thirds of U.S. exports to Japan. Over one-third of U.S. exports to Japan in 1995 were either foodstuffs (21.2 percent) or raw materials (12.4 percent). For Japan, foodstuffs and raw materials exports were only 0.6 percent of its total exports to the United States in 1995.

Note that, in the 1970s, as the Japanese challenge mounted, the United States was even more reliant than it would be in 1995, in relative terms at least, on exports of foodstuffs and raw materials to Japan. In 1979, 55.5 percent of U.S. exports to Japan took the form of these basic materials. In that year 65 percent of Japan's raw materials imports from the United States were soybeans (5.7 percent of total imports), wood (11.2 percent), and coal (5.0 percent). By 1995 Japan imported a somewhat larger quantity of soybeans (but the proportion of total imports fell to 1.5 percent), and absolutely smaller quantities of wood (4.2 percent) and coal (0.9 percent). Hence over the sixteen-year period, the relative importance of foodstuffs for U.S. exports to Japan was maintained, while the relative, and in some cases absolute, importance of raw materials declined.

U.S. agriculture is a case in point for the necessity of in-depth, industry-specific analyses of the sources of sustainable competitive advantage. Looking at the trade data, an economist might conclude that the importance of raw materials and foodstuffs in U.S. exports to Japan is simply a matter of very different land–labor ratios in the two nations' factor endowments. But to draw such a conclusion would be to miss the critical importance of collective and cumulative learning on a national scale over the past century in making agriculture the one industrial sector in which the international competitive advantage of the United States is most sustainable. Such a conclusion would neglect a century-long history of organizational learning in agriculture, akin to the managerial revolution that occurred within major U.S. industrial corporations. In the agricultural managerial revolution, the U.S. Department of Agriculture created a national system of research and development that diffused new technology to millions of farmers through the state-based activities of land-grant colleges, experiment stations, and county agents. Indeed, the legacy of this massive investment in organizational learning is not only productive supremacy in agriculture but also the world's foremost structure of industrial research institutions embedded in the U.S. system of higher education (Ferleger and Lazonick 1993, 1994).

Note also that the relative importance of machinery exports from Japan increased substantially in the first eight-year period, while the relative importance of U.S. machinery exports increased from 1979 to 1995, with the major gains being made in the late 1980s and early 1990s. The United States made these gains despite the continuing decline of its machine tool industry in the face of relentless Japanese competition. By 1991, compared with the U.S. machine tool industry, the value of Japanese machine tool production was up 356 percent and machine tool exports 443 percent (Tsuneta Yano Memorial Society 1993, 199). In the 1990s, the Japanese also successfully challenged the German machine tool manufacturers, surpassing them for the first time, in 1992, in the value of production, and in 1993, in the value of exports. Having captured larger and larger shares of export markets through 1996, Japanese companies now completely dominate the mid-range and high-range markets for CNC (computer numerically controlled) machine tools. The low-end markets have been left mainly to Taiwanese companies, and the high-end niches in non-CNC machine tools remain in the hands of the Swiss, the Germans, and, to a more limited extent, the Americans (Forrant 2000).

Between 1987 and 1995, the U.S. gains in machinery were mainly in integrated circuits (up 4.6 percent) and automobiles (up 3.9 percent), these two categories accounting for almost 75 percent of the increase in U.S. machinery exports as a proportion of total exports. Within the category of Japanese transportation-equipment exports, in 1985, 30.2 percent was of automobiles (3,278,724 vehicles), and auto parts made up another 6.2 percent in 1995, these figures were 18.2 percent (2,066,255 vehicles) and 6.6 percent respectively. The decline in Japanese exports reflected the Japanese strategy of foreign direct investment in automobiles, either directly in the United States or in Southeast Asian countries, such as Thailand and Indonesia, that then exported automobiles or parts to the United States. In 1985, Japanese automobile companies produced 254,000 cars and 107,000 trucks in the United States; in 1995, 1,942,000 cars and 414,000 trucks (*Nikkei Weekly* 1996, 151). In 1987, the leading U.S. industry within the transportation-equipment category was aircraft, which represented 5.0 percent of all exports. In 1995, aircraft had declined to 2.6 percent of U.S. exports to Japan, and had been surpassed by automobiles, which constituted 4.2 percent of U.S. exports (294,874 vehicles), up from only 0.3 percent (88,395 vehicles) in 1987.

It was mainly Japanese companies operating in the United States that were doing the exporting. Of just over 100,000 automobiles exported from the United States to Japan in 1994, 53,500 were from Honda, USA, and another 11,300 from Toyota USA. Only about 35 percent of the exports came from GM, Ford, and Chrysler (some of whose cars were produced through joint ventures with Japanese companies). The total number of cars exported to Japan by the three U.S. automakers was less than the number exported by Volkswagen–Audi and only about 60 percent of the combined sales of BMW and Mercedes-Benz in Japan. Each of the U.S. companies was also outsold in Japan by Rover, Opel (owned by GM), and Volvo (*Nikkei Weekly* 1996, 101, 103).

The United States and Japan almost balance trade within the classification "aircraft engines and parts" (Almeida 2001). Parts increasingly dominate the trade in aircraft engines, especially from Japan to the United States. The ability to integrate innovation in advanced materials with precision engineering has been key to Japan's growing success. Building on pioneering investments in the development of polyacryonitric carbon fiber by Toray Industries in the 1970s, three Japanese synthetic fiber producers now dominate 60 percent of the world market (Kodama 1995, 59–60; *Nikkei Weekly* 1997, 210; Suzuki 1994). Finding a market at first as a light and durable material for sports equipment such as tennis rackets and golf clubs, in the 1980s Japanese-made carbon fiber became a primary composite material used in both aircraft and engines. For example, Ishikawajima–Harima Heavy Industries—one of the three major Japanese companies involved in jet engine manufacture—currently produces carbon fiber blades for jet engines made by General Electric (Glain 1997). Japan's competitive advantage in producing these parts that combine advances in chemical and mechanical engineering would seem to derive from its investments in broad and deep skill bases.

Organizational integration also appears important in explaining trade in semiconductors. In 1995, Japanese exports of integrated circuits accounted for 6.2 percent of all Japanese exports to the United States (up from 1.4 percent in 1987), and hence represented one-quarter of the 1995 electrical machinery exports. This bilateral trade in integrated circuits reflects U.S. specialization in microprocessors and Japanese specialization in dynamic random access memories (DRAMs), an international division of labor built on investments in different skill bases in the two nations. Describing the "lagged parallel model" of new product development, pioneered at Toshiba and subsequently diffused to other Japanese enterprises as well as U.S.-based Texas Instruments, Okimoto and Nishi (1994, 197–98) have pointed out that

> the lagged parallel project model is effective for work only on certain types of technology. It works for DRAMS, SRAM [*sic*], and other commodity chips, which share highly predictable linear trajectories of technological advancement. The model is not particularly well suited for products based on

nonlinear, highly volatile technological trajectories, where the parameters of research for the next and successive product-generations cannot be understood ahead of time. Thus, it is not accidental that Japanese companies have dominated in commodity chips but have lagged behind U.S. companies in logic chips, microprocessors, and software for applications and operating systems. The latter may require a different, perhaps less structured, organizational approach.

As for computers, American success in PCs and packaged, standardized software does not mean that the Japanese have not been successful competitors. U.S. government agencies, including the military, have been buying supercomputers from the Japanese. The success of a company like Toshiba in laptop computers reflects Japan's long-standing success at miniaturization, a technological advance that requires the integration of design and manufacturing. Japan also dominates international competition in liquid crystal displays (LCDs), a technology invented by RCA in 1967, but developed from the early 1970s most successfully by the Japanese company Sharp in a growing number of applications. By 1992, Sharp controlled 38 percent of the world's rapidly growing market for LCDs (Johnstone 1999, ch. 3; Kodama 1995, 56–58).

In the United States, there is growing evidence that, even in industries such as jet engines and medical equipment, the trend in the United States is out of manufacturing and even design, and into the low-fixed-cost and highly lucrative business of servicing high-technology equipment (Lazonick and O'Sullivan 2001b). A recent hostile takeover attempt of Giddings & Lewis, the largest machine-tool maker in the United States, by another American company, Harnischfeger, had as its objective the shedding of the target's business of manufacturing machine tools for the automotive industry so that the company could focus on servicing installed machinery (*Wall Street Journal* 1997). In the end, a "white knight," the German company, Thyssen, acquired Giddings, promising to maintain its manufacturing business. But the fact is that considerable money can be made by taking a reputable manufacturing company and turning it into a servicing company.

Precisely because the United States has been a leader in industries such as jet engines, medical equipment, and machine tools, the nation has a huge accumulation of experienced technical specialists, many of whom no longer have as secure employment with equipment producers as they had in the past. Some of these people are finding continued employment servicing the equipment that their companies used to both produce and service. In the past, they acquired these skills through organizational learning. But their utilization of these skills today confines them to narrow and concentrated functions that remove them even farther from the processes of organizational learning that will drive innovation in the future.

In the absence of indigenous manufacturing capability and organizational learning in these industries, where will the next generation of American high-technology service specialists accumulate new state-of-the-art skills? The U.S. economy has a vast accumulation of high-technology skills that derives from the organizational learning that took place in managerial structures over the past century, and on which it can live, and even innovate, for some time into the future. But, if instead of using this organizational learning to build broader and deeper skill bases, American businesses move toward relying on even narrower and more concentrated skill bases, the trends toward income inequality of the last two decades will continue.

The University and Regional Economic Development

If major U.S. corporations do indeed exhibit a pronounced bias toward investing in narrow and concentrated skill bases, such investment behavior poses a problem for regional economic development. Through a combination of education and employment, people who make up a relatively small portion of the population will acquire valuable productive capabilities and be in positions to reap the benefits of economic growth. But a larger and larger proportion of the population will not acquire such capabilities, and the remunerative employment of these large numbers of people will not be sustainable. It also appears that over an ever-widening range of activities and technologies, sustained competitive advantage is going to those enterprises and industries that are investing in broader and deeper skill bases. If so, then narrow and concentrated learning collectivities may find that the array of product markets in which they can remain competitive on the basis of such an investment strategy is shrinking over time. A proliferation of narrow and concentrated skill bases may not remain capable of generating innovation, let alone sustainable prosperity.

Much research remains to be done to identify the sources of the growing inequality of income that Americans have experienced since the 1970s. But the research that has been done strongly suggests an important cause of the phenomenon is the failure of U.S. corporations to invest in broad and deep skill bases in the face of innovative foreign competitors who have. But why should such an investment bias be systematic across U.S. industrial corporations? If, as is the case, strategic decision makers of major corporations have considerable autonomy over how they allocate corporate resources, why is *the nation* a relevant unit of analysis?

An integral determinant of the investment behavior of U.S. corporations, in my view, has been the evolution of U.S. social institutions—particularly financial and educational institutions—in ways that systematize the bias of U.S. business enterprises against broadening and deepening the collective and cumulative learning processes in which they invest. As Mary O'Sullivan and I have documented elsewhere, during the 1980s and 1990s the U.S. financial system has become increasingly powerful in enforcing financial liquidity on U.S. corporations, as manifested by downsizing, high payout ratios, and stock repurchases. These corporations still invest in organizational learning, but given their stated objective of "maximizing shareholder value," they seek to do so by investing in narrow and concentrated skill bases (Lazonick and O'Sullivan 1996, 2001a).

The structure of the U.S. educational system supports a strategy of investing in narrow and concentrated skill bases. The United States possesses a system of higher education that attracts students from around the world. Indeed, foreign competitors send their graduate students and scientists in increasing numbers to American universities to bring back home the wealth of knowledge that these institutions impart through both education and research. At the same time, however, the primary and secondary schooling systems in the United States are not providing large segments of the American population with the capacity to take advantage of the nation's system of higher education. In international comparisons of achievement levels in the mass schooling system, especially in math, the United States has not fared well over the past few decades. The K–12 educational system equips some Americans to take advantage of the knowledge-creation opportunities in higher education. But the public system of primary and secondary education does not adequately prepare most Americans—even many of those who do attend and graduate from institutions of higher education—to make productive use of their capabilities and engage in life-long learning.

Within this institutional structure and competitive environment, a technologically oriented university embedded in a regional industrial economy can be, depending on its own choice of investment strategy, part of the problem or part of the solution. From the perspective that I have presented here, the answer depends on how the university's activities influence organizational segmentation or integration. Do the activities of the university reinforce conditions of organizational segmentation that pose barriers to regional innovation in international competition? Or do its activities foster conditions of organizational integration that support new structures of collective and cumulative learning in the regional economy?

In light of the organizational requirements of international competition that I have outlined in this essay, let us consider which types of activities would make the university a reinforcer of organizational segmentation or a catalyst of organizational integration. To draw out the implications of my analysis of business investment and international competition for the role of the university in generating sustainable prosperity, I discuss these issues briefly in terms of the distinctive dimensions of integration and segmentation, hierarchical, functional, and strategic, which determine the collective capability of an organization.

Hierarchical Integration

In the first decades of this century the system of higher education in the United States was transformed to channel technical and administrative personnel into managerial organizations within both the business and state sector (Ferleger and Lazonick 1994; Noble 1977). This transformation was central to the process that organizationally segmented managerial employees from shop-floor workers. Such organizational segmentation is not inherent in a reliance on higher education as a pre-employment qualification for managerial personnel. In corporate Japan, for example, the requirements that managerial employees have a university education and shop-floor employees a high school education only are observed even more strictly than in the United States (Dore and Sako 1998). In the United States, however, concomitant to the transformation of higher education to generate a supply of managerial employees was the transformation of the production processes of major corporations to exclude shop-floor workers from the processes of organizational learning (Lazonick 1990; chs. 7–10).

It was this widespread corporate strategy to deskill shop-floor labor while investing in managerial skill bases that explains the otherwise paradoxical coexistence in the United States of the world's leading institutions of higher education beside a world-lagging system of K–12 public education (Ferleger and Mandle 1992, 43, 45–46; Ferleger and Mandle 1994, tbls. 3–3, 3–4 on 21–22). Notwithstanding the American rhetoric of equal opportunity, there was no need in the United States to create a system of uniformly high-quality public education when a substantial proportion of the population was expected to spend a lifetime performing routinized jobs devoid of the potential, or even the requirement, for learning. At the same time, the concentration of collective and cumulative learning in managerial organizations that were the foundations of American prosperity made it imperative that resources be allocated at national level to a high-quality system of higher education.

As I have detailed elsewhere (Ferleger and Lazonick 1993), from the late nineteenth century, coterminous with the managerial revolution in American industry, a national system of higher education emerged from a national strategy to develop technologies critical to the productivity of the agricultural sector. Initially, the major investments in the new system of higher education were in the publicly endowed and funded "land-grant" colleges rather than in the elite "Ivy League" universities. But, especially after the turn of the century, when corporate enterprises increasingly recognized the importance of the higher education system for supplying both personnel and research to their organizations, the elite universities transformed themselves to compete for corporate and government support on the land-grant model. In the process, by the 1920s and 1930s, higher education had become a prime route for the social mobility of individuals within the American economy. In the post–World War II decades, the GI Bill, the civil rights movement, and the "War on Poverty" helped open this avenue of mobility to larger and larger proportions of the population. In the 1950s, 1960s, and even the 1970s, there were growing numbers of well-paid and stable blue-collar jobs available in the U.S. economy that did not require any higher education. But if one wanted employment that provided not only good pay and stability but also learning opportunities and social status, one had to get a higher education. The numbers of Americans who went on to higher education expanded enormously in the post–World War II decades. In 1954 (subsequent to a large increase in enrollments under the GI Bill), the number of first degrees granted were 129 per 1,000 persons twenty-three years old; in 1969 (just before a falloff because of the Vietnam War draft), this number was 289 (U.S. Bureau of the Census 1976, 385). Between 1980 and 1994, the percentage of high school graduates who enrolled in college increased from 49.3 percent to 61.9 percent, even as the proportion of high school students who failed to graduate dropped from 12.0 percent to 9.5 percent (U.S. Department of Commerce 1996, 176–80). Over this period, college enrollments in the United States increased from 12.1 million to 14.3 million students, an increase of 18 percent, whereas the number of public high school graduates *declined* by 19 percent (U.S. Department of Commerce 1996, 179, 181).

As it has catered to an ever-growing population of students from diverse backgrounds and with diverse employment prospects, the quality of the U.S. system of higher education has become more unequal. This inequality reflects both the uneven quality of public education across the school districts from which different kinds of institutions of higher education drew their students and the unequal access of students from different socio-economic backgrounds to institutions of different quality. Institutions of higher education that drew students from predominantly working-class school districts had to provide many of these students with "remedial" education—teaching them what they should have learned, but did not or could not, learn in high school. This accommodation of the system of higher education to inequality in high school education has taken some of the pressure off the K–12 system to upgrade and equalize its quality.

Accompanying the increasing rates of enrollment in higher education in the 1980s and 1990s has been the decline in the availability of stable and remunerative blue-collar jobs. If, on the supply side, the quality of higher education was preparing the larger student body for a lifetime of collective and cumulative learning, and if, on the demand side, the employment opportunities to engage in collective and cumulative learning were expanding in the economy, the concomitant shift out of traditional blue-collar work and toward higher education could be laying the foundations for sustainable prosperity. Unfortunately, in the United States in the 1980s and 1990s, neither of these conditions held—leaving the institutions of higher education that are attracting the new enrollees to cope with an untenable situation. These institutions—and UMass Lowell is surely one of them—are expected to ready an ill-prepared student body for a world of innovative international competition that the strategic decision makers in corporate America are trying to avoid. In the absence of a regional, if not national, strategy, to upgrade public education and a regional, if not national, strategy to invest in broader and deeper skill bases, such a university faces an impossible task.

Yet—and here we can see the paradox of the entire U.S. educational system within many individual institutions of higher education—the same university can be simultaneously educating small numbers of students who will be prepared for lifelong learning and large numbers of students for whom neither their time spent in college nor in future employment will be a learning experience. The same university that, in order to maintain its enrollments and income, is forced to allocate most of its resources to remedying the deficiencies of the national K–12 system, may also possess a wealth of knowledge, especially in its graduate faculties. Especially in fields where the university has a history of relevance to industrial innovation, it can be a highly beneficial resource for students who have the high-quality educational backgrounds and business enterprises that have made the investments in innovative capabilities required to absorb this wealth of knowledge.

Not surprisingly, given the quality of the U.S. system of public education, growing numbers of these students come from abroad. Between 1980 and 1994, the number of foreign students ("nonresident aliens") enrolled at U.S. institutions of higher education increased by 50 percent, with increases in 78 percent of those from Asia and 170 percent of those from Europe. In 1980, foreign students represented 2.5 percent of all enrollments; in 1994, 3.1 percent. In 1994, 65 percent of foreign students came from Asia and 14 percent from Europe (U.S. Department of Commerce 1996, 181). And it was predominantly higher degrees that they were after. Of degrees awarded in higher education in 1993, foreign students accounted for 3.8 percent of the bachelor's degrees (up from 2.1 percent in 1981), 12.0 percent of masters' degrees (7.5 percent in 1981), and 27.3 percent of doctoral degrees (12.8 percent in 1981) (U.S. Department of Commerce 1996; 181, 190, 194).

Without this influx of foreign students, it may well be that many if not most graduate programs, particularly in engineering and science, could not survive. Of course, many of these foreign students, upon graduation, remain in the United States as employees. But their availability makes it all the easier, and in some industries, even makes it possible, for U.S. corporations to choose to invest in narrow and concentrated skill bases. And many of those who take up employment in the United States eventually return to their home countries, with high levels of specialist expertise that can then be integrated into broader and deeper skill bases.

Functional Integration

Even in those areas where the university has the resources to contribute to industrial innovation, there is reason to believe that the learning that the university has to impart to the economy has become limited by functional segmentation. Within enterprises, as already described, functional segmentation poses barriers to the collective and cumulative learning that generates higher product quality and more rapid product development.

It is more difficult to measure the impacts of functional segmentation within the university, but it clearly exists. Academic disciplines and subdisciplines have developed their own specialized modes of communication, fundamental assumptions, and realms of inquiry that create barriers to the integration of cross-disciplinary knowledge, let alone interdisciplinary research. Traditional departmental structures and course offerings are vigorously defended, the specialized research focus becomes narrower and narrower over time, and interdisciplinary work is accorded little professional status. Ensuring the insularity of the particular "domain" of research becomes, in effect, a professional goal.

For faculty members in a functionally segmented intellectual environment, the most potent source of individual advancement derives from interuniversity labor mobility, or at least the credible threat of such mobility, rather than from engaging in collective and cumulative learning processes. And the basis of interuniversity mobility is the approval of the individual by a small coterie of fellow subdisciplinarians. In the end, the academic disciplines run the universities; the leading members of their subdisciplines run the academic disciplines. Specialization runs amok. Integration of specialized knowledge fades from the academic scene.

As a result, it is very difficult for a university as an organization to set its own intellectual strategy to engage in organizational learning. But if the members of the university cannot engage in a collective and cumulative learning process, how would they have any unique intellectual resources to contribute to the regional economy to foster sustainable prosperity? Indeed, in the presence of functional segmentation, when the specialized knowledge that is generated at the university is relevant to industrial enterprises, it is most likely because it reinforces the structures of functional segmentation within those enterprises. The university–industry relationship may remain a powerful means of developing highly specialized knowledge, but the very access to this knowledge may make it all the less necessary for enterprises to invest in broader and deeper skill bases in order to

compete. In many cases, moreover, the highly specialized technical and social knowledge generated in the university may simply be irrelevant to the
regional economy because the narrow and concentrated subdisciplines in
which this knowledge develops are entirely cut off from any technical or
social application.

For those academics who seek to influence regional economic development, the challenge is to break down and overcome functional segmentation within their own organizations. In doing so, they cannot simply bypass
the existing bodies of specialized knowledge. Rather they must understand
thoroughly the intellectual foundations on which this knowledge is built
to probe its strengths and weaknesses. For those who want to engage in
collective and cumulative learning within the university, the critical questioning of the existing bodies of specialized knowledge, including those
that manifest functional segmentation, is a prelude to the integration of
their own specialized knowledge into a collective and cumulative learning
process. The challenge then is to bring together people who understand the
foundations of existing specialized knowledge, and who have a commitment to organizational learning about the sources of regional economic
development as a major intellectual goal. To set this process in motion requires that the university have a strategy to invest in organizational learning, which, as in a business enterprise requires the organizational integration of, and financial commitment to, the specialized division of (in this
case academic) labor to achieve common goals.

Organizational learning must involve the learners in the evolution and
transformation of the subject that they are studying. To contribute to the
sustainable prosperity of the region, academics must motivate, engage in,
and assess the relevance of, their own organizational learning through exposure to the "shop floor"—the producers in their region. The danger is
that academics will offer their functionally segmented knowledge to functionally segmented enterprises. If so, there may well exist a university–
industry symbiosis, but it will be one that is part of the problem rather
than part of the solution. To be part of the solution, academics committed
to regional development must themselves become organizationally integrated into a regional strategy to generate sustainable prosperity.

Strategic Integration

Innovative strategy is a process that allocates resources to collective and
cumulative learning processes. To engage in such strategy, the strategists
must be integrated into the learning processes that they are trying to influence (O'Sullivan 2000). Academics committed to regional economic development are in unique positions to mobilize different constituencies in the
region to engage in collective and cumulative learning processes. How we
as academics become integrated into the processes of regional learning to

influence regional strategy is, at this stage, what our own organizational learning is mainly about.

ACKNOWLEDGMENTS

This essay draws heavily on my research on industrial development and international competition carried out under grants from the Jerome Levy Economics Institute, the Center for Global Partnership of the Japan Foundation (in collaboration with the Massachusetts Institute of Technology International Motor Vehicle Program), and the Targeted Socioeconomic Research Programme of the European Commission (in association with STEP Group, Oslo, Norway). In 1996–97, I carried out this research at the University of Tokyo, where I was professor of economics, and at INSEAD (the European Institute for Business Administration), where I was a visiting scholar. I have benefited greatly from the excellent library facilities at both the Faculty of Economics, University of Tokyo and the Euro-Asia Centre, INSEAD, Fontainebleau, France. I also owe a significant debt to Mary O'Sullivan for her comments on this particular study and for the development of the central ideas in our joint work.

REFERENCES

Abe, E. 1996. Shibusawa, Eiichi (1840–1931). In *International encyclopedia of business and management*, ed. M. Warner, 44–51. London: Routledge.

Adams, T., and I. Hoshii. 1972. *A Financial history of the new Japan*. Tokyo: Kodansha International.

Almeida, B. 2001. Good jobs flying away: The U.S. jet engine industry. In *Corporate governance and sustainable prosperity*, ed. W. Lazonick and M. O'Sullivan. London: Palgrave.

Appelbaum, E., and R. Batt. 1994. *The new American workplace: Transforming work systems in the United States*. Ithaca: Cornell University Press.

Atkinson, A., L. Rainwater, and T. Smeedling. 1995. *Income distribution in the OECD countries*. Paris: Organisation for Economic Cooperation and Development.

Bisson, T. 1951. *Zaibatsu dissolution in Japan*. Berkeley: University of California Press.

Clark, K., and T. Fujimoto. 1991. *Product development performance*. Boston: Harvard Business School Press.

Cohen, S., and J. Zysman. 1987. *Manufacturing matters*. New York: Basic Books.

Cole, R. 1989. *Strategies for learning: Small group activities in American, Japanese and Swedish industry*. Berkeley: University of California Press.

Cusumano, M. 1985. *The Japanese automobile industry: Technology and management at Nissan and Toyota*. Cambridge: Harvard University Press.

Dertouzos, M., R. Lester, and R. Solow. 1989. *Made in America: Regaining the productive edge*. Cambridge: MIT Press.

Dore, R., and M. Sako. 1998. *How the Japanese learn to work*. London: Routledge.

Ferleger, L., and W. Lazonick. 1993. The Managerial revolution and the developmental state: The case of U.S. agriculture. *Business and Economic History*, 2d ser. 22:2.

———. 1994. Higher education for an innovative economy: Land-grant colleges and the managerial revolution in America. *Business and Economic History*, 2d ser., 23:1.

Ferleger, L., and J. Mandle. 1992. Co-signs and derivations of America's two-score decline: Poor math skills, poor productivity. *Challenge* 35 (3): 48–50.

———. 1994. *A new mandate: Democratic choices for a prosperous economy*. Columbus: University of Missouri Press.

Forrant, R. 2001. Good jobs and the cutting edge: The U.S. machine tool industry. In *Corporate Governance and Sustainable Prosperity*, ed. W. Lazonick and M. O'Sullivan. London: Palgrave.

Funk, J. 1992. *The teamwork advantage: An inside look at Japanese product and technology development*. Cambridge, Mass.: Productivity Press.

Glain, S. 1997. IHI keeps Japan in the jet-engine race. *Wall Street Journal*, June 17.

Gordon, A. 1985. *The evolution of labor relations in Japan: Heavy industry, 1853–1955*. Cambridge: Harvard University Press.

Hadley, E. 1970. *Antitrust in Japan*. Princeton: Princeton University Press.

Halberstam, D. 1986. *The Reckoning*. New York: Morrow.

Hirschmeier, J., and T. Yui. 1981. *The development of Japanese business*. London: George Allen & Unwin.

Hiwatari, N. 1996. Japanese corporate governance reexamined: The origins and institutional foundations of enterprise unions. Conference on Employees and Corporate Governance, November 22, Columbia University Law School, New York.

Hounshell, D. 1984. *From the American system to mass production, 1800–1932*. Baltimore: Johns Hopkins University Press.

Hunt, D. 1988. *Mechatronics: Japan's newest threat*. London: Chapman and Hall.

Hunter, J. 1984. *A concise dictionary of modern Japan*. Berkeley: University of California Press.

Imai, M. 1986. *Kaizen: The key to Japan's competitive success*. New York: Random House.

Ishikawa, K. 1985. *What is total quality control? The Japanese way*. Englewood Cliffs, N.J.: Prentice-Hall.

Iwauchi, R. 1989. The growth of white-collar employment in relation to the educational system. In *Japanese Management in Historical Perspective*, ed. T. Yui and K. Nakagawa. Tokyo: University of Tokyo Press.

Jaikumar, R. 1989. Japanese flexible manufacturing systems: Impact on the United States. *Japan and the World Economy* 1.

Johnstone, B. 1999. *We were burning: Japanese entrepreneurs and the forging of the electronic age*. New York: Basic Books.

Jürgens, U., T. Malsch, and K. Dohse, 1993. *Breaking from Taylorism: Changing forms of work in the automobile industry*. Cambridge, U.K.: Cambridge University Press.

Kochan, T., H. Katz, and N. Mower. 1984. *Worker participation and American unions: Threat or opportunity?* Kalamazoo, Mich.: W. E. Upjohn Institute for Employment Research.

Kodama, F. 1995. *Emerging patterns of innovation: Sources of Japan's technological edge.* Boston: Harvard Business School Press.

Krugman, P. 1997. *Pop internationalism.* Cambridge: MIT Press.

Krugman, P., and R. Lawrence. 1994. Trade, jobs, and wages. *Scientific American,* April.

Lawler, E., G. Ledford, and S. Mohrman. 1989. *Employee involvement in America: A study of contemporary practice.* San Francisco: American Quality and Productivity Center.

Lazonick, W. 1990. *Competitive advantage on the shop floor.* Cambridge: Harvard University Press.

———. 1991. *Business organization and the myth of the market economy.* Cambridge, U.K.: Cambridge University Press.

———. 1992. *Organization and technology in capitalist development.* Cheltenham, U.K.: Edward Elgar.

———. 1999. The Japanese economy and corporate reform: What path to sustainable prosperity. *Industrial and Corporate Change* 8 (4): 607–33.

———. 2001. Organizational learning and international competition: The skillbase hypothesis. In *Corporate governance and sustainable prosperity,* ed. W. Lazonick and M. O'Sullivan. London: Palgrave.

Lazonick, W., and M. O'Sullivan. 1996. Organization, finance, and international competition. *Industrial and Corporate Change* 5 (1): 1–49.

———. 1997a. Finance and industrial development, Part I: The United States and the United Kingdom. *Financial History Review* 4 (1): 7–29.

———. 1997b. Finance and industrial development, Part II: Japan and Germany. *Financial History Review* 4 (2): 117–38.

———. 1997c. Big business and skill formation in the wealthiest nations: The organizational revolution in the twentieth century. In *Big business and the wealth of nations,* ed. A. Chandler, F. Amatori, and T. Hikino. Cambridge, U.K.: Cambridge University Press.

Lazonick, W., and M. O'Sullivan. 2001a. Maximizing shareholder value: A new ideology for corporate governance. In *Corporate governance and sustainable prosperity,* ed. W. Lazonick and M. O'Sullivan. London: Palgrave. Previously published in *Economy and Society* 29 (1): 18–35.

———, eds. 2001b. *Corporate governance and sustainable prosperity.* London: Palgrave.

Levy, F., and R. J. Murnane. 1992. U.S. earnings levels and earnings inequality: A review of recent trends and proposed explanations. *Journal of Economic Literature* 30 (3): 1333–81.

Lichtenstein, N. 1989. The man in the middle: A social history of automobile industry foremen. In *On the line: Essays in the history of Auto Work,* ed. N. Lichtenstein and S. Meyer. Urbana: University of Illinois Press.

Maddison, A. 1994. Explaining the economic performance of nations, 1820–1989. In *Convergence of productivity: Cross-national studies and historical evidence,* ed. W. Baumol, R. Nelson, and E. Wolff. New York: Oxford University Press.

Marglin, S. 1979. Catching flies with honey: An inquiry into management initiatives to humanize work. *Economic Analysis and Workers' Management* 13.

Mass, W., and A. Robertson. 1996. From textiles to automobiles: Mechanical and organizational innovation in the Toyoda enterprises, 1895–1933. *Business and Economic History* 25:2.

Mizuno, S. 1984. *Company-wide total quality control.* Tokyo: Asian Productivity Organization.

Montgomery, D. 1987. *The fall of the house of labor.* Cambridge, U.K.: Cambridge University Press.

Moore, J. 1983. *Japanese workers and the struggle for power, 1945–1947.* Madison: University of Wisconsin Press.

Morris-Suzuki, T. 1994. *The technological transformation of Japan.* Cambridge, U.K.: Cambridge University Press.

Moss, P. 2001. Earnings inequality and the quality of jobs: Current research and a research agenda. In *Corporate Governance and Sustainable Prosperity,* ed. W. Lazonick and M. O'Sullivan. London: Palgrave.

Nakamura, K. 1997. Worker participation: Collective bargaining and joint consultation. In *Japanese labour and management in transition,* ed. M. Sako and H. Sato. London: Routledge.

Nikkei Weekly. 1996. *Japan economic almanac 1996.* Tokyo: Nihon Keizai Shimbun.

———. 1997. *Japan economic almanac 1997.* Tokyo: Nihon Keizai Shimbun.

Noble, D. 1977. *America by design.* New York: Oxford University Press.

———. 1984. *Forces of production: A social history of industrial automation.* New York: Oxford University Press.

Nonaka, I. 1995. The development of company-wide quality control and quality circles at Toyota Motor Corporation and Nissan Motor Co. Ltd. In *Fordism transformed: The development of production methods in the automobile industry,* ed. H. Shiomi and K. Wada. Oxford: Oxford University Press.

Nonaka, I., and H. Takeuchi. 1995. *The knowledge-creating company.* Oxford: Oxford University Press.

Okimoto, D., and Y. Nishi. 1994. R&D organization in Japanese and American semiconductor firms. In *The Japanese firm: The sources of competitive strength,* ed. M. Aoki and R. Dore. Oxford: Oxford University Press.

O'Sullivan, M. 2000. *Contests for corporate control: Corporate governance and economic performance in the United States and Germany.* Oxford: Oxford University Press.

Sako, M. 1998. Supplier development at Honda, Nissan and Toyota: A historical case study of organizational capability enhancement. Photocopy. Said Business School, Oxford University.

Samuels, R. 1994. *"Rich nation, strong army": National security and the technological transformation of Japan.* Ithaca: Cornell University Press.

Shimokawa, K. 1994. *The Japanese automobile industry.* London: Athlone.

Suzaki, K. 1987. *The new manufacturing challenge.* New York: Free Press.

Suzuki, T. 1994. Toray Corporation: Seeking first-mover advantage. In *Japanese business success: The evolution of a strategy,* ed. T. Yuzawa. London: Routlege.

Thomas, R. 1994. *What machines can't do: Politics and technology in the industrial enterprise.* Berkeley: University of California Press.

Tilly, C. 2001. What prognosis for good jobs? The U.S. medical diagnostic imaging equipment industry. In *Corporate governance and sustainable prosperity,* ed. W. Lazonick and M. O'Sullivan. London: Palgrave.

Tsuneta Yano Memorial Society. 1993. *Nippon: A charted survey of Japan, 1993/ 94.* Tokyo: Kokusei-sha.

Tyson, L. 1992. *Who's bashing whom? Trade conflict in high-technology industries.* Washington, D.C.: Institute for International Economics.

Uchida, H. 1989. Comment (on paper by Iwauchi). In *Japanese management in historical perspective,* ed. T. Yui and K. Nakagawa. Tokyo: University of Tokyo Press.

U.S. Bureau of the Census. 1976. *Historical statistics of the United States from the colonial times to the present.* Washington D.C.: U.S. Government Printing Office.

U.S. Department of Commerce. 1996 and various years. *Statistical abstract of the United States.* Washington D.C.: U.S. Government Printing Office.

U.S. Department of Health, Education and Welfare. 1972. *Work in America.* Cambridge, Mass.: MIT Press.

Urabe, K. 1988. Innovation and the Japanese management system. In *Innovation and management international comparisons,* ed. K. Urabe, J. Child, and T. Kagono. Berlin: Walter de Gruyter.

Wada, K. 1995. The emergence of the "flow production" method in Japan. In *Fordism transformed: The development of production methods in the automobile industry,* ed. H. Shiomi and K. Wada. Oxford: Oxford University Press.

Wall Street Journal. 1997. Giddings accepts buyout offer from Thyssen of $675 million, June 9.

Walton, R. 1975. The diffusion of new work structures: Why success didn't take. *Organizational Dynamics* 3 (Winter).

———. 1979. Work innovations in the United States. *Harvard Business Review* 57 (July–August).

Westney, D. 1987. *Imitation and innovation,* Cambridge: Harvard University Press.

———. 1994. The evolution of Japan's industrial research and development. In *The Japanese firm: The sources of competitive strength,* ed. M. Aoki and R. Dore. Oxford: Oxford University Press.

Yonekawa, S. 1984. University graduates in Japanese enterprises before the Second World War. *Business History* 26 (July): 193–218.

Zimbalist, A. 1975. The limits of work humanization. *Review of Radical Political Economics* 7 (Summer).

Can the Shift to Services Support Sustainable Prosperity?

Louis Ferleger and William Lazonick

OVER the course of the twentieth century, American workers have experienced major sectoral shifts of employment, first out of agriculture and more recently out of manufacturing. These sectoral shifts have left Americans increasingly looking to "services" as a new source of employment opportunity. Over the past two decades or so, the shift out of manufacturing has been characterized by the massive elimination of what were previously relatively well-paid and stable jobs (Lazonick and O'Sullivan 2001). Over the same period, there has been a persistent worsening in the distribution of income among American households (Moss 2001). A major economic and social issue that now faces Americans is whether the quality of the employment opportunities that are available and will be forthcoming in the services sector will help to reverse the worsening income distribution and form the foundation for sustainable prosperity over the next generation.

The rise of services employment began in the early part of the century. Over the period from 1920 to 1970, the proportion of nonmanagerial, nonprofessional services workers increased from 21 percent to 38 percent of the labor force, while the proportion of professional and managerial employees in both services and manufacturing rose from 12 percent to 22 percent. Nonfarm manual workers still represented 37 percent of the U.S. labor force in 1970, down only 3 percent from their share half a century earlier. Indeed the proportion of the labor force classified as craft or operative ("semiskilled") workers, as distinct from unskilled laborers, increased from 29 percent in 1920 to 32 percent in 1970.

From 1970 to 1997 (the most recent figures available), however, manufacturing employment declined from 26.4 percent to 16.1 percent of the labor force. The proportion of employment in all of the goods-producing sectors (agriculture, mining, construction, manufacturing, and utilities) declined from 44.3 percent to 32.7 percent over this period (table 1). During the 1990s, two out of three participants in the labor force were employed in sectors classified as producing services rather than goods (Bureau of the Census 1998, 421).

91

Table 1 shows the changes in the allocation of the U.S. civilian labor force across sectors, as classified by the Bureau of the Census in the *Statistical Abstract of the United States 1998*. We have regrouped the data into a "goods-producing" sector, and a "services-producing" sector. The goods-producing sector includes agriculture, mining, construction, manufacturing, and utilities, while the services-producing sector includes wholesale and retail trade, and finance, insurance, real estate, and public administration, plus the categories of business and repair services, personal services, entertainment and recreation, and professional and related services.

Later, we shall reclassify the categories in the "services-producing" sectors to provide a more useful categorization for understanding the actual and potential contribution of services employment to economic development. But, whatever the classification, services have become more important as a source of employment in the U.S. economy, and that trend will continue. What, then, are the prospects that services can generate high-quality employment opportunities that will contribute to a reversal of the worsening income distribution? Will the expansion of services yield employment opportunities that provide high pay and opportunities for learning and advancement? Or is the shift to services characterized chiefly by low-paid employment with little opportunity for learning and productivity growth, and hence part of the problem of increasing income inequality? The answers to these questions are critical to an assessment of the potential for employment in services to provide a foundation for sustainable prosperity in the United States.

Goods and Services

But what are "services," how do they contribute to productivity, and what quality of employment opportunities do they provide? Economists have long referred to an economy's products as "goods and services," but the nature and importance of the distinction between "goods" and "services" are not usually made clear. A "good" is a product that requires that the user—be he or she a consumer or a producer—interact with a physical object to get access to the desired qualities. A "service," in contrast, requires that the user interact with other people to get access to the desired qualities. Hence, labor as an input into a production process is a service, not a good, because some people—employers as users—have to motivate other people—employees as producers—to supply their skills and efforts to the tasks at hand. A "user-friendly" good is one that supplies the desired qualities to the users of the product without having to call on the services of another human being. In contrast, a "user-friendly" service is one for which the user finds that the producer—another person—is competent, reliable, cooperative, and even innovative.

Table 1. Percentage of U.S. civilian labor force employed in goods and services, 1970–1997

Sector	1970	1980	1990	1994	1997
Total employed (thousands)	78,678	99,303	118,793	123,060	129,558
Goods producing	44.3%	39.3%	34.7%	32.8%	32.7%
Agriculture	4.4	3.4	2.7	2.8	2.6
Mining	0.7	1.0	0.6	0.5	0.5
Construction	6.1	6.3	6.5	6.1	6.4
Manufacturing	26.4	22.1	18.0	16.4	16.1
Transportation, communication, and other public utilities	6.8	6.6	6.9	7.1	7.1
Services producing	55.7	60.7	65.3	67.2	67.3
Finance, insurance, real estate	5.0	6.0	6.8	6.6	6.4
Wholesale and retail trade	19.1	20.3	20.7	20.9	20.7
Wholesale	3.4	3.9	3.9	3.8	3.8
Retail	15.7	16.4	16.8	17.1	16.9
Services	25.9	29.0	33.1	34.9	35.8
Business and repair services	1.8	3.9	6.3	5.9	6.5
Advertising	0.2	0.2	0.2	0.2	0.2
Building and dwellings	n.a.[a]	0.4	0.7	0.7	0.6
Personnel supply	n.a.	0.2	0.6	0.7	0.8
Computer and data processing	n.a.	0.2	0.7	0.8	1.2
Detective/protective	n.a.	0.2	0.3	0.4	0.4
Automobile	0.8	1.0	1.2	1.3	1.3
Other business and repair, n.e.c.[b]	0.8	1.7	2.6	1.9	2.0
Personal services	5.4	3.9	4.0	3.5	3.4
Private households	2.3	1.3	0.9	0.8	0.7
Hotels and lodging places	1.2	1.2	1.5	1.2	1.2
Other personal, n.e.c.	1.9	1.4	1.6	1.5	1.5
Entertainment and recreation	0.9	1.1	1.3	1.7	1.9
Professional and related services	16.4	20.0	21.3	23.6	23.9
Hospitals	3.6	4.1	4.0	4.1	4.0
Health services, exc. Hospitals	2.1	3.4	3.9	4.5	4.9
K–12 schools[c]	7.8	5.6	5.0	5.2	5.3
Colleges and universities	n.a.	2.1	2.2	2.2	2.1
Social services	1.1	1.6	1.9	2.5	2.5
Legal services	0.5	0.8	1.0	1.0	1.0
Other professional, n.e.c.	1.3	2.5	3.3	4.0	4.1
Other services, n.e.c.	1.4	0.2	0.1	0.1	0.1
Public administration	5.7	5.4	4.7	4.7	4.4

Sources: Bureau of the Census, U.S. Department of Commerce. *Statistical abstract of the United States 1998* (Washington, D.C.: U.S. Government Printing Office, 1998), 421; Bureau of the Census, U.S. Department of Commerce, 1996 Current Population Survey (unpublished data), table 1, Employed persons age 16 and older by detailed industry, class of worker, age, and educational attainment.

[a]n.a. = not available
[b]n.e.c. = not elsewhere classified
[c]includes employees of colleges and universities in 1970

"Goods" and "services" are not, however, generally produced independently of one another. With almost any good that involves complex technologies to make it function usefully, there is a complementary element of service that is required to ensure that the good actually delivers the desired qualities. For example, when you buy a new television set, you may require up-to-date (and, it is hoped, accurate) information from a salesperson about the latest features of different makes and models. When you shop at a retail outlet that provides such a service, you are actually purchasing this service along with the good. In addition, when you buy the new set, you automatically receive a warranty for a certain period of time that provides you with servicing should it need repair. For a fee, you can also extend the warranty, and hence the access to service when and where required. When you purchase a television set, therefore, you are in effect purchasing a combination of a good and a service.

But to what extent are you purchasing a good or a service? Thirty or forty years ago, the average consumer who bought a television set might well have been a first-time purchaser, and hence might have been more reliant on the advice of a salesperson than the average consumer is now. Moreover, given the state of television technology, the consumer back then expected that after-sale service would, sooner rather than later, become a necessity. The fee for an extended warranty, if available, was accordingly more expensive. Since that time, the development of the electronics and tube technologies that go into a set, by making the product more reliable, has reduced the service element of the product. Retailers just love to push extended warranties for highly reliable products onto consumers precisely because the probabilities of the need for repair services have become very low. Thus, the development of television technology since the 1960s means that the experience of most people who buy a set today is buying just a good, not a service.

Indeed, through the development of technology, whole categories of products that once counted as services have become goods; for example, automatic household appliances such as washing machines, a substitution of goods for services that goes back to the consumer durable revolution of the 1920s (Olney 1991). As with the television sets, the more reliable these goods, the less the complementary need for repair services (hence Maytag's advertisement that features its repairman as the loneliest man in town). A more recent example is software. When a computer programmer develops a particular software application for a particular customer, he or she is providing a service. But when that particular software application is developed sufficiently so that it can be used by many different users who need never come into contact, directly or indirectly, with the computer programmer or any other software technician or adviser, what was once a service can be sold as a packaged product that includes on-line instructions concerning its features and how to use them. The service has become a good.

A dramatic example is the Microsoft operating system. Back in 1980, when Microsoft was developing an operating system for IBM's soon-to-be-released personal computer, IBM was buying a service from Microsoft (Campbell-Kelly and Aspray 1996). As purchasers of the software, the builders of the IBM PC were in intimate and continuous contact with the Microsoft programmers. But as, through a learning process, the Microsoft system developed into a product that could be used easily and reliably on most brands of computers for an ever-increasing number of applications, it became possible for Microsoft to sell its product with preprogrammed instructions directly to the user as a good that requires very little service.

As these examples illustrate, there is a tendency, through the development and utilization of technology, for services to be transformed into goods. Why then have services become more and more important sources of employment in the economy? The most prevalent demand-side explanation is that, with growth in real incomes, consumers increase the proportion of services that they demand. The most prevalent supply-side explanation is that, insofar as producer services are generated at lower levels of productivity than the manufacture of goods to which they are complementary, an expansion in manufacturing output will increase services employment more rapidly than goods employment. Both explanations help us to comprehend the growth of services employment, but both explanations also take technology as given. Yet technological change is central to the generation of higher real incomes in the economy as a whole and to the structural relations between goods and services in the production processes in which people are employed (Hauknes 1996, 15–17).

An approach to understanding the rise of services employment that places the dynamics of technological change at the center of the analysis looks for an explanation in the demands for and supplies of new services created by the very innovation processes that transform existing services into goods. For example, advances in computer hardware capabilities can create demands for new software services, and hence new types of software engineers. Or advances in medical technology can make it possible to supply wholly new medical services, delivered by highly specialized technicians. In our view, it is this dynamic process of technological change that makes possible the emergence of new services employment, even as technological change, by its very success, eliminates the demand for many types of existing services employment. To what extent can we expect these newly created employment opportunities in services to entail "high-quality jobs"?[1] The central characteristic of a high-quality job is that it involves the job-holder in a learning process that can generate better ("higher quality") and more affordable ("lower cost") products, and hence can provide the job-holder with high pay on a sustainable basis. Such employment opportunities must enable jobholders to contribute to product and process innovation, however incrementally. Without innovation a developmental economy is impossible, and without engaging ever-growing numbers of people in innovation processes, sustainable prosperity—the spreading of the benefits of economic development to more and more people over a prolonged period of time—is not achievable. We leave the crucial issue of what constitutes higher quality, lower cost goods and services for consideration later at our conclusion.

The assessment of the potential for services employment to support sustainable prosperity requires a categorization of the role of services in the economy that is rooted in a theory of economic development.[2] Two questions are relevant, the first having to do with the direct employment opportunities inherent in the production of a particular type of service, and the second having to do with the impact of the provision of that type of service on the process of economic development more generally. What is the quality of employment opportunities, in terms of pay, stability, and promotion that the production of a particular type of service provides? How does that service as an input to or output from the production processes in the economy promote or undermine the process of economic development as it occurs in a particular society? Services employment will best support sustainable prosperity in the society when the production of the service both creates high-quality employment opportunities and generates products that promote the process of economic development. Such a complementarity between employment opportunities and economic development, however, need not necessarily ensue. We argue, for example, that certain types of financial services entail high-paying jobs while undermining the process of economic development in the United States as a whole.

To consider the role of services employment in supporting sustainable prosperity requires, therefore, a theory of economic development that links the production of services to the performance of the economy as a whole over a sustained period of time. Central to a theory of economic development are characterizations of, first, strategic control over allocations of productive resources to innovation processes and, second, the development and utilization of productive resources in innovation processes. We define innovation as the process in the economy that generates higher quality, lower cost products without reducing the standards of living of those people involved in the innovation process. Investments in innovation entail strategic control over the allocation of resources (O'Sullivan 2000a). To contribute to an innovation process, certain types of services, such as education, may require a high degree of government (that is, public-sector) control over the strategic allocation or resources, while other services, such as software programming, might require a high degree of business (that is, private-sector) control. In addition, a particular type of service can be either an input to or an output from an innovation process. Hence, as a first cut at the assessment of the role of services employment in supporting sustainable prosperity, we can, as shown below, categorize services in terms of the prime locus of strategic control over the category of services and the relation to the innovation process of the category of services:

	Relation to innovation process	
Prime locus of strategic control	Inputs	Outputs
Public sector	Institutional	Social
Private sector	Producer	Consumer

High-quality jobs—employment opportunities that entail learning and that enable the jobholder to make productive contributions that can sustain high levels of income—do not just happen. Those who exercise strategic control over the allocation of resources in the economy—be they public-sector governments and institutions or private-sector businesses and individuals—must have a motivation to allocate resources to services that entail high-quality employment opportunities. Moreover, for these employment opportunities to support sustainable prosperity, the products that these services jobs generate must contribute to the development of the economy as either inputs to or outputs from an innovation process.

In the matrix, we categorized services employment as "institutional," "producer," "consumer," and "social" according to the combination of "prime locus of strategic control" and "relation to the innovation process" attributes that can enable these services to contribute to the development of the economy. Institutional services can, through educational systems, develop the capabilities of people to participate in innovation processes and, through financial systems, mobilize the financial resources needed to initiate and sustain innovation processes. Producer services can be integral to the innovation processes that generate higher quality, lower cost goods in the goods-producing sectors. Consumer services can help ensure, and even create, a demand for the goods and services generated by innovation processes. Social services can influence innovation processes to serve the ends of social as well as economic development. Table 2 transforms the sectoral categories in table 1 to provide a breakdown of civilian employment in the U.S. economy from 1970 to 1997 in terms of goods-producing sector and services-producing sectors, and for the latter, in terms of institutional, producer, consumer, and social services.

Table 3 shows gross value added as a percentage of GDP in the same categories for 1970 to 1994 (the latest year currently available). Note that

Table 2. Percentage of U.S. civilian labor force employed in services-producing sectors, 1970–1997

Sector	1970	1980	1990	1994	1997
Total employed (thousands)	78,678	99,303	118,793	123,060	129,558
Goods/producing	44.3	39.3	34.7	32.8	32.7
Services-producing	55.7	60.7	65.3	67.2	67.3
Institutional	12.8	13.7	14.0	14.1	13.8
Education	7.8	7.7	7.3	7.5	7.4
Finance, insurance, real estate	5.0	6.0	6.8	6.6	6.4
Producer	6.4	9.5	12.5	12.2	12.8
Wholesale trade	3.4	3.9	3.9	3.8	3.8
Business and repair[a]	1.4	3.4	5.7	5.3	5.9
Hotels and lodging[b]	0.6	0.6	0.8	0.6	0.6
Legal[b]	0.3	0.4	0.5	0.5	0.5
Other professional, n.e.c.[c]	0.7	1.2	1.6	2.0	2.5
Consumer	23.4	21.8	22.6	23.0	22.9
Retail trade	15.7	16.4	16.8	17.1	17.0
Hotels and lodging[b]	0.6	0.6	0.8	0.6	0.6
Legal[b]	0.3	0.4	0.5	0.5	0.5
Entertainment and recreation	0.9	1.1	1.3	1.7	1.9
Private household	2.3	1.3	0.9	0.8	0.7
Automobile services[a]	0.4	0.5	0.6	0.6	0.6
Other personal	1.9	1.4	1.6	1.5	1.5
Other services, n.e.c.[d]	1.4	0.2	0.1	0.1	0.1
Social	13.1	15.6	16.2	17.8	17.8
Social welfare	1.1	1.6	1.9	2.5	2.5
Health	5.7	7.4	7.9	8.6	8.9
Public administration	5.7	5.4	4.7	4.7	4.4
Other professional, n.e.c.[c]	0.7	1.2	1.6	2.0	2.5

Sources: Bureau of the Census, U.S. Department of Commerce, *Statistical abstract of the United States, 1998* (Washington, D.C.: U.S. Government Printing Office, 1998), 417–20, and various earlier editions of the *Statistical Abstract.*

n.e.c. = not elsewhere classified
[a] automobile services component divided equally between business and consumer services (included in business and repair services under producer services)
[b] hotels and lodging and legal services divided equally between business and consumer services
[c] other professional services divided equally between social and business services
[d] other services n.e.c. allocated entirely to consumer services

these measures of gross value added must be interpreted with caution, especially when comparing across categories. The Organization for Economic Cooperation and Development (OECD), which generates these numbers, uses a variety of measurement methods for different sectors of the U.S. economy (OECD 1996). In table 3, for example, gross value added of "education" is very low, primarily because the measure that the OECD uses is a price index of private consumption expenditure for private education (OECD 1996, 96), thus omitting the vast public component of value added

Table 3. Gross value added as a percentage of GDP in U.S. services-producing sectors, 1970–1994

Sector	1970	1980	1990	1994
GDP (in billions of current $)	$1,035.6	$2,784.2	$5,743.8	$6,931.4
Goods-producing	43.77	43.00	36.89	34.86
Services-producing	56.23	57.00	63.11	65.14
Institutional	14.86	15.60	17.85	19.12
Education	0.69	0.59	0.69	0.74
Finance, insurance, real estate	14.17	15.01	17.16	18.38
Producer	10.74	11.99	14.70	15.51
Wholesale trade	6.97	7.01	6.39	6.66
Business and repair	2.62	3.50	7.20	7.36
Hotels and lodging	0.31	0.36	0.40	0.41
Legal	0.35	0.45	0.70	0.68
Other professional, n.e.c.[a]	0.50	0.67	0.01	0
Consumer	12.34	11.21	11.73	11.82
Retail trade	9.68	8.83	8.77	8.80
Hotels and lodging	0.31	0.36	0.40	0.41
Legal	0.35	0.45	0.70	0.68
Entertainment and recreation	0.68	0.72	1.03	1.11
Private household	0.43	0.22	0.16	0.16
Other personal	0.90	0.63	0.68	0.67
Social	18.30	18.21	18.83	19.09
Social welfare	0.97	0.94	1.12	1.27
Health	3.03	4.00	5.36	5.89
Public administration	13.80	12.60	12.34	11.93
Other professional, n.e.c.[a]	0.50	0.67	0.01	0.00

Source: Organisation for Economic Cooperation and Development (OECD), *Value added in services* (Paris: Statistics Directorate, OECD, 1997).

[a]n.e.c. = not elsewhere classified

from education; for example, in 1994 expenditures on public elementary and secondary schools were about twelve times greater than expenditures on private elementary and secondary schools (Bureau of the Census 1998, 161). In contrast, the measure of gross value added of "public administration," which is very high, is based on an index of the numbers of people employed in public administration (OECD 1996, 114).

Gross value added as a percentage of GDP to sectoral employment as a percentage of total employment in the same year, table 4 provides a measure of the average "productive contribution" of employees in each sector to the economy, with a value of 1.00 as the average productive contribution of all employees across all goods and services sectors in the economy in any given year. Any value greater than 1.00 reflects an above average contribution, and vice versa. Thus, table 4 presents a picture of the relative

Table 4. "Productive contribution" of employment in U.S. services-producing sectors, 1970–1994 (Gross value added as percentage of GDP to sector employment as a percentage of total employment, with 1.00 as the average productive contribution of *all* employees)

Sector	1970	1980	1990	1994
Goods-producing	0.99	1.09	1.06	1.07
Services-producing	1.01	0.94	0.97	0.97
Institutional	1.16	1.13	1.27	1.36
Education	0.09	0.08	0.09	0.10
Finance, insurance, real estate	2.83	2.50	2.52	2.78
Producer	1.68	1.26	1.18	1.27
Wholesale trade	2.05	1.80	1.64	1.75
Business and repair	1.87	1.03	1.26	1.39
Hotels and lodging	0.52	0.60	0.50	0.68
Legal	1.17	1.13	1.40	1.36
Other professional, n.e.c.	0.71	0.56	0.01	0.00
Consumer	0.53	0.51	0.52	0.51
Retail trade	0.62	0.54	0.52	0.51
Hotels and lodging	0.52	0.60	0.50	0.68
Legal	1.17	1.13	1.40	1.36
Entertainment and recreation	0.76	0.65	0.79	0.65
Private household[a]	0.16	0.12	0.11	0.11
Other personal[b]	0.27	0.39	0.40	0.42
Social	1.40	1.17	1.16	1.07
Social	0.88	0.59	0.59	0.51
Health	0.53	0.54	0.68	0.68
Public administration	2.42	2.33	2.63	2.54
Other professional, n.e.c.	0.71	0.56	0.01	0.00

Source: From tables 2 and 3.

[a] Includes consumer automobile services
[b] Includes other services, n.e.c.

importance of the productive contributions of employees across sectors in any given year as well as the trends in the relative importance of particular sectors over the time span indicated.

Recognizing the limitations of the available data and making use of existing supplementary information where warranted, our task is to provide a preliminary assessment of whether and to what extent, from the perspective of a developmental economy, employment in these four services categories has been in fact supporting sustainable development in the 1980s and 1990s in the United States. We describe the general characteristics of each category of services employment in terms of the locus of control and the relation to the innovation process, as well as how these characteristics have changed over time. Within the framework for each category of services, we consider the evidence on the quality of employment opportunities that the production of a category of services generates and the impact of this category on the prospects for economic development in the U.S. economy as a whole.

Institutional Services

Education and finance are two types of institutional services that are key to the performance of the American economy. These institutions are key because they provide innovation processes with capable people and mobilized money, both of which are essential to the development and utilization of technology. During the twentieth century, educational services have been primarily financed and organized by federal, state, and local governments. Despite substantial deregulation of financial services since the late 1970s, the financial system, including the stock markets, the bond markets, banking, and consumer lending, is still subject to substantial public-sector control in terms of the purposes for which and the terms on which different types of financial institutions can channel financial resources into the economy.

Educational Services

The building of the higher education system in the United States reflected a national commitment to develop technological capabilities, with the agricultural sector taking the lead in the late nineteenth century and the manufacturing sector not too far behind (Ferleger and Lazonick 1993, 1994). By the early decades of this century, the U.S. system of higher education channeled technical and administrative personnel into managerial organizations within both the business and government sectors. These managerial organizations became the focal point for organizational learning, the essence of the innovation process.

During most of the twentieth century, however, production workers increasingly found themselves excluded from this organizational learning process as skill development was taken off the shop floor. This transformation created the basis for an organizational segmentation between managerial employees and shop-floor workers in U.S. industrial enterprises. Reflecting this organizational segmentation was the (otherwise paradoxical) coexistence in the United States of a system of higher education that attracts students and researchers from around the world with a system of K–12 public education that for the past three decades has continually ranked among the lowest in average quality among the advanced economies. Notwithstanding the American rhetoric of equal opportunity, there was no need in the United States to create a system of uniformly high-quality public education when a substantial proportion of the population was expected to spend lifetimes performing routinized jobs devoid of the potential, or even the requirement, for learning. At the same time, the concentration of collective and cumulative learning in managerial organizations that provided the foundations for American prosperity in the twentieth century made it imperative that resources be allocated at a national level to a high-quality system of higher education (Lazonick and O'Sullivan 1997c).

During the past two decades, the system of higher education has continued to play an ever more important role in the U.S. economy. Between 1980 and 1995, the percentage of high school graduates who enrolled in college increased from 49.3 percent to 61.9 percent, while the proportion of high school students who failed to graduate dropped from 12.0 percent to 9.9 percent. Over this period, college enrollments in the United States increased from 12.1 million to 14.3 million students, an increase of 18 percent, even though, because of the demographic shift, the number of high school graduates declined by 16 percent (Bureau of the Census 1997, 176, 179–81).

As it has catered to an ever-growing population of students from diverse backgrounds and with diverse employment prospects, the quality of the U.S. system of higher education has become increasingly unequal. This inequality reflects both the uneven quality of public education across the school districts from which different types of institutions of higher education draw their students and the unequal access of students from different socioeconomic backgrounds to institutions of different quality. Moreover, growing numbers of students at American institutions of higher education come from abroad, with, it would appear, better quality public educations on average than their U.S.-bred counterparts.[3] Many of these foreign students, upon graduation, remain in the United States as highly skilled employees, thus taking the pressure off the U.S. public sector to upgrade the average quality of K–12 education.

Hence, as an input into the innovation process, the U.S. educational system both creates essential pre-employment foundations for a proportion of the population to hold high-quality jobs, while denying these same foundations to a substantial, and perhaps growing, proportion of people raised in the United States. A general upgrading of the U.S. educational system is a basic precondition for sustainable prosperity. But, as much as Americans might recognize the need for higher quality mass education, such an upgrading will not occur unless employers in both the business and government sectors pursue strategies that integrate more and more people into processes of collective and cumulative learning.

As for employment within the educational system itself, these are jobs that in general entail considerable knowledge and skill. Employing 7.4 percent of the civilian population in 1997 (see table 1), the educational services sector provides an important source of high-quality services employment. Most of these people work in the public sector, although the proportion has been declining. In 1980, 85.2 percent of those employed within the educational system worked for public-sector institutions; by 1996, this proportion had declined to 78.6 percent. The average annual salaries of public school teachers in the United States rose from $15,913 in 1980 to $36,531 in 1994, an increase of 128 percent, compared with an increase of 90 percent in nominal annual earnings of all full-time employees in the United States over the same period (Bureau of the Census 1991, 413; 1997, 168, 429). As for higher education, from 1980 to 1994, the average salaries of faculty members at public colleges and universities rose by 111 percent for assistant professors ($18,000 to $38,000) and 108 percent for full professors ($28,800 to $59,800), and at private institutions by 136 percent for assistant professors ($17,000 to $40,100) and 137 percent for full professors ($30,100 to $71,200) (Bureau of the Census 1997, 169, 187).[4] During this period, therefore, the salary advantage of employment in private over public higher educational institution, has increased substantially.

Between 1980 and 1996, pupil–teacher ratios—one measure of teachers' working conditions—declined by 7.5 percent from 18.7 to an estimated 17.4 (Bureau of the Census 1997, 164, 168). Over this period, the proportion of the civilian labor force employed in K–12 education has declined somewhat, by 0.3 percent, reflecting to some extent the demographic shift that has reduced the numbers of K–12 students. This demographic shift, however, is in the process of reversing itself. What will happen to the quality of jobs in K–12 education as the numbers of enrolled students increase depends on the extent of the social commitment, exercised through local, state, and federal governments, to deliver high-quality mass education to all young people in the United States. At the higher education level, the proportion of the civilian labor force employed has barely increased over the past two decades, despite the growing demands on the system of higher education to enhance the capabilities of the future labor force. Employment in educational services still provides substantial numbers of high-quality jobs in the U.S. economy. But without a widespread commitment to upgrade the quality of the mass education system, one would not expect employment in educational services to offer much scope for growth over the next generation.

Financial Services

Finance plays a central role in the development of an innovative economy. Those who control finance can determine the strategies for the development and utilization of productive resources. The key to the role of the financial system in the innovation process is that it transfers control over the productive resources from those economic actors who lack the abilities and incentives to develop and utilize these productive resources to those who do have such abilities and incentives. The latter are generally organizations—business enterprises but also government agencies as distinct from households—because the innovation process entails the integration of the abilities and incentives of large numbers of people with different functional specializations and hierarchical responsibilities. In general, the role of the financial system in economic development is to transfer control over finance from households as savers to businesses and governments as investors. This transfer of control is accomplished by means of both government taxation—an overtly political process—and financial intermediation—a process that relies on the payment of financial returns to savers. Coming into the 1970s, these returns were regulated to favor business and government investors over the household savers, in terms of control of the financial resources. Beginning in the late 1970s, however, the deregulation of U.S. financial institutions swung the balance in favor of household savers over business and government investors.

In making this transfer of finance from savers to investors, the financial system plays an essential, and potentially, productive role from the perspective of the development of the economy as a whole. Contrary to the traditional importance given bank-centered industrial finance, however, the financial enterprises that make up the financial system do not have to be particularly innovative to perform this developmental role, mainly because it is primarily the industrial enterprises, not the financial enterprises, that must possess the organizational capabilities to generate higher quality, lower cost goods and services (Lazonick and O'Sullivan 1997a, 1997b). Hence we have the role of a highly specialized and highly regulated financial system in the process of economic development, of the type, for example, that existed in the United States from the early twentieth century until the onset of financial deregulation in the mid-1970s.

Since the 1970s, as a consequence of financial deregulation, financial enterprises have been very innovative as generators of financial services. But there is good reason to doubt that financial innovation has been beneficial to the development of the U.S. economy as a whole. Rather than transfer control over productive resources from passive savers to active investors, the financial system has been increasingly geared to securing higher returns on financial wealth for passive savers (see Lazonick and O'Sullivan 2000). The result of this shift of the U.S. financial system from what might be called its "value-creating" role to its "value-extracting" role has been to make it more expensive and difficult for the producers of other goods and services to control the productive resources required to generate higher quality, lower cost products.

At the same time, while a small proportion of the American population is reaping the higher returns on financial securities—particularly on corporate securities—a large proportion of the population is using the U.S. financial system to get ever more deeply in debt. Consumer credit as a percentage of household debt rose from just over 6 percent in 1950 to about 16 percent by the late 1970s, and in the first half of the 1990s averaged about 17 percent. As the U.S. household savings rate plummeted to less than 5 percent in 1995, a record 52 percent of growth in consumption was financed by consumer credit (Henwood 1997, 65; U.S. Congress 1998, 319). Consumer spending has contributed to the boom in the economy in the last half of the 1990s, but is that boom sustainable? In the decades ahead, low levels of savings, and hence low levels of accumulated financial assets, will place all the more pressure on financial institutions to extract high rates of returns on financial securities to fund the retirement incomes of Americans. The economy will become more focused on value extraction and less focused on value creation. Even if the financial transformation of America over the past two decades may be undermining the foundations for sustainable prosperity over the coming decades, one might, nevertheless, be tempted to argue that, in the present, financial services is a sector with expanding numbers of well-paid jobs for more highly educated people. On the surface, employment in financial services has been an important source of the generation of high-quality employment opportunities in the U.S. economy over the past two decades. In 1980, average nominal hourly earnings of employees in FIRE (finance, insurance, and real estate) services were $5.79, compared with average hourly earnings of $6.66 in total nonfarm private employment and $7.27 in manufacturing. In 1997, average hourly earnings in FIRE were $13.33, compared with $12.28 in all nonfarm private employment and $13.17 in manufacturing (see table 5). Over the 1980 to 1996 period, (table 5). Over the period 1980–1996 period, the proportion of the FIRE labor force with at least a four-year undergraduate

Table 5. Average hourly nominal earnings in private-sector industry, 1980–1997, in U.S. $

Industry	1980	1985	1990	1994	1995	1997
Total nonfarm private	6.66	8.57	10.01	11.12	11.43	12.28
Mining	9.17	11.98	13.68	14.88	15.30	16.17
Construction	9.94	12.32	13.77	14.73	15.09	16.03
Manufacturing	7.27	9.54	10.83	12.07	12.37	13.17
Durable goods	7.75	10.09	11.35	12.68	12.94	13.73
Nondurable goods	6.56	8.72	10.12	11.24	11.58	12.33
Transportation and public utilities	8.87	11.40	12.92	13.78	14.13	14.93
Wholesale trade	6.95	9.15	10.79	12.06	12.43	13.44
Retail trade	4.88	5.94	6.75	7.49	7.69	8.34
Finance, insurance, real estate	5.79	7.94	9.97	11.83	12.32	13.33
Other private services	5.85	7.90	9.83	11.04	11.39	12.28

Source: Bureau of the Census, U.S. Department of Commerce, *Statistical abstract of the United States 1998.* (Washington, D.C.: U.S. Government Printing Office, 1998), 429.

degree increased from 24.1 percent to 33.9 percent (see table 6). The comparable proportions for manufacturing in 1980 and 1996 were 12.2 percent and 19.0 percent. The proportion of the civilian labor force employed in FIRE increased from 6.0 percent in 1980 to 6.8 percent in 1990, and then declined somewhat to 6.4 percent in 1997 (table 1). With the high level of educational attainment of FIRE employees, this services sector par-

Table 6. Percentages of educational attainment of employees by industry, 1980, 1994, 1996 (employees at least sixteen years old)

Industry	Year	Educational attainment					
		8 yrs or less	hs,nd[a]	hs,nc[b]	1–3 yrs college	B.A. degree	B.A.+
Construction	1980	12.9	18.3	41.8	15.3	6.0	8.3
	1994	7.8	15.7	45.5	23.6	6.4	7.4
	1996	6.6	16.3	44.4	24.3	7.3	8.3
Manufacturing	1980	11.2	16.3	45.1	15.1	8.5	12.2
	1994	5.5	10.5	41.3	24.4	14.0	18.4
	1996	5.2	10.2	40.4	25.2	14.2	19.0
Transportation and	1980	6.4	12.3	48.4	21.0	9.3	11.9
public utilities	1994	1.8	6.3	38.8	32.9	16.4	20.2
	1996	1.6	6.5	37.9	33.2	16.9	20.8
Wholesale	1980	6.5	11.3	42.8	21.3	14.1	18.1
trade	1994	3.4	6.5	37.0	30.0	19.9	23.1
	1996	3.2	8.3	36.8	29.7	19.1	22.1
Retail trade	1980	6.8	23.1	43.0	18.3	6.6	8.8
	1994	4.1	9.2	43.6	28.5	12.5	14.6
	1996	3.4	18.7	37.2	29.9	9.5	10.8
Eating and	1980	9.6	32.4	42.8	16.1	4.6	6.0
drinking places	1994	8.4	12.6	42.4	25.8	9.4	10.7
	1996	6.2	28.1	33.2	25.6	6.3	7.0
Finance, insurance,	1980	2.3	5.6	44.9	22.8	17.5	24.1
and real estate	1994	0.9	2.1	30.3	31.5	27.7	35.3
	1996	0.9	3.1	28.9	33.2	26.7	33.9
Services	1980	6.0	11.0	30.7	18.7	15.3	33.8
	1994	3.1	5.6	27.8	29.9	20.9	33.6
	1996	3.0	8.0	26.6	31.1	20.0	31.3

Sources: Bureau of the Census, U.S. Department of Commerce, 1980 Current Population Survey, table 15, Industry of employed persons 16 years old and over, by age and years of school completed; 1994 Current Population Survey (unpublished data), table 1, Employed persons age 16 and older by detailed industry, class of worker, age, and educational attainment; 1996 Current Population Survey (unpublished data), table 1, Employed persons age 16 and older by detailed industry, class of worker, age, and educational attainment.

[a] 1–4 years of high school, no diploma
[b] high school diploma, no college

ticularly holds out prime possibilities for an expansion of high-quality services jobs in the U.S. economy.

From the perspective of sustainable prosperity, the irony is that financial services that are run for the sake of value extraction generally require the employment of more highly educated and motivated people than financial services that support value creation. That is, it requires innovation to be a successful value extractor, but it is a type of innovation that tends to undermine sustainable prosperity in the economy as a whole. Hence, in terms of educational qualifications and employment remuneration, the transformation of financial services from value creation to value extraction over the past two decades may have entailed an expansion of "high-quality" employment opportunities. But reliance on this sector to generate high-quality services employment may well be a detriment to the creation of high-quality jobs that can contribute to sustainable prosperity in the economy as a whole.

Producer Services

Producer services are, by definition, inputs into the production of goods and services. As such, producer services hold out the potential that those employed in their delivery will be directly involved in the processes of innovation. Producer services that demand highly skilled and highly educated people can provide high-quality employment opportunities. As we have already determined, substantial amounts of producer services are carried out within enterprises that are classified as goods-producing. For example, when members of a development team are employed as consultants, they are classified as members of the services-sector labor force, but should these same people be employed as members of a goods-producing enterprise, they are classified as participants in the "goods-producing sector"— that is, as members of the manufacturing labor force. If we assume, as is undoubtedly the case, that such producer–services employment is more prevalent in enterprises that are classified as goods producing than is goods-producing employment in enterprises that are classified as services producing, then there is an underestimation, perhaps substantial, of the actual amount of producer-services employment in the economy.

A services-producing enterprise that successfully develops its products so that they can be used by purchasers as "goods"—that is, without the aid of services providers—may find itself, and its employees, reclassified as a producer of goods. Individuals can, however, have a working life apart from the enterprises that have employed them. Once, through their involvement in a process of organizational learning, employees of a goods-producing enterprise have acquired unique capabilities, it is possible that they may choose to work in enterprises that produce services rather than goods. Indeed, in an era such as the 1980s and 1990s, when the downsizing of the economy had affected not just shop-floor workers but also managerial employees, involuntary terminations may have provided the impetus for highly educated and trained people in turn to enter the producer-services sector.

Given the character of corporate strategies over which they have no control, it may be perfectly rational for highly educated and trained people to turn from being organization men (or women) to become independent consultants or subcontractors. These people can make their accumulated knowledge available to other enterprises at a fraction of what it would cost these enterprises to develop this knowledge within their own organizations. The potential problem for the development of the economy as a whole is that the innovation process that transforms services into goods requires new organizational learning. Reliance on the knowledge of consultants may permit enterprises to compete in product markets today, but they are not making the investments in cumulative and collective learning processes that will enable them to take on the new competition tomorrow.

Indeed, an attractive strategic option for well-established, goods-producing enterprises can be a prime, or even exclusive focus, on producing services instead of goods. In industries such as aircraft and machine tools, for example, there is evidence that the trend in the United States is out of manufacturing and even design, and into the low fixed cost and highly lucrative business of servicing high-technology equipment (Almeida 2001; Lazonick 1998). Considerable money can be made by taking a reputable goods-producing company, with an installed base of brand-name machinery, and turning it into a services-producing company. Precisely because the United States has been a leader in industries such as aircraft and machine tools, the nation has a huge roster of experienced technical specialists, many of whom no longer have as secure employment with equipment producers as they had in the past. Some of these people are finding continued employment servicing the equipment that the companies for which they worked formerly manufactured and maintained after sale. In the past, these employees had acquired these skills through organizational learning. But their utilization of these skills now may be confining them to narrow and concentrated functions that remove them even farther from the processes of organizational learning that will drive the development of the economy in the future.

As shown in table 2, producer services are both important to employment in the U.S. economy and grew substantially as a proportion of employment in the 1970s. This proportion almost doubled from 6.4 percent in 1970 to 12.8 percent in 1997, with the increase occurring almost equally in the 1970s and 1980s. The biggest single source of increase within producer services was the (still-broad) category of business and repair services.

Within business and repair services (table 1), significant growth occurred in the 1980s and 1990s in personnel supply (0.2 percent of the labor force in 1980, 0.8 percent in 1997) and computer and data processing (0.2 percent in 1980, 1.2 percent in 1997). The expansion of this sector is related to the computer revolution that has occurred over the last two decades and to some extent reflects the complementary development of the production of goods and the production of services in an innovative sector of the economy. The other important source of employment in producer services is wholesale trade, which has remained just under 4 percent of the civilian labor force throughout the 1980s and 1990s.

Earnings in producer services vary widely across categories, and certain categories of producer services have done better than other categories over the course of the 1980s and 1990s. For example, from 1980 to 1996, average real hourly earnings in 1992 dollars rose from $12.18 to $15.00 in legal services, while they barely increased (from $11.52 to $11.65) in wholesale trade, and from $7.38 to only $7.39 for those working in hotels and motels. In business services, excluding auto repair, for which earnings data are unavailable for most of the 1980s, real average hourly earnings in 1992 dollars rose slightly from $10.13 in 1990 to $10.16 in 1996 (Bureau of the Census 1997, 426; U.S. Congress 1998, 284).

These averages of course, conceal varying degrees of within-category variation in earnings. Indeed, an evaluation of the extent to which this employment in producer services has been contributing, and will continue to contribute, to the expansion of high-quality jobs cannot be made with the aggregate data that we are using here. Rather such an evaluation will require much more disaggregated data and an accumulation of case-study research on the role of specific producer services in the innovation process.

What we do know from aggregate data is that in the 1980s and 1990s there has been a substantial increase in investment in information technology in the business services sector. As can be seen in table 7, in 1980 total private investment in fixed, nonfarm, nonresidential information technology was $58.5 billion; in 1996, $193.2 billion. Of these amounts, the proportion invested in enterprises classified as supplying business services was 2.87 percent in 1980 but 9.20 percent in 1996, the biggest percentage shift in any of the classifications in table 7. The classification with the next largest percentage shift in the proportion of total information technology investment was another producer-services sector, wholesale trade, where the proportion of investment increased from 9.20 percent in 1980 to 15.06 percent in 1996. What effect this investment in information technology in business services is having on the quantity and quality of employment opportunities is part of a much larger—and thus far unresolved—debate on the employment impacts of investment in information technology more generally (Moss 2001).

Table 7. Percentages of fixed nonfarm nonresidential private investment in information technologies, 1980–1996

Industry	1980	1985	1990	1994	1995	1996
Total invested (= 100 percent)	$58.55	103.35	122.50	150.16	170.92	193.19
Mining	3.09	2.37	1.21	1.04	1.01	0.77
Construction	0.20	0.15	0.14	0.12	0.10	0.10
Manufacturing	19.66	18.71	19.58	16.14	17.41	15.58
Durable goods	11.78	12.05	10.35	8.45	9.01	8.26
Nondurable goods	7.89	6.66	9.23	7.69	8.40	7.32
Transportation	3.33	3.34	2.70	3.02	3.40	3.15
Communication	29.31	18.47	19.94	18.64	18.11	17.95
Electric, gas, and sanitary services	6.24	8.12	7.09	5.05	5.13	4.46
Wholesale trade	9.20	12.32	9.25	13.29	13.11	15.06
Retail trade	2.98	3.41	3.19	3.55	3.65	3.59
Finance, insurance, and real estate	16.61	21.17	23.18	21.67	20.99	22.17
Other services	9.37	11.95	13.70	17.49	17.08	17.18
Hotels and other lodging places	0.35	0.37	0.38	0.15	0.16	0.23
Personal services	0.42	0.36	0.38	0.28	0.27	0.23
Business services	2.87	4.99	5.41	8.79	9.05	9.20
Auto repair, auto services, and parking	0.65	1.26	0.92	0.69	0.62	0.64
Miscellaneous repair services	0.08	0.11	0.09	0.11	0.10	0.09
Motion pictures	0.57	0.56	0.91	1.33	1.13	1.14
Amusement and recreation services	0.35	.013	0.12	0.16	0.15	0.17
Health services	1.60	1.66	2.70	3.43	2.81	2.66
Legal services	0.49	0.69	0.83	0.48	0.46	0.41
Educational services	0.15	0.10	0.07	0.06	0.06	0.05
Other services, n.e.c.	1.83	1.71	1.89	2.01	2.29	2.36

Source: U.S. Department of Commerce, Fixed nonresidential private capital, by industry and by type of equipment or structure, *Fixed reproducible tangible wealth of the United States, 1925–1996.* CD-ROM (Washington, D.C.: U.S. Government Printing Office, 1997).

Note: Information technology consists of mainframe computers, personal computers, direct-access storage devices, computer printers, computer terminals, computer tape drives, computer storage devices, other office equipment, communications equipment, instruments, and photocopying and related equipment.

Consumer Services

Throughout the 1980s and 1990s, about three-quarters of all employment in consumer services has been in retail trade. In 1997 retail trade employed 16.9 percent of the civilian labor force, up from 16.4 percent in 1980 (table 2). The retail trade sector is characterized by low wage levels, low levels of educational attainment, and low levels of technological investment (see tables 5, 6, and 7).

Retail services may well continue to provide more employment to Americans. With a net addition of 9.5 million jobs to retail trade between 1970 and 1997, this sector has been by far the most important source of new jobs in the U.S. economy over the past few decades. At the same time, with over 21 million people selling the rest of us goods, there is a question whether the growth rate of employment in this sector is reaching its limits.

One of these limits is the amount of disposable income that Americans have to spend. In general, the incomes of most Americans have been stagnating since the 1970s. One cause of this stagnation has been the disappearance of the high-paid, even if relatively low-skill, blue-collar jobs that bolstered the incomes of many Americans in the post–World War II decades. These jobs have disappeared because American companies have been outperformed by foreign companies in a whole range of goods in which the United States had previously been the world leader. These goods include a vast array of household appliances and consumer electronics products. Americans are still buying these foreign-made goods in immense quantities, thus providing ample employment opportunities for U.S. workers in retail services even as the opportunities in manufacturing have shrunk.

The new retail services jobs are not replacing the good jobs that once characterized the manufacturing sector. Americans can buy more consumer goods when they cost less, and one thing that keeps the cost down is the low wages of the people who sell them the goods. In 1982 dollars, the average weekly wage in retail trade employment was $172 in 1980, while the average weekly wage of all American employees was (in 1982 dollars) $275. By 1990, real wages (in 1982 dollars) in retail trade employment had fallen to $146, and in 1996 to $145 (Bureau of the Census 1997, 429). By the mid-1990s, both absolutely and relative to the U.S. average, the pay for work in the retail-trade sector had worsened. Lower wages in retail services can make the consumer dollars of Americans go farther and can help increase the demand for employment in retail services, but if sustainable prosperity is what Americans are after, don't look to the retail sector for the services solution.

Social Services

The purpose of social services is to maintain the well-being of the population insofar as certain standards of well-being are not otherwise achieved through the operation of the economy. Like consumer services, the ability of the economy to provide social services is an outcome of the developmental economy. The difference between the two types of services is that in a market-oriented economy the decision to allocate resources to consumer services is a private-sector (household) decision, whereas the decision to allocate resources to social services is a public-sector (government) decision. When the economy performs poorly in creating high-quality employment opportunities, households have less disposable income for consumer services, but all the more pressure is put on many types of social services to maintain the physical and mental health of the population. At the same time, the demand for certain types of social services, particularly those related to health care, can be positively correlated with the successful performance of the economy insofar as it enables people to live longer and remain healthier. Nevertheless, even then, the supply of social services is constrained by the willingness of the society to allocate resources to uses that are economic expenditures (outputs), not economic investments (inputs).

In terms of employment, the two most important types of social services are health and public administration (see table 2). In 1970 both of these categories employed about the same proportion of the civilian labor force—5.7 percent, or about 4.5 million people. Since then, to 1997, employment in public administration has fallen to 4.4 percent of the civilian labor force, reflecting both a reaction of many Americans to the costs of big government and a rationalization of administrative services in the public sector. Meanwhile, from 1970 to 1997, employment in health rose from 5.7 percent of the labor force to 8.9 percent—an absolute net increase of 7.0 million jobs. In 1997, the health sector employed over 11.5 million people.

The increase in health services employment has gone hand in hand with the demands of an aging population to keep these services alive and functioning. Older Americans make this demand because their employment had provided them with good health coverage—something that has become increasingly unavailable for younger Americans. And where employment-based coverage is not available, older Americans have been the prime recipients of government's funding of health care services. In addition, by cashing in on the assets—mainly pensions and housing—that they have accumulated over lifetimes, older Americans have been able to find resources for meeting critical health care needs.

But if demand for, and employment in, health services has increased, so too has the overall cost. In 1988, 5.01 percent of all expenditures by American households was on health services. Since then, that proportion has been creeping up, to 5.21 percent in 1990, to 5.47 percent in 1992, and to 5.53 percent in 1994 (the latest year available). Many of the additions to the health services labor force are lower-wage jobs, but the earnings of different groups of people, from doctors to hospital janitors, in the sector vary widely. With the quality of health service a prime concern, there are strict limits on using wage costs to keep total costs down in this sector.

Health services will remain expensive in the United States and they are likely to command higher proportions of household expenditures, thus pressing the limits of the health services Americans can afford.

As the present generation of younger workers grows older, moreover, the funds that Americans households will have available to spend on health services is likely to decline. Year by year, for all categories of American employees, the proportion of those who have employment that includes health coverage has been declining. In 1987, almost 13 percent of Americans did not have health insurance from any source; in 1995, this figure had risen to 15.4 percent (Bureau of the Census 1997, 120). So, too, the proportion of workers who have pension coverage has declined, thus leaving the next generation without an asset that can help pay for, among other things, health services in old age. It is highly unlikely, furthermore, that the appreciation of real estate values that occurred during the post–World War II decades—a rise that left homeowners of that generation with extremely valuable assets in the 1980s and 1990s—will occur again on the same scale. Thus, the younger generations, as they age, will not have available an asset that can be used, among other things, to pay for more health services. The high cost of health services and the diminished household resources available to pay for them mean that health services will not continue to provide a services solution to the generation of large numbers of high-quality employment opportunities.

The Future of Services

We live in a world of change, and we should expect the sectoral structure of employment to change over time. Services have become and will become more important to employment in the American economy because the development of the goods-producing economy generates both the demand for new services employment and the possibility for paying for such employment. If there is to be a "services solution" to the problem of sustainable prosperity in the American economy, it will be because the new services employment opportunities are high value added and can be sustained over time.

But what adds value in the economy? The key to adding value over a sustained period of time is organizational learning. Whether in agriculture or manufacturing, and whether through the organization of the innovative enterprise or the developmental state, the ability of the economy to pay its participants high returns over prolonged periods of time comes from a process in which people learn, typically in a cumulative and collective manner, how to produce higher quality, lower cost goods and services, while at least maintaining their existing standards of living. The key question is, therefore, under what conditions different types of services employment contribute to a process of economic development that can be sustained over a prolonged period of time.

If services are to contribute to economic development, they will have to do so by providing employment that, as in the computer revolution, is complementary to the production of goods. When such a complementary development of goods and services occurs, the development and utilization of services, as we have seen in the evolution of a software company like Microsoft, transforms these services into goods that can be mass produced and distributed both at home and abroad. A "services solution," that is, cannot occur separately from a "goods solution." The production of goods and the production of services are complementary activities in the dynamic process that can generate economic development.

Such a dynamic process continues to occur in certain industrial sectors in the American economy—with the development of personal computers and Internet communications being the most important examples of the 1990s. But even in these new innovative sectors, American companies, with a few exceptions that prove the rule, tend to excel in those activities, technologies, and markets in which they can compete by investing in narrow and concentrated skill bases (Lazonick and O'Sullivan 2001). Without some dramatic changes in the institutional conditions for economic development in the United States—conditions that affect both the integration of people in organizational learning processes and the commitment of finance to building broad and deep skill bases—the prognosis for sustainable prosperity in the United States is not good. Where will the jobs of the future be found? The U.S. Bureau of Labor Statistics (BLS) projects those categories of civilian employment that will grow and those that will decline most quickly over the next decade. From year to year, the specific job categories that are projected to grow most rapidly can change dramatically (compare, for example, the data presented in Bureau of the Census 1996, 408; 1997, 414; 1998, 420). Nevertheless, as shown in table 8, the BLS projections for the thirty job categories in which the largest number of jobs would be added in the U.S. economy between 1996 and 2006, no fewer than seventeen of them require only short-term, on-the-job training. Of the total number of 8.6 million additional jobs that are projected across these thirty categories, 4.9 million, or 57 percent, are low-paid jobs that require little training. Moreover, the vast majority of these low-paid jobs are in services employment.

Although the projections also show large numbers of high-paid jobs being created in occupations such as "systems analysts" and "general managers and top executives," yet in our so-called knowledge economy, low-paid services jobs seem to be the wave of the future. Sustainable prosperity requires a commitment from both the business and government sectors to invest in broad and deep skill bases that can engage in organization learn-

Table 8. Job categories with the largest projected job growth, in thousands, 1996 to 2006

Occupation	Employment		Change		Education/
	1996	2006	No.	%	Training
Cashiers	3,146	3,677	530	17	STOJT
Systems analysts	506	1,025	520	103	BA
General managers and top executives	3,210	3,677	467	15	WE+BA(min.)
Registered nurses	1,971	2,382	411	21	AD
Salespersons, retail	4,072	4,481	408	10	STOJT
Truck drivers, light and heavy	2,719	3,123	404	15	STOJT
Home health aides	495	873	378	76	STOJT
Teacher aides and educational assistants	981	1,352	370	38	STOJT
Nursing aides, orderlies, and attendants	1,312	1,645	333	25	STOJT
Receptionists and information clerks	1,074	1,392	318	30	STOJT
Teachers, secondary school	1,406	1,718	312	22	BA
Child care workers	830	1,129	299	36	STOJT
Clerical supervisors and managers	1,369	1,630	262	19	WERO
Database administrators, computer support specialists	212	461	249	118	BA
Marketing and sales worker supervisors	2,316	2,562	246	11	WERO
Maintenance repairers, general utility	1,362	1,608	246	18	LTOJT
Food counter, fountain, and related workers	1,720	1,963	243	14	STOJT
Teachers, special education	407	648	241	59	BA
Computer engineers	216	451	235	109	BA
Food preparation workers	1,253	1,487	234	19	STOJT
Hand packers and packagers	986	1,208	222	23	STOJT
Guards	955	1,175	221	23	STOJT
General office clerks	3,111	3,326	215	7	STOJT
Waiters and waitresses	1,957	2,163	206	11	STOJT
Social workers	585	772	188	32	BA
Adjustment clerks	401	584	183	46	STOJT
Cooks, short order and fast food	804	978	174	22	STOJT
Personal and home care aides	202	374	171	85	STOJT
Food service and lodging managers	589	757	168	28	WERO
Medical assistants	225	391	166	74	MTOJT

Source: Bureau of the Census, U.S. Department of Commerce, *Statistical abstract of the United States 1998* (Washington, D.C.: U.S. Government Printing Office, 1998), 420.

STOJT = short-term on-the-job training
MTOJT = moderate-term on-the-job training
LTOJT = long-term on-the-job training
WERO = work experience in a related occupation

AD = associate degree
BA = bachelor's degree
WE+BA(min.) = work experience plus BA or higher degree

ing. The four different categories of social services—institutional, producer, consumer, and social—all have distinct roles that they can play in this process. Among institutional services, educational services at the K–12 level as well as the higher education level, can generate increasing numbers of workers who are capable of engaging in organizational learning, while the public regulation of financial services can seek to ensure that this learning focuses on creating value in the economy rather than extracting value from it. Through innovative strategies of business enterprises, producer services can become integral to the development of productive resources and part of an innovative dynamic that transforms services into high-quality, low-cost goods. Consumer services can potentially help households become more knowledgeable and discerning so that they can allocate their resources to goods and services that create new challenges for learning that can contribute to the process of economic development. Social services can contribute to a developmental economy, when the quest for a better society, rather than the quest to reduce expenditures at all costs, becomes embedded in the public services that employ people whose mandate is to learn about and solve society's problems and to take leadership roles in initiatives that address new social challenges. The quality of employment opportunities that will emerge in the United States over the next generation, and their contributions to sustainable prosperity, will not just happen, but will require strategic direction by both business and government.

NOTES

We gratefully acknowledge the comments of Mary O'Sullivan, as well as the research assistance of Andreea Balan, and financial support from the Committee for Industrial Theory and Assessment, University of Massachusetts Lowell.

1. For an attempt to answer this question through case studies of generally low-wage retail services, see Bailey and Bernhardt 1997.

2. For an introduction to the perspective on economic development that underlies these arguments, see Lazonick 1999; Lazonick and O'Sullivan 1996; O'Sullivan 2001.

3. Between 1980 and 1994, the number of foreign students ("nonresident aliens") enrolled at U.S. institutions of higher education increased by 50 percent, with increases of 78 percent among those from Asia and 170 percent among those from Europe. In 1980, foreign students represented 2.5 percent of all enrollments; in 1994, 3.1 percent. In 1994, 65 percent of foreign students came from Asia and 14 percent from Europe. And it was predominantly higher degrees that they came for. Of higher education degrees awarded in 1993, foreign students accounted for 3.8 percent of bachelor's degrees (up from 2.1 percent in 1981), 12.0 percent of master's degrees (7.5 percent in 1981), and 27.3 percent of doctoral degrees (12.8 percent in 1981) (Bureau of the Census 1997, 181, 190, 194).

4. Ibid., 169, 187.

REFERENCES

Almeida, Beth. 2001. Are good jobs flying away? In *Corporate Governance and Sustainable Prosperity,* ed. W. Lazonick and M. O'Sullivan. London: Palgrave.

Bailey, Thomas R., and Annette D. Bernhardt. 1997. In search of the high road in a low-wage industry. *Politics and Society* 25 (2): 179–201.

Bureau of the Census, U.S. Department of Commerce. 1991, 1996, 1997, 1998. *Statistical abstract of the United States.* Washington, D.C.: U.S. Government Printing Office.

Campbell-Kelly, Martin, and William Aspray. 1996. *Computer: A history of the information machine.* New York: Basic Books.

Ferleger, Louis, and William Lazonick. 1993. The managerial revolution and the developmental state: The case of U.S. agriculture. *Business and Economic History* 22 (2).

———. 1994. Higher education for an innovative economy: Land-Grant colleges and the managerial revolution in America. *Business and Economic History* 23 (1).

Hauknes, Johan. 1996. Innovation in the service economy. Oslo: STEP Group Report, December.

Henwood, Doug. 1997. *Wall Street.* London: Verso.

Lazonick, William. 1998. Organizational learning and international competition. In *Growth, globalization, and governance,* ed. Jonathan Michie and John Grieve Smith. Oxford: Oxford University Press.

———. 1999. Innovative enterprise in a national economy: The theory and method of historical transformation (May). University of Massachusetts Lowell. Photocopy.

Lazonick, William, and Mary O'Sullivan. 1996. Organization, finance, and international competition. *Industrial and Corporate Change* 5 (1).

———. 1997a. Finance and industrial development, Part I: The United States and the United Kingdom. *Financial History Review* 4 (1).

———. 1997b. Finance and industrial development, Part II: Japan and Germany. *Financial History Review* 4 (2).

———. 1997c. Big business and skill formation in the wealthiest nations: The organizational revolution in the twentieth century. In *Big business and the wealth of nations,* ed. Alfred D. Chandler, Jr., Franco Amatori, and Takashi Hikino. Cambridge, U.K.: Cambridge University Press.

———. 2000. Maximising shareholder value: A new ideology of corporate governance. *Economy and Society* 29 (1).

———, eds. 2001. *Corporate governance and sustainable prosperity.* London: Palgrave.

Moss, Philip. 2001. Earnings inequality and the quality of jobs. In *Corporate governance and sustainable prosperity,* ed. William Lazonick and Mary O'Sullivan. London: Palgrave.

Olney, Martha. 1991. *Buy now, pay later: Advertising, credit, and consumer durables in the 1920s.* Chapel Hill: University of North Carolina Press.

Organisation for Economic Cooperation and Development. 1996. *Services: Measuring real annual value added.* Paris: OECD.

———. 1997. *Value added in services.* Paris: Statistics Directorate. OECD.

O'Sullivan, Mary. 2000a. The innovative enterprise and corporate governance. *Cambridge Journal of Economics* 24 (4).

———. 2000b. *Contests for corporate control: Corporate governance and economic performance in the United States and Germany.* Oxford: Oxford University Press.

U.S. Congress. 1998. *Economic report of the President 1998.* Washington, D.C.: U.S. Government Printing Office.

Diversity Dilemmas at Work

Meg A. Bond and Jean L. Pyle

AS MORE women and people of color have entered the United States workforce, the management of diversity in organizations has become both a focus of academic inquiry and central to the consulting industry (e.g., Arredondo 1996; Chemers, Oskamp, and Costanzo 1995; Cox 1993; Fine 1995; Jackson 1992; Jamison and O'Mara 1991). Recent reports, however, about the racism in several major corporations and the ongoing harassment and discrimination against women make it clear that much racism and sexism still exist in the United States. There is also evidence that when the diversity among workers is addressed, it is approached passively, as if "responding chiefly to government mandates rather than responding proactively to create value for diversity in . . . organizations" (Miller 1994, 1). Thus, even though an increasing number of organizations are articulating a value for diversity and developing diversity initiatives, the overall commitment is at best unstable and ambivalent.

This chapter addresses some critical dilemmas that have emerged as the workplace diversity movement has gained more visibility. We examine some underlying reasons why the incorporation and retention of diverse groups throughout the workforce are fraught with complexities that have constrained progress. To set a context and lay the groundwork for understanding how and why the dilemmas have arisen, we examine the origins of the sudden shift to managing diversity from a focus on equal employment opportunity (EEO) legislation and affirmative action (AA). We briefly describe the current status of diverse groups in the United States workforce. The nation has made surprisingly little progress toward promoting amicable, productive working relationships across differences, given the changing demographics of the workforce and the changing demands on organizations. This lack of progress, along with all the forces that make effective diversity management a clear imperative for the survival of today's organizations, compels us to look deeper.

We argue that current efforts to promote diversity in the workplace are complicated by countervailing forces rooted in economic and social trends, organizational traditions, and interpersonal dynamics. As a result, at least three central dilemmas feed the lack of resolve around workplace diversity. They involve the themes of (a) underlying assumptions regarding equality and difference, (b) backlash dynamics, and (c) the resistance of organizational cultures to incorporating diversity. We conclude by arguing that a multileveled, contextualist approach that attends to both policy-level and more informal interpersonal processes is necessary to manage the tensions inherent in these three dilemmas.

Diversity Policy Trends

Throughout the history of the United States, there have been alternating points of view about whether the United States will be a melting pot or a multicultural society. On the one hand, the melting-pot point of view assumes peoples of different races and ethnicities blend together or assimilate into a common culture. On the other hand, in a multicultural society different ethnic groups retain their cultures and coexist in a smoothly functioning society (Triandis 1995). Through most of the last century, the prevailing point of view had been that the United States was a melting pot and that different ethnic groups and immigrants should assimilate to an "American" way of life. The worker of early managerial approaches such as Taylorism and Fordism was a homogeneous worker. Workers did not have diverse identities. Similarly, citizens were conceptualized as homogeneous, abstract individuals who have equal rights. In spite of contrary realities, political theory denied or ignored most differences, projecting a color-blind or difference-blind point of view (Skrentny 1996).

The United States, however, has always been diverse. There has been a strong current or undercurrent of multiculturalism throughout our history; in reality, workers and citizens have never been homogeneous. In spite of the melting-pot ideology and the notion that people would assimilate, by the 1950s and early 1960s it was clear that all groups were not equally integrated into the workforce. Many women and minorities occupied low-wage jobs in an occupationally segregated labor force (Amott and Matthaei 1996; Blau, Ferber, and Winkler 1998; Jacobsen 1994). White males occupied the preferred, higher wage positions. Blacks and women became acutely aware of the pervasive racism, sexism, and discrimination they faced in education, in the workplace, as consumers, and in the community. They developed strategies for resistance designed to improve their status, which culminated in the civil rights movement, the women's movement, and much unrest in the 1960s. It was at this time that politicians began formulating approaches to redress inequalities through EEO and AA policies.

The ideas about fairness that underlie EEO and AA existed long before their prominence in the United States in the 1960s to 1980s. Skrentny states that "the basic idea [of AA] comes from the centuries-old English legal concept of equity, or the administration of justice according to what was fair in a particular situation, as opposed to rigidly following legal rules, which may have a harsh result" (1996, 6).

Affirmative action first surfaced in the United States in the 1935 National Labor Relations Act, specifying that an employer who was found to discriminate against an employee must take affirmative action to restore the victim to where that worker would have been in the absence of discrimination. In spite of these earlier occurrences of affirmative action policies and practices, EEO and AA are widely thought to date from the 1960s in the United States because of the proliferation of policies that were established beginning in 1961. Policies ranged from those that encouraged more equal treatment to ones that mandated the adoption of goals and timetables. They typically addressed the situation of minorities first, then added gender.[1]

Although values for fairness and equality underlie EEO and AA, the motivations behind the establishment of these policies were more complex. Skrentny (1996) traces the history of the development of EEO and AA policies in the United States in the 1960s, examining why AA policies were established then, when advocacy of racial preferences was a "third rail" of American politics, and there was no major constituency lobbying for them in the early 1960s. He argues that a major motivation for these policies was the need to manage growing urban and racial tensions. He shows that the process of obtaining the support of elites for such policies (that these white men otherwise might be expected to oppose) began with President Kennedy's convincing business leaders that more aggressive hiring of blacks would be beneficial in reducing the threats of urban unrest and rioting. Berman (1996) argues that Nixon aggressively promoted AA to bring blacks into the mainstream and break up the Democratic coalition. Those in power began to recognize that to maintain their positions, they would have to offer more equitable opportunities to minorities.

Even though EEO and Affirmative Action moved some women and minorities into the workforce (Badgett and Hartmann 1995; National Council for Research on Women 1996), the prevailing view of the worker as homogeneous remained unchanged. Employers continued to take an assimilationist view of employees (Loden and Rosener 1991; Thomas 1992). Loden and Rosener (1991) maintain that most organizations and managers in the United States believe "members of all diverse groups want to become and should be more like the dominant group" (28), and that, accordingly, organizations often counsel others to invest in training that will make them more like the dominant group (assertiveness training, public speaking, etc.). Thomas (1992) adds that because it has been assumed that women and minorities would assimilate to the dominant culture, organizations have not been challenged to undergo fundamental change in integrating them.

These policies have generated controversy and backlash from the late 1970s. EEO policies are passive policies that forbid discrimination, whereas AA involves active policies that require employers to make efforts to balance their workforces. It is AA, an often misunderstood term, that has aroused the most controversy. In actuality, AA consists of policies, laws, executive orders, and court-ordered practices as well as voluntary practices that are designed to promote equity. To assert that it requires quotas or the hiring and promotion of unqualified candidates is erroneous (National Council for Research on Women 1996; Simms 1995). The attack on AA so intensified in the years of Ronald Reagan's presidency that federal enforcement of AA was virtually abandoned.

The shift from AA to "managing diversity" began in the late 1980s, during the second term of President Reagan. During the Reagan years, there was a concerted effort to liberalize the economy—that is, to reduce the role of government in the economy and increase the role of the market—and to become more competitive internationally by curtailing production costs. A major component of the deliberate shift to so-called free markets was deregulation; other components involved reductions in federal government expenditures (particularly cutbacks in the provision of social services), and efforts to curtail the already-limited power of labor unions. The reduction in regulation involved diminished emphasis on monitoring and enforcing EEO and AA policies as well as decreased attention to worker health and safety concerns and environmental conditions.

It is within this context, during the second half of the 1980s, that several broad trends made managing diversity appear to be a strategic necessity for organizations seeking to maintain competitiveness (Arredondo 1996; Cox 1993; Jackson and Alvarez 1992). One major impetus was the publication of *Workforce 2000* (Johnson and Packard 1987), which projected that U.S.-born white males would make up only 15 percent of the net new entrants to the labor force (entrants minus leavers) during the period from 1985 to 2000 (Arredondo 1996; Jackson & Ruderman 1995).[2] Also influential were other changes in the structure and the composition of the internal labor force: the movement of some women and minorities up from the lower levels of the hierarchies, the flattening of hierarchies with the elimination of layers of middle managers during the corporate restructuring of the 1980s, and the increasing age diversity of the workforce.

In addition, the ability to manage diversity was thought to lead to better decision making and increased productivity. Organizations that foster good working relationships among diverse people can theoretically draw more widely on the available labor force and more fully develop workers' capabilities. Studies have shown that employee morale is related to identity groups (gender, racio-ethnicity, nationality) and, in turn, is linked to work quality, productivity, absenteeism, and turnover, all of which affect profitability (Cox 1993). In addition, creativity, innovation, and problem solving can potentially be enhanced by pooling the perspectives of diverse people (Thomas 1992). Therefore, if diversity management facilitates people working well together, it can reduce costs and increase productivity (Cox and Blake 1991). Furthermore, organizations were seeking to compete more effectively in domestic and international markets that involved different ethnic or racial groups, and recognized the usefulness of having diverse employees on the front lines (Arredondo 1996; Jackson and Ruderman 1995). Given this wide range of incentives, many corporations have established diversity programs (see Jackson et al. 1992; National Council for Research on Women 1992).

However, the underlying trends and issues were much more complex, resulting eventually in the dilemmas impeding progress in achieving diversity that we discuss in the rest of this chapter. First, the increasing emphasis on managing diversity as a way to incorporate different peoples into the workplace occurred at a time when AA was being challenged. Because both managing diversity and AA focus on diversity in the workplace, it might appear contradictory that managing diversity was rising in prominence when AA was under attack. However, understanding the overall context of the liberalization of the U.S. economy helps explain why this happened. The objective of EEO/AA was equity and fairness; the main rationales used to promote diversity management, however, have centered on improving corporate efficiency and profitability. As Thomas puts it, diversity management is strictly about "enhancing the manager's capability to tap the potential of a diverse group of employees" (1992, 311). He reasons that it is not about civil rights, women's rights, redressing past wrongs, or eliminating racism or sexism. Thus, it was only the rationale of managing diversity that was compatible with overall ideological shifts in the 1980s toward more market solutions and fewer government interventions. This has led to considerable debate over what is fair and to what extent people should be treated the same or differently.

Second, another underlying complexity was that managing diversity became popular during a period when many organizations were reducing the size of their workforces. Downsizing and the shift to managing diversity were both particular manifestations of broader socioeconomic trends during the period. The liberalization of the economy in the 1980s gave corporations increased freedom to restructure and lay off employees as well as "relief" from regulations such as AA. The fact that diversity-management programs and their reassuring language became more prevalent at the same time that organizations were actively downsizing was potentially misleading for employees. The ideology of managing diversity, which suggests that management is focusing positively on the value of different workers, can obscure the full impact of the downsizing and the increasing inequality between top management and workers. However, some workers have recognized the problems with managing diversity programs in the context of shrinking workforces and economic stress and have resisted.

So where are we now? We have witnessed considerable pressure to liberalize the U.S. economy since the early 1980s, the origin and growth of the managing diversity movement since the late 1980s, and the revitalized attack on AA since 1995. Although it may appear that AA is waning in importance while managing diversity is becoming the preferred approach, the societal debate over what should be done with respect to diverse people in the workforce is lively, heated, and controversial.

Both approaches, EEO/AA and managing diversity, are very much part of the way our society is currently addressing diversity issues.[3] However, both types of policies are also now under fire: AA as discussed above and managing diversity more recently, because of, first, high-profile cases in which firms that had implemented diversity programs were subsequently revealed to have blatantly racist attitudes and practices and, second, controversy over whether the consulting industry was providing carefully developed, individually tailored diversity-management programs in return for high fees. The current environment is one in which organizations are struggling to determine how these two approaches can best be utilized and how they can coexist.

In addition, even though there has been considerable recognition that attention to diversity is a strategic necessity, the attention it has received has not translated into significant shifts in the distribution of jobs and resources, into smooth working relationships across differences, or into the necessary cultural transformation of organizations (Bond and Pyle 1998; Cox 1993; Jackson 1992; Pyle and Bond 1997). Our current crossroads represents, at best, an ambivalent or unstable commitment to supporting diverse groups in the workplace. The overarching context for this unsure commitment has been shaped by broad economic and political trends. This ambivalence is evidenced both in the shifting goals that have driven diversity policies and by the lack of real numbers in the workforce. We believe the limited commitment is also fed by some critical, emerging dilemmas that play out at the organizational level. We examine these dilemmas below and explore how they can help explain why the incorporation and retention of diverse groups is fraught with complexities that constrain progress.

Emerging Dilemmas

As is evidenced by historical trends and the current status of women and people of color in the United States workforce, there remains a lack of clear resolve around fully supporting and promoting diversity in our current workforce. Underlying the lack of progress—in very quiet and generally unarticulated ways—are three dilemmas. First, assumptions about the relationship of similarity and difference to equality can give rise to contradictory paradigms for approaching diversity. Second, initiatives on behalf of workplace diversity have triggered serious unintended consequences and backlash reactions at multiple levels of analysis. Third, diversity efforts have been hampered by limited changes in team cultures and organizational values. These dilemmas make progress toward achieving an effective and diverse workforce the particularly complex challenge that it is today.

Dilemma No. 1: False Dualisms

Assumptions about sameness, difference, and equality can give rise to seemingly contradictory paradigms for diversity work.

Policymakers rarely stop to reflect on the paradigmatic assumptions about sameness and difference underlying their EEO/AA or diversity initiatives, yet conflicts at this level can easily derail even the most earnest of organizational efforts. As previously described, the seemingly compatible policies of EEO/AA and managing diversity have actually been driven by very different economic and political priorities. The rationale behind EEO/AA emphasizes equity in terms of giving all people the same access to opportunities. Underlying the managing diversity movement are promises of increased efficiency when differences among employees are respected. In a sense both purport to strive for fairness; however, the rationales, and thereby the assumptions about what constitutes fairness, are quite different. Similarly, when organizations adopt diversity programs (even when they include some elements of both AA and diversity management), they often do so without reconciling the guiding paradigms. As a result, approaches to diversity in the workplace can vary dramatically in terms of the assumptions made about the relative importance of sameness and difference among workers.

In the simplest form, sameness paradigms emphasize that all people should be treated the same and that inequity results when groups are treated differently. This perspective is akin to Ryan's (1981) fair play perspective, which stresses that all individuals should have the freedom to pursue opportunities. However, the perspective also assumes that there is no guarantee of success and that natural processes "will insure that the ablest, most meritorious, ambitious, hardworking, and talented individuals will acquire the most, achieve the most, and become the leaders of society. The relative inequality that this implies is seen as not only tolerable, but as fair and just" (9). In fact, by extension, equality of results is seen "as unjust, artificial, and incompatible with the more basic principle of equal opportunity" (9). This has been the dominant perspective in the United States almost from the beginning.

Difference paradigms emphasize the existence of diversity among people and the need to design strategies for equity that attend to those differences. From this worldview, inequality results from ignoring basic differences in power, opportunities, and access to resources. Similarly, Ryan describes a fair-shares conception of equality that "emphasizes the right of access to resources as a necessary condition for equal rights to life, liberty and happiness" and is "committed to the principle that all members of the society obtain a reasonable portion of the goods that society produces" (9). This is an entirely different view of what constitutes fairness and justice; and, as Ryan very eloquently argues, "the conflict between Fair Play and Fair Shares is real, deep, and serious, and it cannot be easily resolved" (9).

These differing paradigms have major implications for organizations seeking to foster workplace diversity. The sameness versus difference distinction gives rise to questions about whether diversity initiatives should strive for equality of access or move more toward equality of results. Many legal mandates operate from an emphasis on access and define equity as identical treatment. However, this denial of difference conflicts with most organizational development perspectives about how actually to manage a diverse group that emphasize the need to treat people differently based on their varied and unique cultural backgrounds (i.e., tailoring approaches to people based upon their gender, race, culture, etc.).

Although the sameness and difference paradigms operate from different basic assumptions, characterizing them as opposites is problematic. Other dualisms, such as posing "difference" as the antithesis of "equality," have similarly distracted diversity efforts. Scott (1994) describes the way the "equality versus difference" distinction plays out in feminist writings. From the equality perspective, "sexual difference ought to be an irrelevant consideration in schools, employment, the courts and the legislature" (362). The differences perspective believes "that appeals on behalf of women ought to be made in terms of the needs, interests, and characteristics common to women as a group" (362). However, as Scott argues, the posing of these two positions as mutually exclusive does us a disservice—"when equality and difference are paired dichotomously, they structure an impossible choice" (365).

As we illustrate in the following subsections, these various ways of posing sameness and difference as antithetical are the essence of the first dilemma that makes progress on workplace diversity difficult. When such perspectives on diversity are framed as if in opposition to one another, discussions become debates about deeply held beliefs regarding fairness and justice; common ground can be difficult to find.

Policy approaches and assumptions of same versus different. In a manner that parallels the sameness and difference paradigms, legal and policy approaches to rectifying inequities vary, based on whether the objective is equality of opportunity or equality of outcomes. On the one hand, some employment policies are designed to ensure everyone has equal access to opportunities by eliminating discriminatory behavior in hiring, job assignments, training, promotion, and firing. Here the emphasis is on treating everyone the same at least in terms of access to jobs. However, whereas approaches based on the assumption that everyone should be treated the same might address the issues of equal opportunities and due process, they do not necessarily address more subtle dynamics internal to the organization that affect the equality of outcomes. On the other hand, laws and policies that focus on the potential for disparate impact—that is, assess the impact of particular organizational practices on different types of individuals and attempt to make accommodations that redress differential outcomes—recognize that groups may be differentially affected and may, therefore, need to be treated differently. From this perspective, organizations are challenged to adapt to groups that differ with respect to basic assumptions about how individuals should approach the tasks and relationships involved in accomplishing their work.

EEO and AA both primarily focus on equality of opportunity; however, as mentioned earlier, EEO does so in a more passive manner than does AA. The overarching rationales for establishing them were to prohibit discrimination (EEO legislation) or to take steps to ensure and monitor movement toward parity (AA). By redressing differences created by historical inequities and by requiring employers to monitor progress, AA also begins to attend to equality of outcomes. Even though driven by pragmatics, diversity management efforts move in the direction of addressing equality of outcomes. Managing diversity is a major shift from EEO/AA in many ways. The worker is no longer considered homogeneous. In fact, quite the opposite occurs. As the dimensions of managing diversity unfold, each worker becomes a combination of many diverse identities. Managing diversity is taken to mean valuing the uniqueness of each individual. Thus, rather than treating all people the same (a focus of EEO) or providing some preferential treatment for selected groups (AA in its special efforts to recruit, promote, and retain women and minorities), this approach moves toward treating individuals differently. This view assumes people will be integrated but not necessarily assimilated.

Although the paradigms of fair play and fair shares or equality of opportunity and equality of outcomes represent different worldviews that shape and reflect deeply held beliefs about equality and justice, posing them as opposites can serve primarily to "lampoon [Fair Shares] as the recipe for a ridiculous, bizarre, and inhuman world . . . as though it were an either/or proposition. Logically of course, the terms are not at all contradictory; more important, defining the issue in this oversimplified fashion obscures and dismisses most of the relevant questions" (Ryan 1981, 26–27).

In similar fashion, Berry (1995) shows how opponents currently distort and oversimplify the meaning of AA, making it seem incompatible with their notions of equity and fairness in order to bias public opinion and to distort and control the widespread debate over the continuation of the policies.

Another dynamic that hinders thoughtful attention to diversity dynamics and intensifies the polarization of perspectives is the focus on legal definitions of inequity. The enforcement of policies is driven by whatever has become the formal definition of what is (and then, by extension, what is not) inequitable treatment. The law and many organizational policies essentially require one to set up a dichotomy between guilty and not guilty. "A little bit guilty" is not typically an option. This definitional process poses "equitable" as the opposite of "illegal" (i.e., if it is not illegal, it must be just). However, there are many subtle shades of unequal treatment and much harassment and discrimination that would not hold up in a court of law but that is nonetheless based in some sort of biased actions and involves some level of unfair treatment. The rules of the courtroom favor arguments that are unidirectional and unconvoluted by "it depends" or other conditional statements. The requirements are for logical consistency, and "truth" is determined by the tightness of an argument. Ironically, a myopic focus on what is considered justice under the policies of organizations or laws of our nation can hinder the way we think about equality and the diversity dilemmas. Although laws and policies play an essential role in holding people accountable for their actions, they also provide a language and a decision-making framework that can constrain the development of creative solutions.

In short, policy debates are often inappropriately framed as a choice between dualisms or dichotomies (i.e., sameness vs. difference, difference vs. equality; equal opportunity vs. equal outcomes, fair play vs. fair shares, illegal vs. equitable). In addition, there are some complex side effects of adopting a singular focus on either similarities or differences—effects that contradict the original intent of the stance.

Side effects of the focus on sameness. The intent of the equal opportunity and fair play perspectives is ostensibly to create an even playing field and move forward from there. Adopting such a position based on the sameness of treatment, however, is in essence rooted in an individualistic analysis. This stance, in effect, de-emphasizes the importance of group membership and represents an individualizing of experience even when the original intent of a policy (e.g., AA) might have been to address prior inequities experienced by a class of people.

Emphasizing sameness of treatment is also an essentially acontextual perspective. It ignores the fact that the meaning of an individual's behavior can vary with race, gender, and power (e.g., jokes that might be considered harassing behavior by a woman when told to her by a male supervisor in her office, would not necessarily constitute harassment when told by a lone woman coworker in an otherwise all-male shop). The interpretation of a person's behavior is affected by race and gender dynamics (e.g., the classic scenario in which the woman who leaves early to pick up children is seen as lacking commitment whereas the man who leaves for the same reason is considered heroic). The implications of a policy can clearly differ based on race and gender. For example, given common differences in African American and Anglo family structures, an African American man living in the north who comes from a close, extended Southern black family may need more than one day off to travel to an uncle's funeral, whereas his white coworker might not even need the full day he would be allowed because attending a distant funeral is less likely. Another, perhaps more classic, example involves family leave policies, which generally have more impact on women since they tend to be the family caretakers. Thus, as Scott (1994) argues, if we ignore difference in the case of subordinated groups, it "leaves in place a faulty neutrality" (362) and totally decontextualizes our analysis of organizational behavior.

If sameness is equated with equality, we reinforce and recreate the problems associated with assuming a homogeneous worker, we run the risk of ignoring critical contextual determinants of organizational behavior, and we diminish our capacity to think creatively about how to build productive work teams.

Side effects of a focus on differences. From the differences perspective, individual behavior can be best understood, interpreted, and assessed when considered in the context of one's unique culture, current status, and history. However, this stance also has some potential contradictory consequences. First, in our culture, difference is often equated with wrong, pathological, unknown, and scary. When work on race and gender focuses on comparisons and differences, generalizations about white men remain the normative standard against which others are compared (Crawford 1995; Unger 1992). In reaction, some attempts to appreciate cultural uniqueness have simply replaced a "difference as deficit" perspective with a "difference as better" perspective (Fine, Johnson, and Ryan 1990, 306). For example, celebrating diversity has sometimes taken the form of placing a higher value on so-called women's perspectives or inclinations (e.g., empathy or collaborative decision making) or considering minority cultures morally superior to our dominant culture. The key issues here are: Who gets labeled as different? Who gets to define "other"? and Why do differences make a difference in who gets access to job resources and power in organizations? Although valuing less dominant perspectives and approaches might be a welcome relief, the problem is that the "difference as deficit" and "difference as better" perspectives both "assume some hierarchy of skills and behaviors based on comparing white men with women and with minorities" (Fine et al. 1990, 306). A focus on whose approach to the work is better can derail discussions about how organizational power, structures, roles, and/or expected outcomes need to be restructured to accommodate and incorporate diverse persons. Alternatively, if work settings begin with an assumption of heterogeneity, diversity becomes the norm and variations in approach are expected as part of the daily fabric.

Another contradictory consequence is that a focus on differences among groups can reinforce stereotypes and assumptions of homogeneity within the groups, and thereby become a basis for discrimination based on group membership—even if the original intention was to foster an appreciation of the ways people might differ. That is, the focus on differences (among groups) can actually inhibit the appreciation of differences (within groups) (cf. Bond and Keys 1993, 2000). For example, there is considerable diversity both among Asian Americans (Kim and Lewis 1994; Silka and Tip 1994) and among Latinos/as (Bernal and Enchautegui-de-Jesus 1994) that is masked by looking at them as cultural groups. No matter what the original intention, contrasting groups with one another can easily perpetuate occupational segregation by leading back to the argument that one group is better suited for a particular type of work or role than another.

When taken to an extreme, the stance that everyone is different implies that all differences are essentially equivalent and thereby renders diversity "nothing more than a benign, meaningless concept" (Nkomo 1995, 248). This universalizing of difference essentially renders it equivalent to sameness (i.e., we all have our differences in common), and this homogenizing of experience dilutes, downplays, and/or ignores the influence of marginalization and oppression based on group membership (i.e., race, class, gender) in shaping people's experiences in organizations. In addition, the language of managing diversity with its focus on the individual and the myriad ways people are different fails to recognize that all differences are not created equal in terms of their impact on people and their access to power and other resources (Sessa and Jackson 1995). There is a hierarchy within the category of minority groups. The "everyone-is-different" perspective obscures the ways people are disadvantaged based on particular group memberships. Certain diverse groups are particularly devalued and delegitimized and have less power.

A focus on distinctiveness can also result in erroneous or misguided attributions about the sources of difference. Mednick (1989) argues that constructs used to describe the distinctiveness of women (whether fear of success, androgyny, or the notion of a different voice) attribute the uniqueness primarily to gender and thereby ignore the influence of "cultural, socioeconomic, structural, or contemporaneous situational factors that may affect behavior" (1120). This emphasis on gender as the primary correlate of a particular behavior pattern or work outcome for women stands in direct contradiction to research that indicates that some previously assumed gender differences no longer emerge when researchers control for power and status (e.g., Crawford 1995; Holloway 1994; Lott 1987; Unger 1992). Similarly, it is important to avoid confounding cultural distinctiveness among racial groups with adaptive responses to oppression (Ogbu 1993; Watts 1992). A focus on intrinsic differences between women and men or among different racial groups leads us almost inevitably to individual-level change prescriptions and to ignoring sociopolitical analyses (Crawford 1995; Mednick 1989). Thereby, as Mednick argues, what might have begun as genuine efforts to appreciate uniqueness are transformed into support for "conservative policies that, in fact, could do little else but maintain the status quo" (1122).

Summary. It is essential that we address the dilemma that emerges when diversity initiatives pay attention only to difference or only to similarity. The positions of same versus different or equality of opportunity versus equality of outcomes are essentially extreme positions and can result in nonproductive debates if allowed to remain dualisms. Both focusing on and ignoring difference risk recreating it, and, as Scott argues, "the antithesis itself hides the interdependence of the two terms, for equality is not the elimination of difference, and difference does not preclude equality" (1994, 362). To counter the problems that emerge from a focus on sameness versus differences, we need to adopt a perspective that is founded on the belief that each group organizes and defines experience within its own set of cultural assumptions and experiences. However, we need to work from models that both acknowledge differences and refuse to accept them as full explanations for employment patterns, that is, adopt models that challenge dualistic thinking. We need conceptual approaches that are ecologically based and that keep the discussion of distinctiveness rooted in awareness of social, political, and cultural forces such as racism and sexism that shape the unequal distribution of power and resources. In essence, managers, consultants and researchers all need to work from models that consider sameness and difference simultaneously and contextually—that place differences in the context of similarities and similarities in the context of differences.

Dilemma No. 2: The Unanticipated Fallout

Diversity policies and programs have generated a set of reactions that work in opposite direction of the original policy intent.

Both the externally driven policies and the internally designed diversity management programs have generated fallout and side effects that are barriers to further progress toward a fully integrated and effective workforce. First, diversity policies have triggered an intense public debate about whose civil rights are at issue. Second, the use of EEO legislation and AA policies to address the problems of workplace inequities and discrimination has resulted in a conundrum of legislation to address still more inequities, additional guidelines and rulings to interpret the legislation, and more opportunities for litigation. Third, the more recent focus on managing diversity has generated direct reactions to diversity programs. In a sense, we see all three of these phenomena as evidence of the disruption people feel as a result of the increased attention being paid to diversity in the workforce. Each of them is indicative of the fears and the potential for redistribution of resources associated with the deep economic and psychological changes that are facing us.

Questioning of whose civil rights? AA policies have been controversial since the late 1970s and have recently been under widespread attack on many levels. As discussed earlier, much of the controversy is over the actual meaning of AA and perceptions of what constitutes fairness. Statements have become politically charged, as campaigns reduce these issues to hyperbole and sound bites. Opponents often shape the debate in misleading ways that make people fear that AA involves quotas, hiring of less qualified candidates, and reverse discrimination. However, the National Council for Research on Women (1996) reports the results of a survey that shows that when people understand that AA does not specify quotas, they overwhelmingly support it. Plous (1996) reports the results of five public opinion polls in 1995 regarding AA. When surveys offer intermediate choices regarding AA (rather than a dichotomous choice between AA as it currently exists versus no AA at all), most people favor maintaining some form of AA.

Debates regarding EEO and AA policies increasingly focus on "Whose civil rights?" with people on each side indicating they believe that their rights have been infringed upon. A backlash has developed, as some people who were denied positions (particularly males and/or Caucasians) assert that such policies result in reverse discrimination (i.e., the males or whites argue they have been discriminated against because they feel women and minorities are given preference). Considerable press and political attention are given to such allegations. A 1995 Department of Labor study, however, disputes the notion of widespread reverse discrimination against whites. It reports that only a hundred of three thousand discrimination cases filed involved reverse discrimination, and only six of these had claims that could be substantiated (Wilson 1995). The Equal Employment Opportunity Commission (EEOC) reports that of ten thousand reverse discrimination cases filed during the period from 1987 to 1994, only 10 percent had merit (National Council for Research on Women 1996).

This backlash is at least partially rooted in a comfort with what has long been the status quo and a sense that changes in this arrangement are a violation of workers' rights. The reaction that changes are unfair is further fueled by the belief among white men that women and minorities do not experience discrimination in the form of unequal resource support and do have equal, if not better, chances of receiving organizationally based rewards (Fine et al. 1990; Kossek and Zonia 1993). The backlash is like a homeostatic mechanism pulling against change—pulling back to a situation that is believed by those who benefit most from it (i.e., white males) to be stable. Much evidence exists, however, to suggest that the current situation is neither stable nor perceived by many women or people of color as fair.

Such backlashes tend to erupt more in periods of economic distress. The recession of 1990 to 1991 and the long, slow recovery period were characterized by continued corporate downsizing, increased fears of job loss, and stagnation in real wages. Data for the period 1989 to 1995 show that job loss in the 1991 to 1993 recovery was as high as in the 1981 to 1983 recessions. Although losses were higher for blue-collar workers, they increased substantially for white-collar workers as well. In addition, wages stagnated (growing at one-fifth of a percent a year), median family income fell, and the gap between the respective shares going to the top fifth of income groups versus the bottom fifth widened (Mishel, Bernstein, and Schmitt 1997). In such times people can fear even the possibility of the advancement of others, such as minorities or women, even though the data do not support such fears. Statistics on what has actually happened show that the earnings of African Americans as a percentage of whites' incomes fell from 1973 to the present (Wessel and Schlesinger 1997).[4]

In another form, backlash has erupted among workers without children who increasingly resent what they perceive as special policies (i.e., family-friendly policies such as flexible scheduling, family leave, and child care assistance) tailored to employees with children. Although more than 20 percent of those surveyed in a study indicated they had to accept added responsibilities to cover for parents with whom they worked (Williams 1994), the starting assumption is that the current arrangement (with special family policies) is fair or neutral. Workers with children, however, can argue that the current workday structure itself (nine to five) is a biased arrangement that is convenient only for the childless or those with a home-based partner or helper. In response, some organizations have established worklife programs that offer all employees, with or without children, free time from work to attend to needs in their personal lives. More recently, organizations are being encouraged to define family to encompass all an employee's personal concerns for a more balanced life and to restructure jobs to accommodate these concerns (Shellenbarger 1996).

Proliferating legislation. Complications around EEO/AA-driven policies have arisen as new legislation is continually being passed or existing legislation is being interpreted to address additional potentially discriminatory situations. In the 1990s, new legislation has included the Americans with Disabilities Act (1990), the Glass Ceiling Act (1991), and the Family and Medical Leave Act (1993). The number of different groups seeking legal redress under existing or new laws has grown. For example, people who are readily distracted by noise at work have sought protection under the Americans with Disabilities Act (ADA), requesting more isolation, new work assignments, or changes in managerial styles (Pollock and Lublin 1997). As the number of laws rises, it becomes clear that it is impossible to write legislation to specifically address all situations. Meanwhile, as legislation increases, possibilities for litigation multiply. The increased litigation has resulted in conflicting rulings in the courts in different states and at different levels of the judicial system, adding further complexity to implementing "the law."

Furthermore, there is confusion over the meaning of some EEO laws and how they are to be operationalized, particularly in times of rapid technological change. Guidelines multiply in the effort to clarify how organizations can operate in compliance with the law. These, in turn, add to the confusion and increase the danger of a litigious gridlock. For example, EEOC guidelines were designed to assist organizations in job interviewing in a way that does not violate the ADA. The guidelines have led instead to enormous confusion in organizations regarding how to proceed with application questions and interviews. The challenge is how organizations can obtain information needed to accommodate a disability without being able to use it as a basis for discriminatory actions. In addition, the dramatic change in communications caused by widespread use of the Internet has posed further compliance issues for employers. With the rapid rise of the use of the Internet for job postings and searches, companies are faced with collecting race and gender data on everyone who clicks about a job opening in order to comply with Office of Federal Contract Compliance Programs (OFCCP).

In reaction, organizations have adopted a variety of defensive strategies as they seek to minimize the adverse impact of litigation or to avoid legal action. Recent high-profile cases and the increasing network of legislation have fueled organizational fears of litigation, particularly class-action discrimination lawsuits. Many organizations agree to out-of-court settlements to avoid further adverse publicity. Moreover, several sectors have moved to mandatory arbitration of employee complaints to avoid lawsuits. Employee complaints in the securities industry have been submitted to mandatory arbitration, with the three-person panel chosen and paid for by the industry itself. Other sectors such as insurance, banking, aerospace, and construction adopted this model to avoid employee lawsuits. Many legislators, however, fear that employees are being forced as a condition of employment to sign away their right to a trial. Opponents are further concerned that, although arbitration may work when participants have equal power, this balance does not exist in the employer–employee relationship.

While EEO/AA laws were originally passed to enforce more equitable policies, the legal reactions to these policies have brought home the fact that every act of inequality cannot be legislated away. The proliferation of legislation has, at best, created a mass of laws that are difficult to navigate and, at worst, reduce the effort to absurdity. Rather than instilling a value for diversity, which ultimately is a desired outcome of EEO/AA policies, they risk being seen more as a hindrance (even by corporations that believe in AA) and/or as something to be maneuvered around. Ironically, the rise of the managing-diversity movement itself, where the differences that are to be valued are much more widely defined than under the EEO/AA legislation of the 1960s and 1970s, may have contributed to this burgeoning legislative tangle.

Reactions to in-house diversity programs. The prevailing idea behind most in-house diversity programs is that support for diversity will be enhanced if understanding is increased and the sexism/racism of individual employees is confronted, challenged, and changed. Diversity initiatives range from short one-shot orientations or showing of films to extensive training and development that is integrated into other organizational development efforts.

Numerous unanticipated problems have ensued from both the one-shot and the more intense programs. First, many employees perceive their organization's interest in diversity as simply a superficial gesture that will not result in any fundamental change in the organizational culture. This can be particularly problematic when workshops on the value of diversity are also perceived as condescending and/or disconnected from the realities of everyday life at work. In addition, at times, employees see diversity training as a tool to support a strictly management agenda, particularly when the initiatives lack inclusiveness and an understanding of the power differentials at work. In such cases, participants resist the diversity programs by not taking them seriously. Second, when workers feel unfairly accused or blamed for inequalities in the workplace, or when discussion of differences results in the generation or reinforcement of stereotypes, what was designed to be a set of exercises in coming together can result in increased divisiveness and rising animosities. The increased resentment is obviously the opposite of what was originally intended.

A significant part of the reactance seems rooted in how participants in diversity programs understand the meaning of increased diversity. They are being asked to become more sensitive and understanding of others. Requirements to work effectively with others who are different, however, can lead to fear of reduced power, confusion about how to behave, and potential loss of job security. For example, mixed groups seem to signal diminishing status and loss of security for white men (Gruber and Bjorn 1982; Tsui, Egan, and O'Reilly 1992). These feelings may be intensified by evidence that, when the proportion of women in a job classification goes up, the salary level for both men and women in that job class declines (Tomaskovic-Devey 1993). Furthermore, the greatest backlash against diversity initiatives and minority employees tends to occur in organizations experiencing the greatest flux, particularly the type of uncertainty that comes with reorganization and downsizing—that is, when the fear of job loss is the most intense. There is perhaps an inescapable irony here: The economic and political conditions that have made the managing-diversity movement appealing as a national trend (i.e., promises of increased efficiency in times when maintaining national economic competitiveness is a major concern) are exactly those conditions that set the stage for resistance within the organization.

When in-house diversity programs are well integrated into other organizational development efforts and have top-level support, people are challenged to change deeply set patterns. Diversity programs signal not just a possible new world order (i.e., who has what positions of power), but they also signal the need to adopt a new worldview (i.e., basic values and expectations about how people should view relations). As the informal rules or codes of behavior change, the uncertainty about what are acceptable ways to relate on the job can be quite unsettling. This uncertainty about how to deal with the anticipated effects can become a driving force behind much of the resentment of these new in-house programs.

Summary. Any attempts to increase support for diversity must consider the variety of problems and backlashes that can result. From a macroperspective, much of the controversy and backlash can be understood as a recurring phenomena that emerges in response to limited resources and societal power shifts. Whether instituting organizationwide EEO-type policies or mandating diversity training programs, organizations must also understand more microlevel changes in the support systems that people need to enable them to adapt to dramatic shifts in expectations. Any efforts to address the diversity challenge need to incorporate an appreciation for the interplay between the meaning that increased diversity holds for the individual (e.g., threat versus opportunity), the capacity of the organization to provide support for change (both formally and informally), and the options or constraints provided by broader policy and legislative trends.

Dilemma No. 3: Culture Change

Efforts to support workforce diversity have been hampered by limited changes in organizational values.

In addition to the problems posed by diversity paradigms and various backlashes is the equally troublesome dilemma that organizational cultures have not changed dramatically enough to fully support diversity. EEO, AA, and internal diversity programs simply have not been the successful antidotes proponents had hoped for, partly because underlying community and organizational attitudes and values have not changed significantly. In fact, the hostility among groups in society at large is mirrored by workplace conflict involving such issues as race, gender, age, sexual orientation, class, and disability.

A study conducted by the Center for the New American Workforce (Miller 1994) found that, whereas many organizations have implemented some sort of policy and/or training initiative to address the unique needs of different types of people, the initiatives often do not translate into changes in the quality of worklife for employees. Just responding to legislative mandates clearly does not automatically lead to changes in organizational culture and values. We contend that efforts to fully support diverse groups in work settings are hindered by two dynamics that surround the limited changes in organizational cultures. First, as a result of the kinds of backlash described above, much bias and discrimination have gone underground. Most discrimination is of such a subtle and indirect nature (e.g., aversive racism/sexism) that it is difficult to document or even articulate. As a result, although formal policies can open up some previously unavailable options, these policies are only effective if they are backed up by transformation of informal interpersonal values and processes within the organization. The example of family-friendly policies is a good illustration of this dynamic where the organizational values and culture often prevail over the formal policies. Second, there are some inherent value contradictions among the popular organizational trends of employee empowerment, team development, and diversity management that grow out of the varied foci of these development efforts (i.e., the individual focus of empowerment and the group focus of team-building versus the necessity that diversity efforts cut across levels). The fact that these approaches are embedded in our individualistic society (rather than a society that places more value on the group) further complicates diversity efforts.

Underground racism and sexism. As policies have made blatant discrimination illegal and unacceptable, biases are expressed in more subtle forms. Even though this may not be a conscious or deliberate change on the part of any particular individual or organization, it is, nonetheless, insidious. This more subtle form of racism, termed aversive racism, is being increasingly recognized (Dovidio and Gaertner 1996; Gaertner and Dovidio 1986). Aversive racism involves underlying racially biased attitudes and behaviors of people who claim that they are not prejudiced. Parallel to this, gender discrimination perpetuated by people who believe themselves not sexist could be termed aversive sexism. What is often expressed is a subtle racism and sexism communicated through eye contact, who is credited with good ideas and good work, how company celebrations are framed, where and when important decisions are made, and which stated points of view are actually acted upon.

Research indicates that aversive racism works in some unexpected ways. First, it may not lead people to express more negative feelings about minorities or have lower expectations of minority group members. It may, rather, lead them to express fewer positive reactions. That is, there may be little direct expression of negative sentiments, but white men continue to receive more positive evaluations when all else is considered equal (Messick and Mackie 1989). In addition, there is the observed difference between what people indicate as their values and what they are willing to do about them (Messick and Mackie 1989). As a result, much of the discrimination felt by members of underrepresented groups is not the subject of formal complaints and is almost impossible to address through legal and policy-based mechanisms (cf. Bond and Pyle 1998). Organizational policies and diversity initiatives may inhibit direct expressions of hostility, but they leave underlying biases untouched. These more subtle forms of discriminatory behavior have serious consequences in the workplace as well as in society.

Many would hope that a younger generation has grown up with more awareness of diversity and thus would be more tolerant. Studies reveal, however, that even younger generations prefer to work with others like themselves. Contrary to the expectation that younger workers would have backgrounds equipping them to work with diverse peoples, workers younger than twenty-five showed no greater preference for diversity. Employees having a greater tolerance for working with people different from themselves are people who have been exposed to different cultures in their home community or neighborhoods. Most people younger than twenty-five have not had such exposure (Families and Work Institute 1993).

The subtle forms of discrimination are difficult to recognize and, therefore, to address. The move from the blatant to the more subtle expressions is a manifestation of the fact that underlying values cannot be mandated away. Furthermore, the less blatant nature of the biases renders the problem fairly inaccessible to additional policy efforts. Thus, what often emerges is a gap between the formally stated goals of diversity initiatives and the values actually adopted during day-to-day organizational life. Ultimately, the informal values that coalesce into an organizational culture will dictate what supports really exist for diverse groups of employees.

The example of family-friendly policies. Family-friendly policies are not as common as the popular press might lead one to expect. Even in those organizations that offer options for working at home, part-time schedules, flextime, or job sharing, the reality is that limited numbers of employees take advantage of them. Some organizations adopt such policies as public relations and recruitment tools, but the benefits are often not available to all employees. Options provided for top-level managers are less available to those in lower-level jobs, particularly women in nontraditional work.

In addition, some policies are more token than real. The highly-debated Family and Medical Leave Act of 1993 has provided employees in companies with fifty or more employees the option for a three-month unpaid leave when needed for family care. Since its passage, few have made use of the option. An employee survey found that losing three months of income is simply not an option for most working families, particularly when faced with a new and/or ill family member (McGonagle et al. 1995). There is also evidence of widespread organizational noncompliance with the law. A year after it became law, four companies in ten affected by the law were not adhering to it (O'Gara 1995).

When people are able to utilize family-friendly benefits, their actual impact is unclear. On the one hand, some studies show that benefits such as child care have increased both annual hours worked by women and their attachment to their employers (e.g., Lehrer, Santero, and Mohan-Neill 1991). On the other hand, recent work based on an analysis of more than thirty-eight thousand companies indicates that there is a low correlation between policies and the representation of women at higher levels of the organization (Sharpe 1994). Some of the companies with the most progressive and comprehensive family-oriented benefits have some of the worst records for promoting women.

In addition, family-friendly policies have actually had adverse effects for some women. Unwritten rules of the organization often punish women who make use of such policies (e.g., by relegating them to a no-growth "mommy track"). The very existence of the policies can emphasize differences between women and men, which, as discussed under Dilemma No. 1, can make it harder to overcome stereotypes. The use and impact of such policies is highly dependent upon the supportiveness of individual supervisors and the overall organizational culture (Shinn et al. 1989). For example, policies may allow a woman to rearrange her schedule to care for a sick child, but if her supervisor then judges her less committed, gives her a poor performance evaluation, or reassigns her important tasks to others, the policy is virtually useless.

Research in this area suggests that an individual's use of policies will be interpreted in a way that is consonant with the prevailing organizational culture. If family-friendly policies are made available without establishing organizational norms regarding the legitimacy of addressing the relationship between work and family, they will, at best, be an empty effort and, at worst, hurt women.

Empowerment, teams, and diversity. There are several current trends in organization development circles: employee empowerment, team development, and diversity management. Although theoretically more broad-based, empowerment and team-building efforts generally focus on the individual and/or the small work group. Effective diversity efforts need to be operationalized as cutting across levels (individual, group, and organizational). When diversity initiatives are implemented in organizations that are also promoting empowerment and self-directed teams, there can be conflicting messages regarding what core ideology really underlies management priorities and, thus, what values actually guide the diversity initiative.

To be consistent with the goal of increasing tolerance, diversity efforts need to be anchored in a culture of interdependence with sensitivity to the interconnections among individual, group, and organizational factors. For example, sensitivity to the ways coworkers vary based on race or gender requires one to think about how the other might react differently from oneself to a situation and/or communication. Diversity efforts need to be backed up by a sense of accountability for one's impact on others and on the group. Yet the predominant discourse in our society is one of independence. Solo action and swift individual decision making are usually reinforced as superior to responsiveness to the collectivity.

Employee-empowerment and team-development initiatives can be interpreted from a more collective value base; however, implementation of such initiatives in the United States is embedded in a society that places a high value on individualism. Given this context, empowerment is often taken to mean an emphasis on independence, and employee-empowerment efforts are typically very individually focused. The idea is generally to enable employees to have more control over their work as they are given the responsibility to make more decisions without consulting with others. The guiding rationale is that delegating decision making to workers moves it closer to the ones with most knowledge about the task and that more independence, in turn, will improve workers' investment in the work and commitment to the organization. Empowerment initiatives rarely emphasize accountability for the impact of one's actions on other people.

There is a particular irony in developing self-directed teams that give work groups the responsibility to make many decisions about their worklife—ultimately including decisions about performance evaluations, hiring, and firing. Similar to the goals of individual empowerment, self-directed teams are promoted as improving decision making, enhancing the quality of worklife, and thereby increasing productivity. However, as dramatic a power shift as this might appear, it can actually serve to reinforce a status quo with respect to discriminatory practices. Whereas the team focus potentially broadens accountability beyond oneself, the dynamics of member selection and the development of informal norms can create and enforce a lack of support for differences. For example, social-psychological research has revealed a strong preference for and attraction to others who are perceived to be more like oneself. At the team level, this unfolds as a preference for homogeneous work groups. In spite of research that indicates that heterogeneous groups can develop more creative solutions to problems, homogeneous groups tend to be more familiar, predictable, and comfortable (e.g., Jackson and Ruderman 1995). Thus, teams tend to select and reinforce members who are most like the dominant group. Without concerted efforts to counter that tendency, teams can so emphasize the characteristics that members have in common that a monolithic team culture develops with strong norms for conformity and little tolerance for those who move outside those norms (cf. Kelly et al. 1994).

In addition, it is difficult to develop effective teams in organizations that themselves exist in a society that fiercely values individualism. Because an important goal of both empowerment and team development is to give over decision-making power to employees, there is a strong likelihood that policies and procedures will be interpreted within that value framework of American culture. That is, if control is conceded to employees—particularly if the group is dominated by majority group members (i.e., white men)—the way is open for the ethos of the group to be defined more in line with the dominant individualistic culture of our society.

Thus, diversity efforts can be quietly undermined by other simultaneous organizational development initiatives that reinforce organizational and team cultures that are less tolerant (rather than more accepting) of differences. Whereas empowerment and team building can theoretically be operationalized in a manner consistent with support for pluralism, the historic lack of diversity combined with the current inequitable distribution of resources among diverse race and gender groups makes this a rare occurrence. Ultimately, the dilemma here is that, to be effective, diversity efforts need to be anchored in both a value for interdependence and an accountability for one's impact on others, which are both notions rooted in an ethos contrary to the American ideology of independence and individualism.

Summary. Efforts to address diversity through policy and/or other formal means will be ineffective unless supported by deeper changes in attitudes, values, and the culture of the organization. Thus, to create organizations that are truly supportive of diversity, the more subtle forms of discrimination (aversive racism/sexism) must be addressed and reduced. All initiatives, whether they are designed to combat aversive discrimination or whether they are to implement EEO/AA or family-friendly policies, diversity programs, and/or team building, must be accompanied by changes in core organizational and community values. Within organizations, people's preferences are often taken as givens (particularly by economists). They are indeed difficult to change. However, to provide the backdrop for important changes in attitudes, we need to develop supports for communal attitudes, foster a sense of community responsibility, and challenge the rampant individualistic approaches to solving problems. We will need to address the organizational processes that maintain segregation and unequal access to resources. Although we must work to develop such basic attitudes within our current organizations, we also need to challenge the community and social processes that reinforce segregation and produce a workforce unprepared to face the expanding diversity.

Challenge for the Future

Diversity in the workplace has not been adequately addressed. There are serious barriers that preclude what is now considered a strategic imperative for organizations—the full utilization of a diverse workforce. Existing approaches to supporting diversity in the workplace have not worked and, in many cases, have resulted in new sets of problems or dilemmas, often involving difficult theoretical issues. The historical, economic, political, and social setting is critical for understanding current barriers to effective workplace diversity.

Whether considering EEO/AA or diversity-management programs or some combination of both, we must avoid being seduced into a false choice between adopting a sameness versus a difference paradigm as the guide for diversity work. Adopting a perspective whereby context at every level of analysis becomes central to understanding diversity emphasizes the interdependence of cultural history, current circumstances, and future aspirations (Trickett 1996). We suggest moving toward focusing on the interdependence of equality of access and equality of outcomes. In other words, differences need to be understood in the context of similarities and, even more important, similarities need to be considered in the context of differences. For example, when the sameness of treatment has differential effects on different cultural or gender groups, we have to redefine our definition of sameness to incorporate those varied contextual influences. When differences are observed, we need to look at how those differences might be similar across individuals or be embedded in some level of sameness (e.g., what has been attributed to race might also be influenced by class; what is attributed to gender might cut across other groups with less organizational power and status). Our conceptions of difference and sameness also must be more fluid. For example, what is an issue of differential outcomes today may translate into differential access tomorrow.

Whereas managing diversity may in theory be an initiative that is seemingly less dramatic (or at least less imposed) than EEO or AA, it nonetheless also disrupts the status quo and creates anxiety about change. The emerging challenge for organizations is to find creative ways to harness and maximize the benefits of diversity and also to acknowledge the loss of comfort and sense of disruption that come from dealing with the unfamiliar. Thus, it is imperative that we pay more attention to the meaning (e.g., both real and anticipated impact) diversity initiatives have for all employees as well as to the types of organizational mechanisms that will integrate diversity into daily expectations and operations. Diversity-specific efforts need to be paralleled by general quality of worklife initiatives that also incorporate expectations for diverse representation and smooth working relationships across differences. The values that undergird both the diversity and more general organization development efforts need to be consciously institutionalized through integration into organizational practices ranging from job descriptions to performance expectations to organizational rituals and community celebrations.

In short, our analysis of the current dilemmas impeding progress toward a diverse workforce leads us to argue that effectively supporting diversity requires fundamental shifts in organizational processes. First, we need management practices and organizational models that acknowledge that people exist in the context of profound and dynamic cultural, economic, social and political forces. Second, organizations must understand that diversity initiatives signal shifts in power dynamics. They thereby require attention to individual fears and the creation of appropriate organizational supports along with solid resolve to address inequities and to affirm support for integrated work groups. Third, we must recognize that all efforts toward creating an effective diverse workforce must be anchored in deeply rooted organizational values for diverse approaches and contributions.

Clearly, we need to develop new theoretical and practical perspectives on how to incorporate diverse groups successfully. We must draw from multiple disciplines and construct a multilevel analysis that includes individual, work group, organizational, and societal factors. We need to consider varied organizational contexts and how they shape differing adaptive challenges for diverse individuals, and we need to be cognizant of influences ranging from broad historical, economic, and political trends to organizational, team, and personal dynamics. At the larger socioeconomic level, we must attend to both equity considerations and to making efficient use of resources in profitable businesses. We need to heed Trickett's (1996) call to be attentive to both the "contexts of diversity" (i.e., the cultural contexts within which people develop and are socialized along with particular adaptive challenges presented by each present situation) and the "diversity of contexts" (e.g., work settings, teams, family, community) that affect the adaptation of diverse groups within organizations.

NOTES

This essay was published in 1998 under the title "Diversity dilemmas at work," in *Journal of Management Inquiry* 7 (3): 252–69, and is reprinted here with permission. The contributions of both authors were equal. The order of names is alphabetical. This research was funded in part by CITA (Committee on Industrial Theory and Assessment) at the University of Massachusetts Lowell.

   1. Key workplace-related legislation during the 1960s and 1970s consisted of the 1961 Executive Order 10925 (which, under President Kennedy, encouraged federal contractors to hire more minorities); the Equal Pay Act of 1963 (which addressed equal pay for men and women doing work requiring equal skill, effort, and responsibilities); Title VII of the Civil Rights Act in 1964 (which, under President Johnson, barred discrimination because of race, color, religion, sex, or national origin in hiring, firing, promotion, compensation, and other terms and conditions of employment); the Executive Order 11246 of 1965 (which, under Johnson, required federal contractors to adopt goals and timetables to achieve proportional racial representation and placed affirmative action under the jurisdiction of the Department of Labor); Executive Order 11375 in 1967 (which, under Johnson, added sex to the categories protected by AA); Executive Order 11478 in 1969 (which, under President Nixon, imposed specific goals and timetables on federal contractors regarding African Americans); Revised Order #4 in 1971 (which, under Nixon, required federal contractors to establish AA programs for women); and the EEO Act in 1972 (which, under Nixon, empowered EEOC to take legal action in federal courts to enforce Title VII of the Civil Rights Act). A detailed outline of all major legislation and cases is available from the National Council for Research on Women (1996).

   2. Women, minorities, and immigrants would constitute 85%. Updated projections for the period 1990 to 2005 confirmed this trend. White non-Hispanic males would represent only 14.5% of the net increase in the workforce, whereas women would make up 57% of the net increase and minorities, 53.7%. Non-Hispanic white males would represent only 38.2% of the workforce by the year 2005, down from 43.1% in 1990 (based upon data in Fullerton 1991). From these projections, it became clear to many organizations that the ability to incorporate diverse workers productively would be essential for survival.

   3. Approximately 30% of companies surveyed by the Center for the New American Workforce address diversity issues through some sort of in-house diversity program to enable people to work together better, such as sensitivity training or team-building activities (Miller 1994).

   4. What workers might more legitimately fear is the growth in the ratio of CEO pay to worker pay. It rose from 122:1 in 1989 to 172.5:1 in 1995, meaning CEOs in major United States companies earned 172.5 times more than the average worker in 1995 (Mishel et al. 1997).

REFERENCES

Amott, T., and J. Matthaei. 1996. *Race, gender, and work,* rev. ed. Boston: South End Press.

Arredondo, P. 1996. *Successful diversity management initiatives.* Thousand Oaks, Calif.: Sage.

Badgett, M. V. L., and H. Hartmann. 1995. The effectiveness of equal employment opportunity policies. In *Economic perspectives on affirmative action*, ed. M. C. Simms. Washington, D.C.: Joint Center for Political and Economic Studies.

Berman, P. 1996. Redefining fairness. *New York Times,* April 14, 16–17.

Bernal, G., and N. Enchautegui-de-Jesus. 1994. Latinos and Latinas in community psychology: A review of the literature. Special issue ed. I. Serrano-Garcia and M. Bond. *American Journal of Community Psychology* 22:531–57.

Berry, M. F. 1995. The case for affirmative action. *Emerge* 6 (7): 28–30, 34–37, 40, 42–43, 48.

Blau, F. D., M. A. Ferber, and A. Winkler. 1998. *The economics of women, men and work,* 3d ed. Upper Saddle River, N.J.: Prentice-Hall.

Bond, M. A., and C. Keys. 1993. Empowerment, diversity and collaboration: Promoting synergy on community boards. *American Journal of Community Psychology* 21 (1): 37–58.

———. 2000. Strengthening parent–community member relations on agency boards: A comparative case study. *Mental Retardation* 38 (5): 422–35.

Bond, M. A., and J. L. Pyle. 1998. The ecology of diversity in organizational settings: Lessons from a case study. *Human Relations* 51 (5): 589–623.

Chemers, M. M., S. Oskamp, and M. A. Costanzo. 1995. An introduction to diversity in organizations. In *Diversity in organizations,* ed. M. M. Chemers, S. Oskamp, and M. A. Costanzo. Thousand Oaks, Calif.: Sage.

Cox, T. 1993. *Cultural diversity in organizations: Theory, research, and practice.* San Francisco: Berrett-Koehler Publishers.

Cox, T., and S. Blake. 1991. Managing cultural diversity: Implications for organizational competitiveness. *Academy of Management Executive* 5:45–56.

Crawford, M. 1995. *Talking difference: On gender and language.* Thousand Oaks, Calif.: Sage.

Dovidio, J. F., and S. L. Gaertner. 1996. Affirmative action, unintentional racial biases, and intergroup relations. *Journal of Social Issues* 52 (4): 51–75.

Families and Work Institute. 1993. *The national study of the changing workforce.* New York: Families and Work Institute.

Fine, M. G. 1995. *Building successful multicultural organizations: Challenges and opportunities.* Westport, Conn.: Quorum Books.

Fine, M. G., F. L. Johnson, and M. S. Ryan. 1990. Cultural diversity in the workplace. *Public Personnel Management* 19 (3): 305–19.

Fullerton, H. N. 1991. Labor force projections: The baby boom moves on. *Monthly Labor Review* 114 (11): 31–44.

Gaertner, S. L., and J. F. Dovidio. 1986. The aversive form of racism. In *Prejudice, discrimination and racism,* ed. J. F. Dovidio and S. L. Gaertner. Orlando, Fla.: Academic Press.

Gruber, J. E., and L. Bjorn. 1982. Blue-collar blues: The sexual harassment of women autoworkers. *Work and Occupations* 9:271–98.

Holloway, W. 1994. Beyond sex differences: A project for feminist psychology. *Feminism and Psychology* 4:538–46.

Jackson, B. W., F. LaFasto, H. G. Schultz, and D. Kelly. 1992. Diversity. *Human Resource Management* 31 (1, 2): 21–34.

Jackson, S. E., ed., 1992. *Diversity in the workplace: Human resources initiatives.* New York: Guilford Press.

Jackson, S. E., and E. B. Alvarez. 1992. Working through diversity as a strategic imperative. In *Diversity in the workplace: Human resources initiatives,* ed. S. E. Jackson. New York: Guilford Press.

Jackson, S. E., and M. N. Ruderman. 1995. *Diversity in work teams.* Washington, D.C.: American Psychological Association.

Jacobsen, J. P. 1994. *The economics of gender.* Cambridge, Mass.: Blackwell.

Jamison, D., and J. O'Mara. 1991. *Managing workforce 2000: Gaining the diversity advantage.* San Francisco: Jossey-Bass.

Johnson, W. B., and A. H. Packard. 1987. *Workforce 2000: Work and workers for the 21st century.* Indianapolis: Hudson Institute.

Kelly, J. G., S. Azelton, R. Burzette, and L. Mock. 1994. Creating social settings for diversity: An ecological thesis. In *Human diversity: Perspectives on people in context,* ed. T. Trickett, D. Birman, and R. Watts, 424–51. San Francisco: Jossey-Bass.

Kim, P. S., and G. B. Lewis. 1994. Asian Americans in the public service: Success, diversity, and discrimination. *Public Administration Review* 54 (3): 285–90.

Kossek, E., and S. Zonia. 1993. Assessing diversity climate: A field study of reactions to employer efforts to promote diversity. *Journal of Organizational Behavior* 14:61–81.

Lehrer, E. L., T. Santero, and S. Mohan-Neill. 1991. The impact of employer-sponsored child care on female labor supply behavior: Evidence from the nursing profession. *Population Research and Policy Review* 10:197–212.

Loden, M., and J. B. Rosener. 1991. *Workforce America!* Burr Ridge, Ill.: Irwin.

Lott, B. 1987. Feminist, masculine, androgynous or human. Paper presented at the American Psychological Association Annual Meeting, August, New York City.

McGonagle, K. A., J. Connor, S. Heeringa, P. Veerkamp, and R. Groves. 1995. Commission on leave survey of employees on the impact of the family and medical leave act. Ann Arbor, Mich.: Institute for Social Research.

Mednick, M. 1989. On the politics of psychological constructs: Stop the bandwagon—I want to get off. *American Psychologist* 44:1118–23.

Messick, D., and D. M. Mackie. 1989. Intergroup relations. *Annual Review of Psychology* 40:45–81.

Miller, J. 1994. Corporate responses to diversity: A benchmark study. Report from the Center for the New American Workforce. Queens College, New York.

Mishel, L., J. Bernstein, and J. Schmitt. 1997. *The state of working America, 1996–97.* Armonk, N.Y.: M. E. Sharpe.

National Council for Research on Women. 1996. Affirmative action: Beyond the glass ceiling and the sticky floor. *Issues Quarterly* 1 (4): whole issue.

Nkomo, S. M. 1995. Identities and the complexities of diversity. In *Diversity in work teams,* ed. S. E. Jackson and M. N. Ruderman. Washington, D.C.: American Psychological Association.

O'Gara, J. 1995. Making workplaces work: Quality work policies for small business. Washington, D.C.: Business and Professional Women's Foundation. Pamphlet.

Ogbu, J. U. 1993. Difference in cultural frame of reference. *International Journal of Behavior Development* 16 (3): 483–506.

Plous, S. 1996. Ten myths about affirmative action. *Journal of Social Issues* 52 (4): 25–31.

Pollock, E. J., and J. S. Lublin. 1997. Employers are wary of rules on mentally ill. *Wall Street Journal,* May 1, B1.

Pyle, J. L., and M. A. Bond. 1997. Workforce diversity: Emerging interdisciplinary challenges. *New Solutions* 7 (2): 41–57.

Ryan, W. 1981. *Equality.* New York: Pantheon.

Scott, J. 1994. Deconstructing equality-versus-difference: Or, the uses of poststructuralist theory for feminism. In *Theorizing feminism: Parallel trends in the humanities and social sciences,* ed. A. C. Herman and A. J. Stewart. Boulder, Colo.: Westview Press.

Sessa, V., and S. Jackson. 1995. Diversity in decision-making teams: All differences are not created equal. In *Diversity in organizations,* ed. M. M. Chemers, S. Oskemp, and M. A. Costanzo, 133–56. Thousand Oaks, Calif.: Sage.

Sharpe, R. 1994. Family friendly firms don't always promote women. *Wall Street Journal,* March 29, B1, B5.

Shellenbarger, S. 1996. Family-friendly jobs are the first step to efficient workplace. *Wall Street Journal,* May 15, B1.

Shinn, M., N. Wong, P. Simko, and B. Ortiz-Torres. 1989. Promoting the well-being of working parents: Coping, social support, and flexible job schedules. *American Journal of Community Psychology* 17 (1): 31–56.

Silka, L., and J. Tip. 1994. Empowering the silent ranks: The Southeast Asian experience. Special issue, ed. I. Serrano-Garcia and M. Bond. *American Journal of Community Psychology* 22:497–529.

Simms, M. C. 1995. Introduction to *Economic perspectives on affirmative action,* ed. M. C. Simms. Washington, D.C.: Joint Center for Political and Economic Studies.

Skrentny, J. D. 1996. *The ironies of affirmative action.* Chicago: University of Chicago Press.

Thomas, R. 1992. Managing diversity: A conceptual framework. In *Diversity in the workplace: Human resources initiatives,* ed. S. E. Jackson, 13–35. New York: Guilford Press.

Tomaskovic-Devey, D. 1993. *Gender and racial inequality at work: The sources and consequences of job segregation.* Ithaca, N.Y.: ILR Press.

Triandis, H. C. 1995. A theoretical framework for the study of diversity. In *Diversity in organizations,* ed. M. M. Chemers, S. Oskamp, and M. A. Costanzo, 11–36. Thousand Oaks, Calif.: Sage.

Trickett, E. 1996. A future for community psychology: The contexts of diversity and the diversity of contexts. *American Journal of Community Psychology* 24 (2): 209–34.

Tsui, A. S., T. D. Egan, and C. A. O'Reilly. 1992. Being different: Relational demography and organizational attachment. *Administrative Science Quarterly* 37: 549–79.

Unger, R. 1992. Will the real sex differences please stand up? *Feminism and Psychology* 2:231–38.

Watts, R. J. 1992. Elements of a psychology of human diversity. *Journal of Community Psychology* 20:116–30.

Wessel, D., and J. M. Schlesinger. 1997. United States economy's report card: Not all A's. *Wall Street Journal,* May 5, A2.

Williams, L. 1994. Childless workers demanding equity in corporate world. *New York Times,* May 29, 1, 11.

Wilson, R. 1995. Affirmative action: Yesterday, today, and beyond. Report. Washington, D.C.: American Council on Education. May.

Understanding the
Political Geography of the Region
Sustainable Development and the University

Jeffrey Gerson and John Wooding

THERE is currently much discussion of regional development and regional development strategies. Areas within countries are increasingly being defined in terms of their regional characteristics, and regions are often viewed as playing a significant role in the world economy. Indeed, "regionalism" is considered a potent force both in political identification and, particularly, in economic development. In the United States, much is made of regional development strategies, and state and local authorities, universities, and private and public foundations often engage in "region-centered" activities. Abroad, supranational institutions such as the European Community have long engaged in regional development policies (Applebome 1999, 1, 17).

Recently, at the University of Massachusetts Lowell (UML) we created a new interdisciplinary department, the Department of Regional Economic and Social Development to promote the stated mission of the university to engage in regional economic development. Behind all this emphasis on the regional, there appears to be a set of assumptions about regions that too often go unquestioned, not least of which is the assumption that it is meaningful to identify any area as a "region."

In this chapter we ask whether we can discuss the development of a region without fully understanding what it means. At UML, the identification of a regional development strategy is predicated on a set of assumptions about the area usually referred to as the Merrimack Valley. We maintain that the absence of a well-defined political border and the fragmentation of political authority pose a significant barrier to successful social and economic development conceived on a regional basis. Therefore, we explore some of those barriers and suggest that much work needs to be done before the resources of a public university such as UML can be deployed effectively in promoting regional development.

Regions have become increasingly important in both academic and policy circles (Cigler et al. 1994; Noponene, Graham, and Markusun 1993,

149

Ohmae 1995; Swanstrom 1995, 309–14; Cisneros and Weiss 1997, 42–45). At UML, the assumption of regionalism and of a sense of regional identity infiltrates the discussions of economic and social development in such cities as Lowell, Lawerence, Haverhill, and the older industrial towns along the Merrimack Valley. It is also frequently applied to the urban and suburban belt defined by the Massachusetts Turnpike to the south, Interstate 495 and New Hampshire to the north and west, and Route 128 to the south and east. In this belt lie cities and towns as diverse as Lowell and Andover, Marlborough and Lawrence, and Nashua and Newburyport.

Most economists are inclined to think about a region by the location of major industries and transportation routes. Determined in this fashion, the "region" is defined largely by the character of its economic enterprises: in this instance, a high-tech economy consisting of many hundreds of defense contractors, computer and telecommunications companies, and software developers located along the major roadways. It is a legitimate characterization and makes sense to those of us who work in and study this area. But much is being assumed in this delineation: that industrial structures are the most important way of thinking about a region; that local policymakers think this way; and that the people who live here consider themselves to be part of something called a "region," or of belonging, in our case, to the Merrimack Valley.[1]

In the literature (and dependent upon the discipline), academics and others have suggested that a region may be defined variously:

As a discrete economic area characterized by certain specialties of production, distribution, or markets (including labor markets and particular skills).

As a demographic area characterized by particular ethnic groupings, historically determined religious communities, or discrete patterns of immigration.

As a particular set of geographic or physical features characterized by proximity to ports or central roads or rail links; or in relation to river valleys or mineral or other natural resources; and, as a bioregion determined by the peculiarities of the indigenous flora and fauna.

As a regulatory region for various state and federal bureaucracies.[2]

Some questions arise about the utility of the term "region" for the Merrimack Valley as we examine what we term the political geography of the region. In this context, we report some preliminary research on the definition of regional political boundaries and how state and local bodies interact in the region. Given the recent dissolution of some county governments in Massachusetts, it is essential to join the current debate about regional entities, and particularly so, if the university and the community are going to work effectively to encourage sustainable economic development.

If we are to undertake meaningful research and an analysis of regional development, first we have to ask some difficult questions about how we conceive of the region:

How does the discussion of the "new" regionalism help us understand regional development?

What contribution does community-power analysis and regime theory make to our understanding of regionalism?

Which institutions, groups, political and economic structures, and authorities have a "regional" interest?

Can we think of the region as being bounded by a common political, social, and economic culture?

Can a university foster a regional development strategy without a clear definition of the region?

What future research strategies might provide a better understanding of the politics of regions?

Where economists see industrial concentration, markets, labor pools, and the like, geographers see valleys and streams, demographers fluid or fixed populations, immigration and emigration—political scientists, on the other hand, tend to focus on political frameworks, concentrations of political power, and institutional forms of government and authority. Of course, no single perspective can possibly capture the astounding variety of complex social and economic forms that may make up a region. It is possible, however, to try to identify ways in which political authority may be defined as regional.

The New Regionalism

The literature on regions is just beginning to develop. In the 1990s, a number of writers began to stress the symbiotic relationship between the core cities and the regions in which they are located. In particular, the economic and environmental health of the suburbs and surrounding towns is seen as profoundly tied to the health of the central city. As Allan Wallis has pointed out: "Current appeals for regionalism based on economic interdependency often include discussion of the globalizing economy, which reduces the significance of nation states and focuses greater attention on regions based as geographic units of competition. Likewise, arguments for regionalism to protect the environment are now based not on evidence of polluted rivers and water sheds alone but on concerns over global warming and acid rains, which cross national boundaries" (Wallis 1998, 97).

Many of the "new" regionalists stress that the region is now of central concern but also note that a significant vacuum in regional government is apparent. For example, Myron Orfield suggests (building on his experience as a state legislator for the southwestern district of Minneapolis) that what is needed is a metropolitan council able to develop and enforce regulations requiring such things as the building of low- and moderate-income housing in the suburbs, the creation of a fund for housing reinvestment, and strict regulation of suburban sprawl (1997). Politically, it is clear (given the challenges of such initiatives) that coalitions are what is needed to help promote the link between the central city and both the inner- and outer-ring suburbs. Indeed, as others have pointed out, this linking can be facilitated by state governments if they pass legislation to empower county governments to act in the name of the municipality or create such things as a "metrowide" tax-sharing agreement, thus providing an alternative tax base (Rusk 1996). Given the politics of regions, this is not an insignificant task. Indeed, the whole idea of creating "metro" governments is problematic in view of the typical class and ethnic differences between town and county, and the political structures that represent them. Many observers agree that political conflicts render the possibilities for meaningful regional government remote. The only other choice is a system of "mutual adjustment" that address issues of regional concern through coordination and public–private partnerships—but is clearly short of formal regional governance (Savitch and Vogel 1997).

There are other versions of region-level governance that also pull back from any form of centralized political authority. William Dodge, for example, stresses the importance of a vital grassroots movement identified on a regional basis. Such a movement, he argues, builds the necessary links across a region but avoids the pitfalls of regional government. It does so by focusing on issues and community problem solving, linked to the delivery of services (1996). Often, however, the instances he cites of successful regional organization have emerged from some common threat to a number of different constituencies that provides the rationale for their collective action. Our own study of the regional movement that developed to defend Hanscom Air Force Base is a case in point (Gerson and Wooding 2000).

As Wallis points out, what is really missing in efforts to create regional governance is a set of shared values and culture that is essentially "regional." At the level of neighborhoods, community members must begin to recognize the identity of the region to which they belong and the gains to be gotten from such identification. At the same time, limited institutional arrangements must be developed that garner the trust of the community and embody a regional vision. Such a vision (both cultural and institutional) might be served, as Wallis suggests, by concern for regional environmental protection or the value of controlled and carefully zoned economic development (1998, 103).

Community Power and Regime Theory

The literature on urban development and urban politics is vast (Elkin 1987; Fainstein 1983; Mollenkopf 1983; Peterson 1981; Shefter 1985; Whitt 1982). Studies of regional development and politics, however, are rather scarce. Analysts of urban development are mostly concerned with the structures and forces active in cities and towns. This is hardly surprising. Though significant, regional forms of government are dwarfed by the numbers of local government in the United States (Stephenson et al. 1992, 53).[3] Regional forms of government are rare, and it is notoriously difficult to define a region in political terms, in view of the fragmented nature of metropolitan areas where responsibility for governing is often divided among counties, towns, townships, cities, special districts, and school districts (Christensen 1995). The political boundaries of a region are virtually nonexistent. They are not, however, irrelevant (Keating and Loughlin 1997; Savitch and Vogel 1997).

In this essay we argue that, on the surface, no regional regime is apparent that can unify the different political jurisdictions in Northeastern Massachusetts. In fact, as Todd Swanstrom (1995) points out, regional studies detect a trend—by suburbs—to exhibit greater economic independence from central cities, not interdependence. Our preliminary research suggests that, although regional regimes do exist, they are only manifest in critical situations that threaten powerful interests on a regional basis.

The study of local and urban politics is typically focused on the power structures within cities and towns themselves: the power of business elites, the authority structures of local governments, the impact and effectiveness of community groups, and the like. Depending on the perspective, the literature tends to stress the efficacy of the fluidity of interest-group politics; the dominance of local business and inability of the poor and minorities to achieve representation; the importance of the market economy in limiting choices for political leaders (Peterson 1981); and the relevance of political action despite market forces (Mollenkopf 1983).

Towns and cities in the United States tend to lack the overarching "command structures" that integrate community, economic, and business power. Where individual institutions might have highly formalized hierarchies and well-defined goals, there is little cross-institutional coordination and much cross-institutional competition and defense of bureaucratic turf. Offsetting this is the growth of informal arrangements to promote cooperation. As Clarence Stone has shown, these informal arrangements can be empowering to the community as a whole (1989). They can also, of course, permit a small group of insiders to wield political and economic power. To borrow Stone's term in this context: An "informal regime" can spring up that is focused on the development of specific goals. A regime thus involves not just any informal group that comes together to make a decision but an informal yet relatively stable group with access to institutional resources that enable it to play a sustained role in making governing decisions. Such regimes have clearly existed in many towns and cities in the Merrimack Valley (Gittell 1992). The interesting question is, in the absence of an over-arching political authority, whether any similar kinds of regimes exist at the regional level. If so, what do they tell us about the politics of the region? Are they important to facilitating regional economic and social development?

Gaylord Burke of the Merrimack Valley Commission (MVC) claims that a region is officially defined by the federal government (largely through the activities of the Office of Management and Budget and the Bureau of the Census) as a Standard Metropolitan Statistical Area (SMSA), which is a central city with 50,000 population or more, with supporting infrastructure of suburbia and a rural hinterland.[4] The federal government, for example, lumps Lawrence and Haverhill, and the dozens of towns surrounding them, into a region. Lowell itself is in a separate SMSA, thus dividing the Merrimack Valley in Massachusetts into at least two areas. Burke also noted that people in the region have definitions that are more chaotic and fragmented than those of the federal government. Every agency at the federal and state level has its own definition of a region (for instance, the EPA has delineated the Sea Coast region, which stretches from Newburyport to Provincetown (on Cape Cod), all in a single region that covers hundreds of square miles).

The MVC may be considered an example of the kind of embryonic informal regime that has potential for a regional perspective. That potential is limited, however. Formed in 1959 under federal legislation defining regions, the MVC is a voluntary organization that deals primarily with questions of zoning and road and rail communication links. Local cities and towns turn to the MVC to help guide them when they have to formulate and make policy that deals with issues that transcend formal political boundaries. Increased state and federal regulation (on health, pollution, etc.), however, tends to define the region and move power away from local communities to the state.

Planners as a group have different notions of what constitutes a region; some focus on the workplace and transportation and commuter patterns; others see the environment as creating natural regions (defined by watersheds, susceptibility to air pollution, etc.). Whatever the perspective, they tend to be forced to work with very weak regional institutions and must still negotiate with urban and local authorities.

How, then, do the people who work around regional initiatives define the region? Most organizations that use "Merrimack Valley" in their names see the region through the lens of their mission.[5] For example, Jim

Wilde of the Merrimack Valley Housing Partnership (a nonprofit organization that sponsors training for low-income, prospective first-time home buyers primarily in Lowell and Lawrence) sees the valley in terms of his client base, which is Lowell and Lawrence's Spanish- and Khmer-speaking communities. Haverhill and all points east are excluded, because there are other housing groups in the eastern valley, such as Haverhill Neighborhoods, in Haverhill. For Wilde (as for many others), the region's most powerful institutions are, not surprisingly, the banks. By providing money for homeowner loans, banks define who gets to live where. There are some twenty banks active in the mortgage business in the area: from large transstate corporations (Fleet, based in Providence) to the small, locally owned banks such as the Washington Savings Bank in Lowell.

However, economic forces do not act alone to decide the fate of low-income home seekers. The federal government is an important actor, too. The Community Reinvestment Act (CRA) of 1993 obliges banks under strict rules and regulations to lend money in areas where they operate, especially in neighborhoods where there had been underinvestment (Gittell 1992, 205). This, of course, also structures the pattern of urban and regional development. Closer to home, city planning departments are essential players because it is through their grant-writing ability that federal money is ensnared to keep the programs afloat (Kenneally 1997; Shannon 1997).[6]

Massachusetts Political Culture: Antiregionalism?

New England as a political region has a history of tenacious self-governance in the form of towns and town meetings. Two recent cases are compelling evidence of antiregionalism's influence. In April of 1997, voters in the town of Framingham, rejected the idea of making this community, and the state's largest town (1990 pop. 65,000) a city, which would do away with their town meeting system of government in place for 297 years. A mayoral–city council government was rejected by a ratio of more than 2 to 1. Voters were so opposed to the notion that voter turnout was the highest in many years. The town has rejected similar measures over the last twenty-five years. Framingham today remains the largest town in the United States run by a New England–style town meeting (Carney 1997).

When twenty-three Massachusetts towns went against their better judgment and allowed themselves to be coaxed by the Commonwealth of Massachusetts in committing to a twenty-year municipal incinerator contract in the mid-1980s to form a regional association to burn their trash, they only regretted doing so. Today, the Northeast Solid Waste Committee (NESWC) is responsible for trash disposal fees for these towns and cities in the committee that are twice as high as those for area towns that declined to join. Tewksbury, for example, paid over $1 million in trash fees in 1997. By 2005, its bill will be $2 million annually. Moreover, dioxin emissions from the incinerator are four times the allowable federal standard, according to area environmentalists who want the plant shut down. The incinerator is an environmental, economic, and political disaster for town governments. When the towns joined, they believed the plant would be self-sufficient, generating money rather than eating it up (Edgar 1997; Kenneally 1997; Shannon 1997).

However, there is a silver lining to this failure at regional governance. In March of 1999, incinerator operators Massachusetts Refusetech (MRI)[7] and the NESWC agreed to split the cost of a $35.5 million upgrade of their North Andover incinerator, required by changes made to the federal Clean Air Act. Moreover, on August 12, 1999, the state of Massachusetts approved $3 million in rate relief for the twenty-three towns of NESWC (Sweeney 1999). The bailout by the Massachusetts governor and state legislature is in keeping with the solutions proposed by many of the "new regionalists" to overcome opposition to regional alliances. Regionalism's proponents, including Minnesota legislator Myron Orfield and Brookings Institution public policy analyst Anthony Downs, foresee a large role for state legislatures in coordinating comprehensive planning among the fragmented political authorities; frameworks for state planning are already underway in Oregon and Florida (Henderson 1994, 35).

In another example, the town of Dracut's plan to build an energy-plant incinerator within its boundaries has created a storm of protest from the border communities to its south and east, Methuen and Andover, that fear toxic emissions will blow their way. In the tradition of town governance, their opposition has been ignored. Dracut's response has been to claim it has but one responsibility—and that is to Dracut—and to save it a considerable sum of money and reduce chances of brownouts due to electricity shortages. Given this kind of response (understandable and typical as it may be), it is obvious that, for the most part, local and urban authorities have little motivation or inclination to view themselves as part of a wider region. Gaylord Burke believes that the Dracut plant is problematic, because "no one is asking the fundamental question whether this is an appropriate use for this land from the perspective of regional health and economics."[8]

Some efforts are underway in Massachusetts to create regionalism, with dubious results. The publisher and editor of the *Standard Times,* a Fall River newspaper, are trying to popularize the name "SouthCoast," and redefine the area of Southeastern Massachusetts, from Acushnet to Wareham and Seekonk to Swansea, all linked by Interstate 195. The idea first arose when a group of CEOs met under the auspices of the University of Massachusetts Dartmouth to discuss a common economic development agenda.

The merger of three area hospitals into the "SouthCoast Health System" furthered it in June 1996. Part of the impetus for the new name is the history of rivalry between Fall River and New Bedford, the area's largest cities, and part the region's grave economic conditions and the state legislature's failure to address them (Jurkowitz 1997). These brief examples serve to illustrate how problematic the concept of a "region" can be.

John Mullin, a professor of urban and regional planning at the University of Massachusetts Amherst, writes that the case of Dracut's energy plant is typical of Massachusetts and the Merrimack Valley, because it has so little history of regional planning regulations, compared with states such as Vermont, Michigan, and Oregon. No state law can compel Dracut to consider the impact its plant will have on bordering towns and cities. A strong tradition of local government means regional planning is rare. When it does occur, it is because overwhelming evidence exists to show that all communities have something to gain or to lose. He points to Cape Cod as an example in Massachusetts where regional planning has worked, because "local communities realized that ground water pollution in one town harmed everyone else" (Mullin 1997).

Robert Flynn, executive director of the Northern Middlesex Council of Governments (NMCG), a regional planning organization that has existed in the Merrimack Valley since 1963, concurs with Mullin.[9] There are three reasons for towns and cities to find regional solutions. One, they can save money; two, they can have a better quality of service; and three, if towns and cities don't find them, they realize they will lose the service completely. For these reasons, the eight towns and one city that constitute the NMCG (Billerica, Chelmsford, Dracut, Dunstable, Lowell, Pepperell, Tewksbury, Tyngsborough, and Westford), cooperate for the joint purchase of equipment, services, and materials, such as fuel oil and road salt, bus transportation for the disabled, and teacher training. In other matters, however, ones that the NMCG calls "hard tasks, harder tasks and hardest tasks," such as preservation of open space, development of compact mixed-use business and commercial centers, and decreasing auto dependence, to cite one example of each, respectively, the likelihood of success is dim (NMCG 1999).

Flynn believes there are two primary reasons for antiregionalism in New England and in particular, in Massachusetts. New England cities and towns are unique because they did not develop by spreading out into unincorporated areas or, as true regional centers, deliver services for hundreds of square miles. In Massachusetts, even the enabling legislation to encourage intermunicipal cooperation, Chapter 40, Section 4A, of the Commonwealth of Massachusetts Laws, itself spells out exactly what towns and cities can do to cooperate by listing each specific allowable activity. In the rest of the country, Flynn says, cities and towns can do whatever they are not specifically prohibited from doing, making more room for creative, collaborative efforts. For a city or town to do more than is listed in Chapter 40, Section 4A, it must have the legislature's approval. Again, the political culture of the state and New England is hostile to regionalism.[10]

On the other hand, there are three examples of triumphant efforts operating at the regional level in the Merrimack Valley. The first was the successful effort to keep Hanscom Air Force Base in Bedford open (the U.S. Air Force operates the Electronic Systems Division of the Air Force Systems Command at Hanscom AFB). In the face of the Pentagon's four rounds of base closings (a hundred major bases were closed in the last four rounds), Hanscom has survived, and even fended off proposed cuts in jobs over the last few years. A unique coalition, consisting of senior Air Force personnel at the base, the Massachusetts Port Authority (Massport is a regional authority that currently operates a civilian airport, L.G. Hanscom Field, on land that was previously part of Hanscom AFB), the twelve members of the Massachusetts congressional delegation, along with three New Hampshire representatives, area businesses and employees most directly affected by closure, as well as political representatives from four bordering towns who are organized into the Hanscom Area Towns Committee—all have managed to work closely to keep the base afloat (A.P. 1997; Bushnell 1997; Cole 1997; Hanscom Field 1997).

The second was the campaign to widen from four to six lanes, a major Merrimack Valley transportation artery, Route 3, which runs twenty-one miles from Burlington all the way to the New Hampshire line. What is unique about the construction project is that the contractor will design the road and finance its construction as well as build it (a design-finance-build plan popularized in several western states in the last several years). The state will then pay back the $185 million to $225 million anticipated cost over thirty years. Even the Massachusetts Building Trades Council agreed to drop its push for a standard-project labor agreement in favor of a less specific "labor harmony" language. The seven corridor-communities will be represented on a local advisory committee chartered to work with the contractor. It will monitor zoning and traffic detour issues. The idea for this committee came out of the towns' participation in the Northern Middlesex Council of Governments (NMCG). The state legislature voted to continue to own the road and keep its oversight role. The project had been in the works for twenty years. It took the leadership of a public–private partnership, led by former Merrimack Valley Congressman Chester Atkins (now working as a consultant) who represented construction firms and financiers, as well as several of the region's state representatives, member towns of the NMCG, and corporations along the Route 3 corridor, such as Wang Global and Bay Networks, to bring the road-widening plan and legislation finally to fruition. Construction began in 2001.

The third example is the effort of five wealthy towns bordering the Merrimack Valley (Carlisle, Concord, Lexington, Lincoln, and Bedford) to prevent the Massachusetts Port Authority (Massport) from beginning commercial passenger jet service at Hanscom Field. Three citizens' groups, Hanscom Area Towns Selectmen, Save Our Heritage, and Safeguarding the Historic Hanscom Area's Irreplaceable Resources, have blocked until very recently the expansion of Hanscom Field through lobbying the legislature, threats of law suits, and participation on the twenty-five member Hanscom Field Advisory Commission, a body set up by the legislature in 1980 to advise Massport. The primary issues for the towns were noise and congestion, in the form of traffic along the route leading to the airport.

The Role of the University in Regional Development

These comments on the issue of political power and regionalism suggest that we face some difficult questions if we are to think about economic and social development in regional terms. For a public university such as UML this is particularly crucial. If the university is to continue to engage with a variety of local and regional organizations (public and private) and to do so in a way that provides access to and support from the university's resources, then it needs to identify and create networks that help define the region.

A couple of examples will serve to illustrate how UML may fill this role. The university has numerous interdisciplinary centers involved in activities ranging from research on biodegradable plastics, leadership training for new immigrants and refugees, and economic analysis for local manufacturing concerns. The Center for Family, Work, and Community (CFWC) researches local social conditions, organizes training, and helps negotiate among local community groups, planners, and local and city officials. The Center for Immigrant and Refugee Community Leadership (CIRCLE) has provided an educational resource for community activists from Lowell, but needs to reach out consciously to other cities and towns in the Merrimack Valley. The provision of such educational services may begin to create a culture of regionalism in the valley.

Similarly, the university was instrumental in obtaining federal funding from the National Institute for Science and Technology (NIST) to establish one of five regional centers operating as manufacturing extension programs. The NIST regional centers bring together small businesses with fewer than five hundred employees from around the region and provide educational and resource information to them. It is an activity that links manufacturing firms, suppliers, and academic specialists around the issues confronting the manufacturing sector in the region. Some of this work has been done through the University's Center for Industrial Competitiveness (CIC). Again, it is an example of how a university can begin to create a regional identity even when the political, economic, and social evidence of such an identity is barely evident.

These interdisciplinary centers work with a wide range of local and regional business, political, and community organizations. They have begun to create a network of regional links around specific issues and projects. Although this may not be a regional strategy, it is indicative of the ways one important institution can fulfill its own role in regional development. Another possible function for a university in developing a sense of regionalism might be as an institutional mediator bringing together in workshops and conferences business owners, city officials, nonprofit agency administrators, professional planners, residents, and academics to talk about the issues that surround regionalism. Topics for discussion and research could include: The consequences of sprawl, such as dependence on the automobile and traffic congestion, locations for development, and the environmental consequences of antiregionalism. Research that shows how local decisions, such as zoning patterns, affect communities beyond their borders could be shared with state, city, and county officials. A university could provide studies that assist the work of regional planning agencies, such as the MVC and the NMCG, which have limited resources. These studies may be able to help area cities and towns recognize the value of mixed-use development, residential development interspersed with small commercial and industrial development, on the old village model, and how such strategies may ease sprawl in the region (Cole 1999).

Conclusion

Our observations suggest that it is problematic to speak of a region in political terms. We can certainly characterize seats of formal authority in cities and towns in the area, we can identify many of the leading players in a variety of urban policy initiatives, and we can map out the formal political boundaries in the region (for example, we have been repeatedly told that there may be a Merrimack Valley, but, in judicial and political terms, it most certainly does not extend into New Hampshire). What we cannot do at this point is locate regional power in any but the most circumscribed way.

Stone's theoretical assumptions about the existence of regimes in urban politics may provide help: "The study of urban regimes is thus a study of who cooperates and how their cooperation is achieved across institutional sectors of community life. Further, it is an examination of how cooperation is maintained when confronted with an ongoing process of social change, a continuing influx of new actors, and potential breakdowns through conflict and indifference" (Stone 1989).

Can we apply such a formulation to the problem of defining regions and regional power? In our view, this has some possibilities and suggests an integrated research agenda. This agenda should capture the insights provided by the economic definition of the region (major industries, employers); the demographic data on the ethnic and class composition of various communities (and patterns of immigration); and the identification of the central locus of political power. It also requires the investigation of how important agents think about the region: Do business or political leaders think in regional terms?

First, if indeed the idea of a "region" is to be meaningful, we should consider how economic, political, and demographic identities might combine to define it as such. A working definition of the area may require a new way of thinking about regionalism.

Second, we should explore the theoretical value of the notion of regimes to understanding the intersection of the political and the economic, and how central actors form connections to achieve both short-term and long-term goals.

Third, the identification of key agents in political and economic decision making is crucial, not only to aid our understanding of how decisions that affect the health of the region in general are arrived at, but also to test out the concept of regime as it might be applied to sometimes ephemeral regional coalitions.

Fourth, the university—especially a public university such as UML—can begin to recognize its own activities as contributing to a regional identity. The more widespread and diverse the involvement, the more likely the university itself will become an actor on a regional basis. This, in itself, helps to develop a sense of regionalism.

These goals could be achieved with an integrated and sustained research agenda that examines:

The university's prior involvement with cooperative ventures and with mobilizing sets of common interests (Center for Family, Work, and Community and the Center for Industrial Competitiveness, and others).

Case studies of regional development issues, such as the widening of Route 3 (a major highway that links metro Boston to the Merrimack Valley and southern New Hampshire; a project that was twenty years in the making).

Case studies of other regions in Massachusetts and New England, where the resistance to regionalism is deeply embedded in the political culture.

Case studies of subregions around the nation and world that may have characteristics similar to the Merrimack Valley (geography, demography, economic conditions, etc.).

If, however, the university is to succeed in encouraging sustainable economic development at the regional level, we need to be more precise about how we define this region. It may be that this is not a terribly useful general term, and that it may only be appropriate in specific cases where projects and industries need to pool resources (local skills, labor, transportation, budgets) to achieve common aims that are mutually beneficial. This, of course, begs the question as to whether such "regionalism" is now more or less necessary for local economies to succeed in domestic and international markets, or to improve the quality of the environment or community life. Much more work needs to be done on what constitutes a "region" before we can confidently assert which specific activities should be undertaken by academic institutions to aid sustainable regional economic development.

NOTES

1. Key employers are Lucent, Hewlett-Packard, GE, Raytheon, and Malden Mills. The most important employment clusters are the knowledge-creation and information-technology industries, environmental remediation and research, military contractors, and health care.

2. State and federal agencies, such as the Environmental Protection Agency (EPA), divide the area into regions in terms of its activities. The Commonwealth's Office of Business Development defines the area as composed of five regional zones (Lowell is in the North-East), some planning agencies think and act regionally, and a number of community and issue-oriented groups conceive of their activities in regional terms (the Merrimack Valley, in particular).

3. Total regional governmental units in 1992 were 47,295: counties (3,042), school districts (14,721), and special districts (29,532). The combined total of local governments (83,186), municipalities (19,200), and towns (16,691) was 119,077 (Stephenson et al. 1992, 53).

4. Interview with Burke by authors, cassette recording, August 21, 1997.

5. Indeed, the local Lowell Yellow Pages contains nearly two columns of organizations that are named "Merrimack" or "Merrimack Valley."

6. Interviews with both Shannon and Kenneally by Gerson, via telephone, August 1, 1997.

7. MRI is owned by Wheelabrator Technologies.

8. Interview by authors, cassette recording, August 21, 1997. In defense of Dracut, Robert Flynn of the Northern Middlesex Council of Governments argues that the town is less responsible for the power plant's location than area towns believe. The town actually has little say over plant location. The Federal Energy Regulatory Commission, the Federal Regulatory Siting Board, and market conditions also dictate whether there is a need for a power plant in a certain location. Robert Flynn, in interview by Gerson, cassette recording, September 8, 1999.

9. As is typical of regional planning associations in the state, the NMCG has no legal authority to mandate changes in the region; however, it has gained some respect just for tackling the controversial issues surrounding regional growth. Robert Flynn interview by Gerson, cassette recording, September 1, 1999.

10. Interview by Gerson, cassette recording, September 8, 1999.

REFERENCES

Appleborne, Peter. 1999. Out from under the nation's shadows: The hot subject today is the culture of regions. *New York Times,* February 20, 1, 17.

Associated Press (A.P.). 1997. Defense study urges cuts to staff, more base closings. *Boston Globe,* November 10, A11.

Bushnell, Davis. 1997. Some towns are wary as Massport advances Hanscom development. *Boston Sunday Globe,* August 31, 1, 2.

Carney, Beth. 1997. Framingham voters reject changing to a city. *Boston Globe,* April 9, B6.

Christensen, Terry. 1995. Local Politics: Governing at the grass roots. Belmont, Calif.: Wadsworth.

Cigler, Beverly, Annica C. Jansen, Vernon D. Ryan, and Jack C. Stabler. 1994. *Towards an understanding of multicommunity collaboration.* Washington, D.C.: U.S. Department of Agriculture, Economic Research Service, Government and Development Policy Section.

Cisneros, Henry, and Marc A. Weiss. 1997. The wealth of regions and the challenge of cities. *The Regionalist* 2 (4): 42–45.

Cole, Caroline L. 1997. Lawmakers go to bat for jobs at Hanscom. *Boston Sunday Globe,* October 12, 3.

———. 1999. Regional planners conjur '2020 vision.' *Boston Globe,* June 27, 10.

Dodge, William. 1996. *Regional excellence: Governing together to compete globally and thrive locally.* Washington, D.C.: National League of Cities.

Edgar, Randal. 1997. NESWC towns may ask state for financial aid. *Lowell Sun,* November 7, 11.

Elkin, Stephen L. 1987. *City and regime in the American Republic.* Chicago: University of Chicago Press.

Fainstein, Susan S., and Norman I. Fainstein. 1987. *Restructuring the city: The political economy of urban redevelopment.* New York: Longman.

Gerson, Jeffrey, and John Wooding. 2000. Regional power and regional regimes: The case of Hanscom Air Force Base in Massachusetts. In C*ommunity power and policy,* ed. Nancy Kleniewski and Gordana Rabrenovic. Stamford, Conn.: JAI Press.

Gittell, Ross J. 1992. *Renewing Cities.* Princeton: Princeton University Press.

Hanscom Field Web page. 1997. [http://www.hanscom.af.mil].

Henderson, Harold. 1994. Review of new visions for metropolitan America. *Planning* 60 (8): 35.

Jurkowitz, Mark. 1997. Renaming the armpit: call it a controversy, should the standard-times be in charge of relabeling the region. *Boston Globe,* June 5, E5.

Keating, Michael, and John Loughlin eds. 1997. The political economy of regionalism. London: Frank Cass.

Kenneally, Rebecca. 1997. NESWC turns focus to its current pact. *Arlington Advocate,* September 25, 1, 6.

Mollenkopf, John H. 1982. *The contested city.* Princeton: Princeton University Press.

Mullin, John. 1997. *Boston Globe,* May 25 (Northeast Weekly), 6.

Noponene, Helzi, Julie Graham, Ann R. Markusun. 1993. *Trading industries, trading regions.* New York: Guilford Press.

Northern Middlesex Council of Governments (NMCG). 1999. 2020 Vision: Planning for growth in the Northern Middlesex Region: A regional growth management action plan. Lowell: NMCG. June.

Ohmae, Kenechi. 1995. The end of the nation-state: *The rise of regional economies.* New York: Free Press.

Orfield, Myron. 1997. *Metropolitics: A regional agenda for community and stability.* Washington, D.C.: Brookings Institution.

Peterson, Paul E. 1981. *City limits.* Chicago: University of Chicago Press.

Rusk, David. 1996. Baltimore unbound. Baltimore: Abell Foundation.

Savitch, H. V., and Ronald K. Vogel. 1997. *Regional Politics: America in a post-city age.* Thousand Oaks, Calif.: Sage.

Shannon, Christine M. 1997. Protestors call for shutdown of trash plant. *Arlington Advocate,* August 27, 2.

Shefter, Martin. 1985. *Political crisis/fiscal crisis.* New York: Basic Books.

Stephenson, D. Grier, Jr., Robert J. Bresler, Robert J. Friedrich, and Joseph J. Karlesky. 1992. *American government,* brief ed. New York: Harper Collins.

Stone, Clarence N. 1989. *Regime politics: Governing Atlanta: 1946–1988.* Lawrence: University of Kansas Press.

Swanstrom, Todd. 1995. Philosopher in the city: The new regionalism debate. *Journal of Urban Affairs* 17 (3): 309–14.

Sweeney, Emily. 1999. Cellucci approves NESWC rate relief. *Arlington Advocate,* August 15, 1, 13.

Wallis, Allan D. 1998. New visions for a metropolitan America. *National Civic Review* 87 (1): 97.

Whitt, J. Allen. 1982. *Urban elites and mass transportation.* Princeton: Princeton University Press.

II RETHINKING SUSTAINABLE DEVELOPMENT

Health, Work, and
the Environment

Introduction

Charles Levenstein

THE discussion of "sustainability" is frequently dominated by concerns, on the one hand, about environmental degradation and, on the other hand, about economic growth. In fact, sustainability was raised by environmentalists as a critique of unidimensional thinking by economists who lauded growth without considering either potentially undesirable consequences, or limits imposed by environmental considerations. Robert Costanza maintains that "economic growth, which is an increase in quantity, cannot be sustainable indefinitely on a finite planet. Economic development, which is an improvement in the quality of life without necessarily causing an increase in quantity of resources consumed, may be sustainable" (Costanza et al. 1991, 7–8).

The importance of healthful employment and work environments is rarely mentioned. Conceiving of healthy cities, culturally rich communities, places where meaningful social production can occur sometimes seems beyond the pale. Improvements in the "quality of life," however, are an essential aspect of sustainability.

The essays in this part of the book are intended to broaden the conversation about sustainability and to raise questions about the well-being of the workforce and communities subject to development strategies, while at the same time bringing the global debates over sustainability to practical local and regional levels. The authors cover a range of ethical concerns, providing a broader notion of sustainability than that frequently informing economic discussions. These studies highlight the difference in perspectives between those interested in short-

165

term profitability and those concerned with the longer term. All of the authors make the presumption that unethical behavior and socially unjust systems are not sustainable. All of the authors argue for or assume social justice as one of the key criteria in defining sustainability.

The essays by Laura Punnett and John MacDougall focus on the costs of unsustainable industrial activity. The World Health Organization, admittedly relying on incomplete data, estimates that the economic costs of occupational disease and injury amount to approximately 4 percent of world production, not to mention the pain and suffering endured. In the United States, Paul Leigh and associates have estimated that between 2 and 3 percent of gross domestic product are lost to industrial injuries and disease. Punnett focuses on the costs of injuries related to the poor design of industrial facilities, the costs of repetitive strain injuries, and presents data collected from her investigations of a number of specific cases.

Punnett's study, however, not only documents the economic costs of occupational injuries, but also, implicitly, offers the approach of the ergonomist as an important element in sustainable development strategies. Designing productive processes so that they "fit" the workers—not damage them—should be a key aspect of development. The enormous social and political dimensions of this problem and its solutions should not escape those observers of the current battle in Congress over a national standard for ergonomics.

Similarly, MacDougall describes the costs of the environmentally destructive use of chemicals and of the "end-of-pipe" solution, incineration. Since the passage of Superfund legislation in the early 1980s and the adoption of the Basel Convention in 1989 that has restricted international trade in hazardous waste, there has been no denying the enormous cost of the toxic waste problem. In the 1990s the environmental regulatory agencies came to accept, perhaps grudgingly, that pollution prevention is essential. In fact, the United Nations Commission on Sustainable Development calls for comprehensive pollution-prevention strategies: "Environmentally Sound Technologies (ESTs) are not just individual technologies, but total systems which include know-how, procedures, goods and services, and equipment as well as organization and managerial procedures" (UNCSD, August 10, 1999). Mac-Dougall adds another element to the criteria for sustainability—democratic process, public participation in societal choices about the uses of technology. In his study of an anti-incinerator movement in the Merrimack Valley, MacDougall links environmental sustainability to politics and to social justice. Regional strategies failing to take inequality and social disparities into account will not be sustainable.

With the next three essays, we turn explicitly to the design of solutions to the problem of sustainability. Margaret Quinn and her fellow authors propose sustainable production as a humane strategy for the work environment, linking occupational and environmental health. Perhaps what should be stressed is that this study represents a major shift in the thinking of industrial hygienists from control to integrated prevention strategies. The precautionary principle is the overriding consideration in this work; from a public health point of view, this is primary prevention.

Kenneth Geiser and Jim Greiner describe innovation in the adoption of cleaner production techniques. They describe three cases: separation and recovery systems and "electroless" nickel by a plating company; a chilled ozone process for the semiconductor industry; and a "radiance" process as a substitute for a standard chemical-intensive wet cleaning process for the semiconductor manufacturing industry. The development of practical cleaner production processes is complex and difficult. Geiser and Greiner show both the importance of scientific and technical research and the need to study the innovation process. In both directions, the ends of sustainability can be served by university-based researchers.

Finally, Vesela Veleva and Cathy Crumbley discuss efforts to develop effective measures of sustainable production, incorporating ethical considerations as well as those of economic viability. Certainly, good decision making about technology requires reasonable indicators of success and failure. In addition, since sustainability frequently involves improvements in quality—of services, of goods, of processes, of life— it is essential to have commonly accepted criteria even for nonquantifiable dimensions of sustainability for judging progress. These essays argue for or assume the social right to intervene in technological decision making well before the "end of the pipe." Cleaner production is an essential element in a sustainable regional development policy.

REFERENCES

Costanza, R., H. E. Daly, and J. A. Bartholomew. 1991. Goals, agenda, and policy recommendations for ecological economics. In *Ecological economics*, ed. R. Costanza. New York: Columbia University Press.
United Nations Commission on Sustainable Development. 1999. New York.

Economic Costs of Ergonomic Problems

An Obstacle to Sustainable Development

Laura Punnett

SUSTAINABLE development is usually conceptualized in terms of avoiding chemical pollution and its effects on our natural habitat. More recently, public health scientists have pressed for incorporating human health concerns into the definition of sustainability. Nevertheless, the focus is still usually on chemical exposures, both within and outside the factory walls, and their effects on workers and community residents.

Our concern about human health and well-being as a fundamental condition of sustainability should not be limited to the prevention of exposure to toxic chemicals. Physical and psychosocial features of the workplace cause many types of injuries, fatal and otherwise, and chronic health conditions. Since these hazards do not spill outside the walls of the workplace, however, they do not typically command the attention of environmentalists, who have largely been the ones to forward the sustainability agenda. Furthermore, these aspects of the workplace appear to many to be unmodifiable, because they usually result directly from the production goals, the level of technology utilized, and the fundamental organization of the workplace.

Occupational ergonomics is the branch of workplace health and safety concerned with "designing the job to fit the worker" (Kroemer 1997). The adverse exposures that workers experience as a consequence of poor ergonomic conditions in the workplace can be characterized broadly as physical and psychosocial. These include the job–person fit both at the micro level— for examples, workstation height, load weight, dimensions of equipment and tools—as well as at the macro level—production quotas and work pace, wage systems, job content, decision-making opportunities, task variability, information-processing demands, software capability and flexibility, production-process organization, and material flow through a facility. The science and practice of ergonomics is largely concerned with developing a knowledge base for correct job design, as well as understanding the processes that facilitate implementation of the science in real workplaces.

Lack of attention to ergonomic principles in the design of the workplace has many adverse consequences for worker health and safety, including musculoskeletal disorders, cardiovascular and cerebrovascular disease,

169

psychological distress, adverse reproductive outcomes, and increased risk of acute injury (Karasek and Theorell 1990; Warren 1997). In particular, there is substantial scientific evidence that some musculoskeletal disorders (MSDs) are caused by occupational ergonomic stressors such as highly repetitive manual work, forceful exertions, and exposure to segmental and whole-body vibration (Bernard 1997). In addition to the resulting human pain and suffering, there are also monetary costs to employers through workers' compensation claims, decreased productivity and production quality, losses to scrap, medical insurance premiums, and, ultimately, the sustainability of a company. The proportion of these injuries and illnesses that are work related are by definition preventable, as are their costs to employers, to workers, and to society.

We focus here on the MSDs associated with ergonomic hazards in the workplace ("strains and sprains" or "repetitive motion injuries"), the frequency of which is high and increasing in the United States, as in many other countries that maintain statistics on occupational injury and illness. Work-related musculoskeletal disorders have accounted for over 50 percent of recognized occupational health injuries and illnesses in the United States since 1992, totaling about 4 million cases in 1995 (U.S. Bureau of Labor Statistics 1992–1996). There is evidence, however, that the frequency of work-related musculoskeletal disorders (MSDs) is severely underestimated, by as much as 60 percent, in traditional administrative data sources such as workers' compensation records and Occupational Safety and Health Administration (OSHA) logs of recordable injury and illness (Fine et al. 1986; Maizlish et al. 1995; Nelson et al. 1992; Silverstein 1997).

In 1989, U.S. MSD costs from workers' compensation alone were $11.4 billion for low back and $563 million for upper extremity (Webster and Snook 1994a, 1994b). Data from such administrative sources, however, underestimate the total economic losses, not only because some work-related MSDs are not compensated but also because of incomplete compensation to workers and uncompensated losses to the employer. The latter include lost production and excessive downtime, difficulty retaining trained and experienced employees, cost of hiring employees because of injury-related turnover, high inspection costs and scrap rates, reduced product quality or market share, and adverse impacts on labor and public relations (Eklund 1995; Oxenburgh 1991, 1997; Simpson and Mason 1990). It has been estimated that the combined direct and indirect economic burden of all musculoskeletal disease in the United States, including arthritis and repetitive trauma disorders, totaled $149 billion in 1992 (Yelin and Callahan 1995). The work-related proportion of the total morbidity in the general population is not known but is likely quite substantial.

The problem is illustrated here with examples from the U.S. automobile manufacturing industry. Vehicle manufacture is a major industry world-wide, accounting for at least 7 percent of the manufacturing workforce in countries as diverse as the United States, Canada, Mexico, Brazil, Sweden, Spain, Japan, India, and Australia (International Labour Office 1995). In general, this sector is characterized by high capital investment in manufacturing facilities and an organized workforce. As in many other industries, musculoskeletal disorders are the single largest group of injuries, whether ranked by frequency or lost workdays (Warner et al. 1998).

Two projects investigating work-related MSDs were carried out by the author and colleagues in U.S. automotive manufacturing plants near Detroit, Michigan, where the regional economy is heavily dependent on this single industry. "Study One" examined back and shoulder disorders in a single automobile assembly plant (Punnett et al. 1986, 1991, 2000). "Study Two" examined all reported musculoskeletal disorders at two automotive stamping plants and two engine-assembly plants, with additional information about disorders affecting the upper extremity (neck, shoulder, arm, elbow, wrist or hand) in two of those four plants (Punnett 1997, 1998). In both investigations, the risk of musculoskeletal disorders was found to be associated with exposure to physical ergonomic stressors at work, after adjusting for numerous covariates. In the assembly plant, about 75 percent of back disorders and over 50 percent of shoulder disorders were attributable to occupational exposures, including postural stress, heavy lifting, and use of power tools. In the stamping and engine plants, at least 60 percent of upper extremity disorders appeared to be similarly work related.

Important evidence was also obtained regarding the extent to which MSDs were not reported to the employer, despite the availability of in-plant medical services. In Study One, a large proportion of back and shoulder disorders identified by the investigators were not found in the plant medical records; these "unreported disorders" were also found to be associated with awkward postures and heavy lifting (Punnett et al. 1986). In Study Two, there was remarkably little overlap among cases identified through the plant medical departments, workers' compensation records, and cases identified by interview and physical examination, implying that many prevalent musculoskeletal disorders had not led to in-plant medical visits or to workers' compensation claims within a several-year period (Punnett 1997).

In both studies, data were also collected on MSD costs that accrued to both employers and workers. These data are presented here and partially evaluated with respect to their impact on employer productivity. The goal was to estimate both the direct and indirect costs of work-related musculoskeletal disorders in a large and economically important industry, in order to illustrate the substantial impact of development on industrial sustainability.

Methods

Study One

In 1984–85, two parallel case-control studies were conducted of back and shoulder disorders in a final assembly automobile plant in Detroit. The potential cases were all workers from the four largest production departments in the plant who reported back or shoulder pain, respectively, to the plant medical department during a ten-month period; they were subsequently interviewed and received a physical examination of the back, neck, and shoulders. These subjects are referred to here as having "reported disorders," regardless of the findings on interview and examination. The potential controls were randomly selected workers in these departments who had not reported back or shoulder disorders to the medical station during the study period. Only those potential controls who were free of back, neck, and shoulder pain on interview and physical examination were considered controls for the case-control analyses, and the workers with unreported pain were excluded from those reports. For this report, however, those workers with pain that would have met the study-case criterion (pain for at least one week or on at least three occasions in the past year) were defined as having "unreported disorders."

Data collected from the medical records and interviews permitted partial estimation of the economic costs of the back and shoulder disorders experienced in the four production departments. The company medical department provided data on in-plant care provided for reported cases only for the current episode, as defined administratively. Work restrictions and absences were recorded but were not coded in a way that permitted enumeration of length of absence or restriction separately, so the medical records could not be used for these items. The standardized interviews provided information on days lost from work since the onset of pain (current episode), any work restrictions since onset, and job transfers sought in the past five years from jobs that were perceived to aggravate pain. These interview data were available for both the reported and the unreported disorders if they met the study-case criterion of minimum duration or frequency. To calculate costs only for the year of the study, data obtained for the period "since onset" were annualized by dividing by the number of years since onset for the group. For example, if median onset time was two years earlier, it was assumed that one-half of the involved costs were incurred in the year preceding the study interview.

Data were not available concerning workers' compensation to the employees or "indirect" costs to the employer, such as decreases in productivity or quality or increases in medical insurance premiums. No future costs could be estimated.

Thus, information on partial costs to the plant was available for medical visits and treatments, lost work time, work restrictions, and voluntary job transfers. Based on discussions with company personnel, the following (conservative) assumptions were made in calculating these costs:

Each in-plant medical visit or treatment cost U.S. $50, on average;
Each work absence of more than five days resulted in training of one replacement worker and $295 per week in non–workers' compensation replacement costs;
One-half of all work restrictions resulted in training for both the injured worker and the worker replacing him/her on their new jobs;
Each job transfer resulted in training for both the injured worker and the worker replacing him/her; and
The cost of training one worker on a new job was sixteen hours of training at an average labor cost of $23 per hour, or $368 total.

Study Two

Administrative data on musculoskeletal disorders at two automobile engine plants and two automobile body stamping plants were abstracted both from records of employee visits to plant medical departments (January 1, 1989–June 30, 1993) and from workers' compensation claims and First Reports of Injury (January 1, 1988–December 31, 1990). Prospective employment figures were collected from each plant for the duration of the study, in order to obtain denominators for the computation of annual rates of new cases. Standardized interviews and physical examinations were also completed on 1,198 workers in selected departments at one engine and one stamping plant.

The medical-visit data were used to compute incidence rates of musculoskeletal injuries and illnesses and to determine costs associated with the reported conditions. All visits for potentially work-related acute or chronic musculoskeletal disorders recorded in the medical database were collected, including incidents that were entered on the OSHA 200 log—company injury record—with a primary code of strain or sprain or of repetitive trauma disorder, as well as nonloggable incidents ("first aid only") with a secondary code of strain or sprain or of repetitive trauma disorder. For enumeration of unique medical-visit cases, repeat visits were excluded from the database, although information on repeat visits was utilized to estimate the total cost of medical visits. Available information from the medical records included whether the case had lost work time or had a work restriction, temporary job transfer, outside medical treatment, or a compensation claim as of the date of the medical department visit.

To estimate costs to the four plants from the medical-visit data, the following assumptions were made, based on the information obtained from the earlier study and an assumed annual inflation rate of about 3 percent since 1985:

Each in-plant medical visit or treatment cost $60 on average;
Each work absence of more than five days resulted in training of one re-
    placement worker;
Each "temporary transfer" of more than five days resulted in training of
    both the injured worker and the worker replacing him/her on their
    new jobs;
The cost of training one worker on a new job was sixteen hours of training
    at an average labor cost of $27.50 per hour, or U.S. $440; and
The cost of outside medical treatment was $500 per case.

At the request of the investigators, the corporate Workers' Compensation (WC) office identified compensation claims or First Reports of Injury
(potential future claims) for musculoskeletal disorders filed from the same
four study plants. All disorders affecting the musculoskeletal system were
sought, including specific diagnoses and nonspecific "strains and sprains"
and "repetitive trauma disorders." Each worker's compensation (WC) record selected by corporate personnel was reviewed and relevant information
was abstracted manually. The WC records did not explicitly distinguish
First Reports of Injury (FRIs), closed, and open cases; however, cases in
which a claim had been accepted for medical costs could be identified. Payment data appeared to be entered only for (closed) lost-time cases.

In order to estimate worker compensation costs more completely, low
back and upper extremity cases with payment for medical treatment, if no
payment amount had been found in the record, were assumed to have an
average dollar cost equivalent to the median cost of closed compensable
cases in Michigan in 1989, according to the Liberty Mutual Insurance
Company. This assumption was considered justified because over 50 percent of closed cases for each body region were claims for medical treatment
only (Webster and Snook 1994a, 1994b). The respective costs were $244
for low back cases and $500 for upper extremity cases. No comparable
estimate was available for musculoskeletal cases involving the neck, lower
extremity, or other body regions.

From the interviews of workers with upper extremity MSDs, information was extracted for the past year on days lost from work because of the
condition, the length of any work restrictions for the same problem, and
whether or not a transfer had been sought from a job perceived to aggravate the problem. Costs were estimated on the basis of assumptions similar
to those used for estimating the costs of medical visits.

All statistical analyses for both studies were performed with Statistical

Analysis Software (SAS Institute 1990) and Microsoft Excel(r) spreadsheet software. Statistical significance was defined at the level of $p = 0.05$.

Results

Study One

In the assembly plant, the four production departments studied had a total workforce of 1,335 employees. A total of 248 potential shoulder or back cases were identified from the reports to the medical department. Of these, 211 workers (118 back and 93 shoulder disorders) were interviewed and examined. Ninety-five (81 percent) of the reported back disorders and 79 (85 percent) of the reported shoulder disorders met the case criterion of minimum duration or frequency of pain for the case-control study.

Among the 310 potential controls randomly selected from the production department rosters, 259 (84 percent) were interviewed and examined. Only 125 (48 percent) were free of back, neck, and shoulder pain as well as by physical examination findings. Ten workers denied persistent pain but had positive findings on examination. Eighty-four (32 percent) had shoulder pain, and 79 (31 percent) had back pain that met the study case criterion including 39 workers with both shoulder and back disorders. These 124 workers, who were excluded from the case-control study because they had pain that met the a priori case definition, are referred to here as "unreported disorders." It was noted, however, that at least some of them had apparently visited the plant medical department at sometime in the past, since there had been some work restrictions for this group, since there had been some work restrictions for this group (see below).

Since the potential controls had been selected at random from the workforce of the four production departments, it could be assumed that they were representative of all workers without back or shoulder medical reports ($n = 1,124$). From the prevalences of back and shoulder pain among the potential controls, it was thus possible to estimate the total prevalence of unreported musculoskeletal pain severe enough to meet the case criterion for the case-control study. Standardized by production department, the prevalence of unreported shoulder pain was 34 percent and of unreported back pain, 31 percent. The medical department records thus identified only 17 percent (79/465) of workers with shoulder pain and 22 percent (95/440) of those with back pain in this workforce.

For both back and shoulder pain, the unreported disorders were twice as long in duration, on average, from onset to interview. Workers with both reported and unreported back disorders had experienced an average of 140 days of pain in the previous year. Both the occurrence and the number of lost workdays per case between the two groups were similar. In contrast, among the shoulder disorders, about one-half of each group had ever lost time from work in relation to the current episode of shoulder pain, but the number of lost workdays was higher among the unreported disorders.

Among both the 118 workers with back disorders and the 93 with shoulder disorders reported to the plant medical department, about one-third received medication or physical therapy. According to the medical record, 52 percent of both back cases ($n = 61$) and shoulder cases ($n = 48$) required up to one month of time off work or restricted duty. A total of 100 back cases (55 percent reported and 44 percent unreported) and 30 shoulder cases (24 percent reported and 10 percent unreported) had had at least one work restriction since the onset of pain. Most of these restrictions had been for less than one work week. Only 24 back cases (10 reported and 14 unreported) and 12 shoulder cases (5 reported and 7 unreported) had been on restricted duty for ten days or more.

Difficulty in performing the current or past job was reported by 73 percent of reported back cases and 62 percent of unreported cases; 60 workers (30 percent) had transferred out of at least one job due to back pain within the five years prior to the study. Similarly, 72 percent reported and 70 percent unreported shoulder cases had difficulty performing the current or past job, and 40 (23 percent) had bid off at least one job due to shoulder pain within the five years prior to the study.

The costs of back disorders totaled $536,058 (table 1), excluding many costs incurred prior to the year of the study as well as workers' compensation and future costs. The average total cost per case was actually lower for reported ($2,290) than unreported ($3,365) cases, reflecting the fact that unreported cases were of longer duration. In contrast, the average cost per case in the past year was about 50 percent higher for reported disorders than unreported ones ($1,209 versus $841).

Both the total cost of shoulder disorders and the average cost per case were about one-half of the corresponding estimates for back disorders,

Table 1. Costs of medical care, lost work time, work restrictions, and voluntary job transfers for back and shoulder disorders on interview: automobile assembly workers, Detroit, 1984–85

	Back disorders			Shoulder disorders		
	Reported ($n = 118$)	Unreported ($n = 79$)	Total	Reported ($n = 93$)	Unreported ($n = 84$)	Total
Total cost ($)	270,221	265,837	536,058	172,131	78,618	250,749
Average cost per case ($)	2,290	3,365	2,721	1,851	936	1,417
Total cost in past year[a] ($)	142,618	66,459	209,077	172,131	39,309	211,440
Average cost per case in past year ($)	1,209	841	1,061	1,851	468	1,195

[a]Estimated from average duration of cases as reported on interview

which involved more and longer work absences than shoulder disorders. The average cost for reported shoulder disorders was about twice as high as for unreported ones ($1,851 versus $936), again primarily due to the difference in number and length of work absences.

Thus, the known costs of these back and shoulder disorders combined, without any information on workers' compensation, totaled $786,807, or $589 per active employee on the payroll, averaged over the entire workforce of the four study departments. For the past year alone, the corresponding figures were $321,822 in total and $241 per employee. These estimates were further adjusted in two ways. First, since the magnitude of underreporting was known in a random sample of the total workforce, the costs of unreported disorders were assumed to be generalizable to the entire workforce. Second, the costs were multiplied by the proportion of back and shoulder disorders that were estimated to be caused by poor ergonomic design, or approximately 75 percent overall (Punnett et al. 1986, 1991, 2000). The resulting figure for work-related back and shoulder disorders combined was $868,804 in total and $425,534 in the past year. This represented a payroll cost per employee of $651 in total ($319 in the past year). Since disorders reported in other years would also have been generating costs during the study period, $319 should be considered the lower bound of the known annual costs for back and shoulder disorders per worker on the plant payroll, excluding any compensation premiums or claims paid.

Study Two

In the administrative data for two stamping and two engine plants, a total of 5,025 first medical visits for "new" musculoskeletal disorders (defined administratively) were recorded during the four-and-one-half-year study period. The majority of these disorders were characterized by the plant medical departments as "strains and sprains" of sudden onset; a smaller number were considered to be illnesses, or "repetitive trauma disorders," presumed to have developed over a longer period of time. The overall rate of medical visits for new musculoskeletal disorders, both OSHA-recordable and nonrecordable, for all body parts combined, was about 120 cases per 1,000-person-years, or 12 percent of the workforce annually (table 2). This rate increased over the time period studied; by 1992–93 it was approximately 150 to 200 cases per 1,000-person-years in each plant.

By body region, the highest rates in each plant were for back disorders (30–60 cases per 1,000-person-years) and the next highest for wrist/hand (20–50 cases per 1,000-person-years). By job classification, skilled trades had higher rates than production employees for back and leg disorders, while production employees had higher rates than the skilled trades for the arm, wrist, and hand.

Table 2. Incidence rates and costs of medical visits (MV) and workers' compensation (WC) claims for musculoskeletal disorders at two automotive stamping and two engine plants, Detroit

	Engine 1	Engine 2	Stamping 1	Stamping 2	Total
Average annualized incidence rate of MV $(\times 10^{-3})$[a]	120.12	163.25	133.03	84.89	121.10
Total cost for MV ($)	100,340	210,360	205,320	173,500	689,520
Average cost per MV case ($)	152	127	145	135	137
Average annualized incidence rate of WC claims $(\times 10^{-3})$[b]	50.23	78.43	47.28	32.47	49.89
Total cost for WC claims ($)[c]	59,643	283,744	243,223	198,652	785,262
Average cost per WC case ($)	324	533	726	604	569

[a] Medical visits, 1989–93
[b] Compensation claims and First Reports of Injury, 1988–1990
[c] Costs of lost work time and medical care, combined

In the four plants combined, less than one percent of medical visits for MSDs were flagged as workers' compensation cases, even though almost 8 percent resulted in lost work time, 17 percent had work restrictions, and 5 percent required outside medical treatment. The immediate cost of these medical visits was estimated at about $137 per case, with little variation by facility.

A total of 1,380 WC claims and First Reports of Injury for musculo-skeletal disorders were identified at the four study plants for the three-year period, 1988–90. However, there was surprisingly little overlap between the medical visit cases and WC claims from the same plants (Punnett 1997), and only 43 (0.9 percent) of all musculoskeletal disorder medical visits were flagged as worker compensation claims.

The WC incidence rates were about one-half as high as the rates of medi-cal visits (table 2). About 20 percent (303) of the workers with WC claims had lost at least one day and the lost-time payments totaled $579,482. About 34 percent (473 cases) had received coverage of medical costs, estimated at $205,780. Thus the total cost of workers' compensation pay-ments for these four plants over the three year period was about $1,474,782, or an average of $1,069 per WC case.

Because of the minimal overlap among cases from medical visits and WC claims, a more complete estimate of administratively recorded costs of upper extremity disorders was obtained by combining them, for a total average annual cost of $414,981 in all four plants. Since the combined workforce of these plants averaged about 9,221 workers, the average annual payroll cost of reported cases was estimated at about $45 per worker per year.

In the interview survey, about one-third of the subjects ($N = 1198$) met the case definition for a disorder of one or more regions of the upper extremity. The wrist/hand and shoulder/upper arm were more likely to be affected than the neck or elbow (table 3). There was very little overlap between these cases and those recorded in the medical visit and compensation data from the same departments; thus, the costs reported by these workers were tabulated separately.

Elbow and wrist/hand cases found on interview were slightly more likely (20 percent) to report having visited the plant medical department within the past twelve months than neck or shoulder cases (13 percent). Wrist/hand disorders were most likely to have prolonged work restrictions (9 percent of cases), and shoulder disorders were least likely (5 percent of cases). About 10 percent to 15 percent of each case group had sought a job change in the past year because of the upper extremity condition. Neck disorders were the most expensive per case, because of one long absence and slightly higher proportions of work restrictions and job transfers. However, wrist/hand disorders generated the highest costs because they were both expensive and the most prevalent. The total payroll cost of all upper extremity disorders combined, based on the interview data, was $152 per worker per year in the six department groups studied (total population, $N = 1550$).

Other information provided in the interviews indicated that nearly one-fourth of each group of cases indicated that they had ever changed or given up leisure or household activities because of their upper extremity problems, some of these many years before the interview took place. In addition, in the past twelve months, about 20 percent had seen their family physician and another 10 percent to 15 percent had seen other medical care providers outside the plant for these conditions.

Table 3. Costs associated with medical care, lost work time, work restrictions, and voluntary job transfers for upper extremity musculoskeletal disorders on interview, by body region: workers in one automotive stamping and one engine plant, Detroit, 1992–93

	Neck ($n = 129$)	Shoulder/ upper arm ($n = 206$)	Elbow/ forearm ($n = 99$)	Wrist/ hand ($n = 296$)	All upper extremity[a] ($n = 434$)
Total cost ($)	52,789	57,081	22,172	103,206	235,248
Average cost per case ($)	409	277	224	349	542

Note: One medical visit per subject assumed
[a] Some cases had more than one upper extremity region affected

Discussion

This chapter describes the partial costs of musculoskeletal disorders incurred by two U.S. automobile manufacturing companies over the past decade. In the production areas studied in these five plants, over one-half of the MSDs identified from clinic records, worker interviews, and physical examinations were explained by physical features of the work environment such as repetitive hand motions, heavy lifting, non-neutral body postures, and vibrating tools.

The direct cost estimates reported here are incomplete because of limitations in the data available from the study plants, especially the inability to follow cases forward in time after the first report or treatment and the incomplete assessment of indirect costs, and because of case underreporting in administrative data systems, especially when workers seek medical attention independent of the workplace facility and/or leave employment because of pain and impaired performance. Nevertheless, the results indicate that work-related musculoskeletal disorders were a major drain on productivity and profits for the five plants studied, representing costs of several hundred dollars or more per employee per year.

Although workers' compensation claims are often thought to be the major expense for MSDs, in-plant medical care and case management here cost at least as much as WC claims. In Study One, these costs for shoulder cases were almost three times higher than the median cost per case of "compensable upper extremity cumulative trauma disorder" in Michigan in 1989, according to published data from Liberty Mutual Insurance Company (Webster and Snook 1994a). This is consistent with other estimates that the real costs to employers of workplace injuries and MSDs can total at least two to three times the amount paid in workers' compensation (Leigh et al. 1997; Oxenburgh 1991, 1997; Yelin and Callahan 1995).

Many investigators have reported musculoskeletal disorders to be associated with the physical demands of jobs in the automobile manufacturing industry (Byström et al. 1995; Fransson-Hall, Byström, and Kilbom 1995, 1996; Johansson et al. 1993; Laflamme, Döös, and Backström 1991; Nelson et al. 1992; Park, Krebs, and Mirer 1996, Park et al. 1992). It has also been documented that there are feasible ergonomic hazard abatement measures for many jobs similar to the ones studied in these five plants. Better design and selection of tools, equipment, and product components, job rotation, and alternative forms of work organization have all been shown to reduce ergonomic hazards and the risk of musculoskeletal injuries (Armstrong et al. 1986; Bullinger, Rally, and Schipfer 1997; Echard, Smolenski, and Zamiska 1987; Habes 1984; Johansson et al. 1993; Kadefors et al. 1996; Karwowski 1987; Keyserling et al. 1993; McAtee 1987; Moore 1994; Schütte and Schüder 1997; VanBergeijk 1987; Yu and Keyserling 1989). Thus, the costs of work-related musculoskeletal disorders in the auto industry are preventable, both in theory and in practice.

In the United States, many ergonomic programs have already been implemented by automotive manufacturers, motivated in part by OSHA citations and by demands from the unionized labor force. Typically such programs include worker training, plant-wide committees, and re-engineering of jobs where workers are injured. There has been little formal evaluation of the success of these efforts, but the data presented here suggest that they have not been sufficient. Many of the facilities are old, however, and are unlikely to be redesigned in a comprehensive manner. Probably more important, the fundamental organization of work around the fixed-pace assembly line is rarely questioned; on the contrary, modifications to increase productivity and profitability, such as just-in-time systems, lean production, and total quality management or continuous improvement, appear to have intensified job demands (Landsbergis, Cahill, and Schnall 1999; Leslie and Butz 1998).

Health Policy Implications

Several important policy issues are raised by these results, including the externalization of occupational disease to the general burden of health care. In the unionized automobile manufacturing workforce in the United States, employees commonly use private medical insurance to cover services for work-related musculoskeletal morbidity (Nelson et al. 1992). Many workers in these two studies reported seeking medical care from outside providers, although the source of payment was not determined. The increased cost to the employers and employees of medical insurance premiums associated with medical treatment for MSDs could not be estimated here, nor has it been evaluated by others.

Another concern is the apparent low overlap between medical care seeking within the workplace and filing of WC claims in Study Two. The WC records had not yet been fully computerized, so it could not be confirmed that all of the relevant records had been identified, and some workers may file WC claims many years after first seeking medical attention. However, others have also called attention to a substantial proportion of work-related MSDs that do not result in WC claims (Michaels 1998; Rosenman et al. 2000). Why some workers in these studies did not file compensation claims could not be determined; possible reasons are lack of knowledge about occupational etiology, concern about job security, anticipated rejection of the claim, and denial of the injury itself because of financial need or a sense of self-worth contingent on providing for oneself and one's family (Maizlish et al. 1995; Michaels 1998; Rosenman et al. 2000; Tanaka et al. 1995; Teschke and van Zwieten 1999).

Other costs of MSDs are externalized to the injured workers and their families and would be compensated incompletely, at best, through the WC system. Financial losses include uncompensated medical care and lost work time, lost future earnings and fringe benefits, reduced job security and career advancement, lost home production and child care, and home care costs provided by family members; non-monetary losses include pain and suffering, decreased quality of family, social and community relations, sense of self-worth and identity, and recreational activities (Leigh et al. 1997; Levenstein 1999; Morse et al. 1998; Yelin and Callahan 1995). Quantifying the costs of diminished health and quality of life is further complicated because musculoskeletal problems may persist for many years, as seen here in Study One, and may have cascading effects not only for the injured workers but also for family members. In Study Two, the workers who filed FRIs and WC claims for musculoskeletal disorders had a median of two dependents each. These dependents were likely to experience both economic losses due to decreased income and psychosocial consequences such as adverse effects on spousal and parent-child relations. In addition, family members who care for an injured worker may also bear a direct burden in the form of decreased employability or social opportunities. The value of home care services provided by family members is likely a large, although unestimated, economic cost.

The Apparent Paradox of Work-related MSDs

If MSDs are both expensive and preventable, it is a dilemma that they continue to occur in such large numbers without large-scale or effective abatement measures being implemented by employers. Perhaps part of the explanation lies in imperfect information; work-related MSDs are underestimated, and a large proportion of these costs often are not specifically identified by traditional accounting methods as due to ergonomic problems in the work process (Oxenburgh 1991, 1997; Simpson and Mason 1990). There is also incomplete technical knowledge about occupational ergonomics among those who are in best position to carry out solutions with the least expense. Costs of design changes are low in the planning stages of products, jobs, work processes, and facilities. However, industrial designers, workplace managers, and engineers typically regard workplace health and safety as the responsibility of safety and medical staff. Management and engineering personnel are often not well educated as to the causal role played by the forms of production technology and work organization for which they are directly responsible. Occupational ergonomics is not taught routinely either to engineering or to business students, despite the fact that they will have the most direct opportunities to achieve prevention, through ergonomic planning in organization of the work process and design of equipment and facilities at both the micro and the macro levels. In fact, lack of understanding of workers' experience may constitute a health hazard in itself; discrepancies in evaluation of ergonomic exposures by Dutch managers and workers were most pronounced in workplaces where the MSD risk was highest (Warren 1997).

Even perfect information, however, with dramatic estimates of the costs of work-related injury and illness, may not always provide sufficient motivation for employers to initiate or sustain effective health and safety programs over the long term. Frick (1997) has argued that monetary losses are insufficient to motivate managers because the problem is rooted more fundamentally in short-term reward systems and issues of control over the organization and management of production, whether of goods or services. Thus, the disciplinary gap in education is paralleled by a gap in accountability for the consequences of technical and managerial decisions, since worker health and safety is rarely a criterion for job performance evaluations of engineers or managers. These issues must be addressed in order to achieve effective primary prevention of work-related morbidity and mortality.

Another problem is that cost analysis, such as this one, values some lives more than others because of wage discrepancies among economic sectors and demographic groups. The workers studied here were union members, mostly male, with good wages, in-plant medical services, medical insurance, and protection against retaliation for seeking redress of health and safety problems. Occupational gender segregation, racial discrimination, obstacles to unionization, and other sociopolitical forces affect the monetary value of workers' labor, although they are irrelevant to the human value of workers' physical and psychosocial well-being. Small companies are even less motivated to implement safety measures by workers' compensation and other costs than are large companies (Sims 1988). In unorganized workplaces, workers may avoid reporting injuries because of fear of retaliation (Michaels 1998). Company size and sector may also influence the likelihood or accuracy of recording MSDs in administrative data systems (Paquette 1998). Thus other workers, with similar health problems to those studied here, might be valued less in monetary terms and might not find the results of cost-benefit analysis to be in their favor, even if managers were always persuaded by such arguments.

Whether they are recognized or not, it is evident that the monetary costs alone of MSDs represent an impressive obstacle to sustainability, at least in automobile manufacturing. As noted above, occupational injuries recognized as related to ergonomic hazards represent about 4 million cases annually, nationwide, and cost about $12 billion in workers' compensation costs alone. The available data argue strongly that such figures are merely the tip of the iceberg. These findings, and those of other authors (e.g.,

Snook, Campanelli, and Hart 1978) indicate that between one-half and three-fourths of the total cost could be avoided by the application of ergonomic principles to workplace design. Even workplace modifications carried out after an injury, by preventing future injuries in other workers, can have payback periods as short as a few months (Kemmlert 1996; Oxenburgh 1991). The resulting productivity losses are tremendous; in the U.S. economy, perhaps $18 billion or more per year of gross national production is being literally wasted.

In conclusion, the costs of ergonomic problems are incompletely measured by traditional accounting methods and are partially externalized, both to employers and to employees. Therefore, the market-based approach to solving workplace health and safety problems is likely to fail, regardless of whether development is motivated by community needs or by a search for export markets. Many MSDs could be prevented by applying known principles of occupational ergonomics to achieve user- and health-centered design of workplaces. If monetary losses alone are insufficient to motivate managers to implement health and safety programs, then perhaps it must be government's role to safeguard the common good by using its legitimate authority to require the primary prevention of work-related musculoskeletal disorders.

"Sustainable development" means economic development that promotes the health and well-being of the population as well as of the natural environment, so protection of the human labor force from injury and illness is clearly an important criterion. The long-term consequences of many injuries impact adversely on people's functional capacity and their ability to obtain and retain jobs. Those concerned to promote sustainability should not focus exclusively on chemical contamination of the environment, whether inside or outside the industrial facility, but should define the issue broadly enough to include the full range of physical hazards affecting members of the community in their places of work.

NOTE

The original studies were funded by contracts from the Ford Motor Company and from the United Auto Workers–Chrysler Joint National Committee on Health and Safety, respectively. Many individuals were instrumental in carrying out these studies. In particular, Drs. Lawrence J. Fine, Don B. Chaffin, W. Monroe Keyserling, and Gary D. Herrin collaborated on Study One; Dr. David H. Wegman and Ms. Jung-Soon Park collaborated on Study Two. Preparation of this manuscript was partially supported by the Committee on Industrial Theory and Assessment of the University of Massachusetts Lowell. An earlier version was published in 1999 in *New Solutions: A Journal of Environmental and Occupational Health Policy* 9 (4): 403–26.

REFERENCES

Armstrong, T. J., B. S. Joseph, R. G. Radwin, and B. A. Silverstein. 1986. Analysis of selected jobs for control of cumulative trauma disorders in automobile plants. Ann Arbor: University of Michigan, Center for Ergonomics.

Bernard, B. P., ed. 1997. *Musculoskeletal disorders and workplace factors: A critical review of epidemiologic evidence for work-related musculoskeletal disorders of the neck, upper extremity, and low back.* Cincinnati: Department of Health and Human Services, National Institute for Occupational Safety and Health.

Bullinger, H.-J., P. J. Rally, and J. Schipfer. 1997. Some aspects of ergonomics in assembly planning. *International Journal of Industrial Ergonomics* 20:389–97.

Byström, S., C. Fransson Hall, T. Welander, and Å. Kilbom. 1995. Clinical disorders and pressure-pain threshold of the forearm and hand among automobile assembly line workers. *Journal of Hand Surgery* (British and European vol.) 20B:782–90.

Echard, M., S. Smolenski, and M. Zamiska. 1987. Ergonomic considerations: Engineering controls at Volkswagen of America. In *Ergonomics interventions to prevent musculoskeletal injuries in industry,* 117–31. Chelsea, Mich.: Lewis Publishers.

Eklund, J. A. E. 1995. Relationships between ergonomics and quality in assembly work. *Applied Ergonomics* 26:15–20.

Fine, L. J., B. A. Silverstein, T. J. Armstrong, C. A. Anderson, and D. S. Sugano. 1986. Detection of cumulative trauma disorders of upper extremities in the workplace. *Journal of Occupational Medicine* 28:674–78.

Fransson Hall, C., S. Byström, and Å. Kilbom. 1995. Self-reported physical exposure and musculoskeletal symptoms of the forearm–hand among automobile assembly line workers. *Journal of Occupational and Environmental Medicine* 37:1136–44.

———. 1996. Characteristics of forearm-hand exposure in relation to symptoms among automobile assembly line workers. *American Journal of Industrial Medicine* 29:15–22.

Frick, K. 1997. Can managers see any profit in health and safety? *New Solutions: A Journal of Environmental and Occupational Health Policy* 7:32–40.

Habes, D. J. 1984. Use of EMG in a kinesiological study in industry. *Applied Ergonomics* 15:297–301.

International Labour Office. 1995. *1995 Yearbook of labour statistics.* Geneva: International Labour Office.

Johansson, J. Å., R. Kadefors, S. Rubenowitz, U. Klingenstierna, I. Lindstrom, T. Engstrom, and M. Johansson. 1993. Musculoskeletal symptoms, ergonomic aspects and psychosocial factors in two different truck assembly concepts. *International Journal of Industrial Ergonomics* 12:35–48.

Kadefors, R., T. Engström, J. Petzäll, and L. Sundström. 1996. Ergonomics in parallelized car assembly: A case study, with reference also to productivity aspects. *Applied Ergonomics* 27:101–10.

Karasek, R. A., and T. Theorell. 1990. *Healthy work: Stress, productivity and the reconstruction of working life.* New York: Basic Books.

Karwowski, W. 1987. Prevention of cumulative trauma disorders of the upper extremity through job redesign: Case studies. In *Trends in Ergonomics/Human Factors* 4, ed. S. S. Asfour, 1021–28. Amsterdam: Elsevier Science Publishers B. V. .

Kemmlert, K. 1996. Economic impact of ergonomic intervention: Four case studies. *Journal of Occupational Rehabilitation* 6 (1): 17–32.

Keyserling, W. M., M. Brouwer, and B. A. Silverstein. 1993. The effectiveness of a joint labor-management program in controlling awkward postures of the trunk, neck, and shoulders: Results of a field study. *International Journal of Industrial Ergonomics* 11:51–65.

Kroemer, K. H. E. 1997. *Fitting the task to the human: A textbook of occupational ergonomics,* 5th ed. Oxford and New York: Taylor & Francis.

LaFlamme, L., M. Döös, and T. Backström. 1991. Identifying accident patterns using the FAC and HAC: Their application to accidents at the engine workshops of an automobile and truck factory. *Safety Science* 14:13–33.

Landsbergis, P. A., J. Cahill, and P. Schnall. 1999. The impact of lean production and related new systems of work organization on worker health. *Journal of Occupational Health Psychology* 4:108–30.

Leigh, J. P., S. B. Markowitz, M. Fahs, C. Shin, and P. J. Landrigan. 1997. Occupational injury and illness in the United States: Estimates of costs, morbidity and mortality. *Archives of Internal Medicine* 157:1557–68.

Leslie, D., and D. Butz. 1998. "GM suicide": Flexibility, space, and the injured body. *Economic Geography* 74:360–78.

Levenstein, C. 1999. Economic losses from repetitive strain injuries. *Occupational Medicine: State of the Art Reviews,* ed. M. Cherniack. Philadelphia: Hanley & Belfus.

Maizlish, N., L. Rudolph, K. Dervin, and M. Sankaranarayan. 1995. Surveillance and prevention of work-related carpal tunnel syndrome: An application of the Sentinel Events Notification System for Occupational Risks. *American Journal of Industrial Medicine* 27:715–29.

McAtee, F. L. 1987. Reducing repetitive motions injuries in overhead assembly. *Seminars in Occupational Medicine* 2:73–74.

Michaels, D. 1998. Fraud in the workers' compensation system: Origin and magnitude. *Occupational Medicine: State of the Art Reviews* 13:439–42.

Moore, J. S. 1994. Flywheel truing: A case study of an ergonomic intervention. *American Industrial Hygiene Association Journal* 55:236–44.

Morse, T. F., C. Dillon, N. Warren, C. Levenstein, and A. Warren. 1998. The economic and social consequences of work-related musculoskeletal disorders: The Connecticut upper-extremity surveillance project (CUSP). *International Journal of Occupational and Environmental Health* 4:209–16.

Nelson, N. A., R. M. Park, M. A. Silverstein, and F. E. Mirer. 1992. Cumulative trauma disorders of the hand and wrist in the auto industry. *American Journal of Public Health* 82:1550–52.

Östlin, P. 1988. Negative health selection into physically light occupations. *Journal of Epidemiology and Community Health* 42:152–56.

Oxenburgh, M. S. 1991. *Increasing productivity and profit through health and safety.* Chicago: Commerce Clearing House.

———. 1997. Cost–benefit analysis of ergonomics programs. *American Industrial Hygiene Association Journal* 58:150–56.

Paquette, P. D. 1998. Occupational carpal tunnel syndrome surveillance in Massachusetts: A comparison of two reporting sources and an analysis of ergonomic exposure variables. Master's thesis. Lowell: University of Massachusetts Lowell, Department of Work Environment.

Park, R. M., J. M. Krebs, and F. E. Mirer. 1996. Occupational disease surveillance using disability insurance at an automotive stamping and assembly complex. *Journal of Occupational and Environmental Medicine* 38:1111–23.

Park, R. M., N. A. Nelson, M. A. Silverstein, and F. E. Mirer. 1992. Use of medical insurance claims for surveillance of occupational disease: An analysis of cumulative trauma in the auto industry. *Journal of Occupational Medicine* 34:731–37.

Punnett, L. 1996. Adjusting for the healthy worker selection effect in cross-sectional studies. *International Journal of Epidemiology* 25:1068–75; 26:914.

Punnett, L. 1997. Ergonomic stressors and surveillance in automotive manufacturing: Final report to UAW–Chrysler Joint National Committee on Health and Safety, rev. Lowell. University of Massachusetts Lowell, Department of Work Environment.

Punnett, L. 1998. Ergonomic stressors and upper extremity disorders in vehicle manufacturing: Cross-sectional exposure–response trends. *Occupational and Environmental Medicine* 55:414–20.

Punnett, L., L. J. Fine, W. M. Keyserling, G. D. Herrin, and D. B. Chaffin. 1986. The injury and cumulative trauma disorder surveillance project at Dearborn Assembly Plant: The health effects of non-neutral trunk and shoulder postures. Ann Arbor: Center for Ergonomics, University of Michigan.

———. 1991. Back disorders and non-neutral trunk postures of automobile assembly workers. *Scandinavian Journal of Work Environment and Health* 17:337–46.

Punnett, L., L. J. Fine, W. M. Keyserling, G. D. Herrin, and D. B. Chaffin. 2000. Shoulder disorders and postural stress in automobile assembly work. *Scandinavian Journal of Work Environment and Health* 26:283–91.

Rosenman, K. D., J. C. Gardiner, J. Wang, J. Biddle, A. Hogan, M. J. Reilly, K. Roberts, and E. Welch. 2000. Why most workers with occupational repetitive trauma do not file for workers' compensation. *Journal of Occupational and Environmental Medicine* 42:25–34.

SAS Institute, Inc. 1990. *SAS Procedures Guide,* version 6, 3rd ed. Cary, N.C.: SAS Institute.

Schütte, M., and D. Schüder. 1997. Case study: Investigation into the subjective strain at two differently designed automobile assembly workplaces. *International Journal of Industrial Ergonomics* 20:413–22.

Silverstein, B. A., L. J. Fine, and D. Stetson. 1987. Hand-wrist disorders among investment casting plant workers. *Journal of Hand Surgery* 12A (2, pt. 2): 838–44.

Silverstein, B. A., D. S. Stetson, W. M. Keyserling, and L. J. Fine. 1997. Work-related musculoskeletal disorders: Comparison of data sources for surveillance. *American Journal of Industrial Medicine* 31:600–608.

Simpson, G., and S. Mason. 1990. Economic analysis in ergonomics. In *Evaluation of human work: A practical ergonomics methodology,* ed. John R. Wilson and E. Nigel Corlett, 798–816. Bristol, Pa.: Taylor & Francis.

Sims, R. H. 1988. Hazard abatement as a function of firm size: The effects of internal firm characteristics and external incentives. Doctoral thesis.

Snook, S. H., R. A. Campanelli, and J. W. Hart. 1978. A study of three preventive approaches to low back injury. *Journal of Occupational Medicine* 20:478–81.

Tanaka, S., D. K. Wild, P. J. Seligman, W. E. Halperin, V. J. Behrens, and V. Putz-Anderson. 1995. Prevalence and work-relatedness of self-reported carpal tunnel syndrome among U.S. workers: Analysis of the Occupational Health Supplement data of 1988 National Health Interview Survey. *American Journal of Industrial Medicine* 27:451–70.

Teschke, K., and L. van Zwieten. 1999. Perceptions of the causes of bladder cancer, nasal cancer, and mesothelioma among cases and population controls. *Applied Occupational and Environmental Hygiene* 14:819–26.

U.S. Bureau of Labor Statistics. 1992, 1993, 1994, 1995, 1996. *Occupational injuries and illnesses in the United States by industry.* Washington D.C.: U.S. Department of Labor.

VanBergeijk, E. 1987. Selection of power tools and mechanical assists for control of occupational hand and wrist injuries. In *Ergonomics interventions to prevent musculoskeletal injuries in industry.* Chelsea, Mich.: Lewis Publishers.

Warner, M., S. P. Baker, G. Li, and G. S. Smith. 1998. Acute traumatic injuries in automotive manufacturing. *American Journal of Industrial Medicine* 34:351–58.

Warren, N. 1997. The organizational and psychosocial bases of cumulative trauma and stress disorders. Doctoral thesis. Lowell: University of Massachusetts Lowell, Department of Work Environment.

Webster, B. S., and S. H. Snook. 1994a. The cost of compensable upper extremity cumulative trauma disorders. *Journal of Occupational Medicine* 36:713–17.

———. 1994b. The cost of 1989 workers' compensation low back pain claims. *Spine* 19:1111–16.

Yelin, E., and L. F. Callahan, for the National Arthritis Data Work Group. 1995. The economic cost and social and psychological impact of musculoskeletal conditions. *Arthritis and Rheumatism* 38:1351–67.

Yu, C.-Y. and W. M. Keyserling. 1989. Evaluation of a new work seat for industrial sewing operations: Results of three field studies. *Applied Ergonomics* 20:17–25.

Waste Incinerators

Strategies for Community Sustainability and Social Justice

John MacDougall

THE primary question addressed here is how can a political base be built to oppose incinerators that burn municipal and medical solid waste? These incinerators have serious environmental and health effects, which is one reason why the earlier rapid expansion of this industry came to an end in the 1990s (Walsh et al. 1997, 7). Polluting facilities like incinerators are an important issue not only for environmentalists, but also for advocates of social justice, because communities of color and low-income neighborhoods are disproportionately harmed by such facilities (Field 1998).

The theoretical framework I use is the analysis of social movements. Movements are very important sources of public policy innovations on such matters as industrial development and social justice. Since the nineteenth century in industrial countries, the state has incorporated many movement organizations into both the process and the substance of policy-making (Tarrow 1994, 46–47, 85–86).

Social movements can be defined as organized attempts to change society, using extrainstitutional means. Movements have to take into consideration what analysts have called the political opportunity structure. This consists of relevant political actors' interests and the major economic and demographic features of the society (Einwohner 1999). Within this context, movement leaders have to make strategic choices, about mobilization, targets, and framing. Mobilization includes informing the public, changing passive supporters into active long-term adherents, and forming alliances (McCarthy and Zald 1977). Targeting decisions cover not only a movement's main adversaries, but also its broad goals, its shorter-term objectives and campaigns (Gamson 1990). Framing involves the choice of language and symbols that will successfully appeal to the general public and neutralize arguments by the movement's adversaries. Effective frames draw on existing popular cultural themes, yet maintain the movement's integrity in the face of a possibly antagonistic mainstream culture (Snow et al. 1986; Tarrow 1994).

Two specific movements are important to the study of incinerator

189

opponents. The first is the environmental movement, whose supporters are mainly middle class and have concentrated on wilderness preservation and national–international issues. The second is the environmental-justice movement and its predecessors, opposed to local pollution sources, mainly backed by low-income people and people of color, and heavily influenced by the civil rights movement. These two movements have often been in conflict, because of the demographic differences in their adherents, and because environmental-justice advocates have frequently used more disruptive tactics and stressed the role of humans in ecosystems more than middle-class environmentalists have (Edwards 1995; Gottlieb 1993).

In these pages I analyze as a case study the movement against incinerators in the Merrimack Valley region of Massachusetts, which has an exceptional concentration of trash-burning facilities. Then I analyze the political opportunity structure in which this movement operates, discussing players at the federal, state, and local levels. In the next section, I briefly describe the movement's history and evaluate its strategic decisions in the areas of mass mobilization, organizing campaigns, and framing the issues, with particular attention to the issues of race and class that are central to the environmental-justice movement. In the concluding section, I offer some suggestions about movement strategies and about the collaborative-advisory, direct and indirect roles of universities in pursuing environmental and social goals.

Incinerators in the Merrimack Valley: The Political Opportunity Structure

The National Context

In the 1970s and early 1980s, incineration of municipal solid waste (MSW) seemed to many to be an environmentally friendly alternative to landfills, at a time when the latter were reaching saturation and were increasingly seen as environmentally unacceptable. Cleaner new incinerator designs had been developed in Europe and were expected to be adopted in the United States. Incinerators also produced electricity, a commodity in short supply at the time because of Middle East oil embargoes. By the mid-1980s, however, the public and government turned against incinerators. Electricity prices fell rather than rose, and the perception of an energy crisis faded. In addition, the harmful environmental consequences of trash-to-energy plants became apparent. By the late 1980s, many local movements arose in the United States that opposed construction and operation of incinerators (Walsh et al. 1997, 6–9).

Despite the widespread opposition to MSW incinerators, the waste-management industry remains very powerful. Before 1970 MSW was handled locally by small firms, but now it is a commodity traded on a global scale, and the industry is dominated by a few giants that handle all kinds of trash including toxic and medical waste, are highly profitable, and wield great political power (Crooks 1993; Field 1998).

The Massachusetts Context

The trends summarized in the preceding two paragraphs apply to Massachusetts. In the early 1980s, state and local officials vigorously promoted incinerators, but they no longer do so. In the 1990s there was a moratorium on new landfills and many were closed. Though the state's priorities now are: first recycling, then incineration, then landfilling, in 1996, 44 percent of the state's MSW was incinerated (LeBlanc 1997; Massachusetts DEP 1997, vol. 1, 3–6).

At this writing, however, these policies are being questioned. Partly because of the closing of landfills and the Lawrence MSW incinerator, it is claimed that there is a shortage of MSW-handling capacity. For a number of years, Massachusetts was a net importer of MSW, partly in order to keep the incinerators running at full capacity, but since 1995 the state has exported waste. In June 1999, the state announced it would lift the landfill moratorium, allowing up to six lined facilities to be reopened (none of them are in the Merrimack Valley). In addition, doubts have been widely expressed as to whether the state can reach its ambitious recycling goal of 46 percent of solid waste—the 1999 level was 34 percent (Kirchofer 1999; MVEC et al. 1997, 17).

The Merrimack Valley: History, Demography, Health

Three neighboring communities in the region are hosts to MSW and medical-waste incinerators. Among these is Lawrence, an old industrial city with an exceptionally high concentration of Latino/Latina residents and poor people—many experts estimate the true proportion of Hispanic residents at 70 percent. The second host community, Haverhill, is also primarily industrial, but less ethnically diverse and somewhat more affluent. By contrast, the neighboring suburbs, such as North Andover, the site of the third incinerator, are almost entirely white and much more affluent. For details see table 1. The table also gives figures for asthma hospitalizations, which are partly attributable to air pollution from incinerators. Measured by its asthma morbidity and almost all other health indicators, Lawrence fares much worse than its neighbors (Declercq 1998).

The first MSW incinerator built in the Merrimack Valley was operated in Lawrence by Ogden-Martin Corp. It was opened in 1980, but closed in 1998 on account of the prohibitive cost of retrofitting to meet more stringent environmental standards. The second facility, located in North Andover and operated by Wheelabrator Corp., started work in 1985. The third, and cleanest, MSW incinerator, also operated by Ogden-Martin, is in

Table 1. Demographic and health statistics for selected Greater Lawrence locations

	Demography 1990				Health (1996)
Location	Median household income ($)	Percentage nonwhite	Percentage Hispanic	Percentage with B.A. and B.S.[b]	Asthma hospitalizations[c]
Andover[a]	40,875	3.4	5.2	47.1	56.4
Haverhill[a]	36,945	5.1	5.0	19.9	257.6
Lawrence[a]	22,183	34.9	41.8	9.0	426.8
North Andover[a]	51,692	3.1	1.2	36.0	115.2
Massachusetts	36,952	10.0	4.6	25.2	179.9

Sources: Cols. 1–4: Bureau of the Census, U.S. Department of Commerce, 1990 Census. (Washington, D.C.: Government Printing Office, 1990). Col. 5: E. Declercq, *The health of the Merrimack Valley* (Lawrence: Lawrence Prevention Center, 1998), 30.

[a] Site of MSW incinerator
[b] Among population age 18 or over
[c] Discharges per 100,000

Haverhill and opened in 1989. The latter two plants are among the largest in Massachusetts, with permitted capacities of 1,500 and 1,650 tons a day respectively.[1] There is also a medical-waste incinerator in Lawrence, operated since 1980 by Browning-Ferris Industries and permitted to burn 24 tons a day. This is the largest medical-waste burner in New England and takes waste from the whole region. Though its capacity is far below that of MSW plants, emissions of toxics per ton from medical incinerators are much greater.[2]

Even after the closure of the Lawrence MSW facility in 1998, the region accounted for 31.3 percent of all the incinerator waste accepted in Massachusetts (Lawaroff 1998a; Massachusetts DEP 1997, vol. 2, I-1). When these incinerators were built in the 1980s, decision makers in business and government saw Lawrence as a poor, largely nonwhite city that would not object to the clustering of dangerous facilities. In addition, as indicated, Haverhill is not a very affluent city, and the particular section of North Andover that hosts the Wheelabrator incinerator is poorer than the rest of the town, and very close to the Lawrence line. Municipal officials in all three communities saw incinerators as an economic boon and actively courted them at the time.

The North Andover incinerator was built largely in response to the creation by the state government of the Northeast Solid Waste Committee (NESWC), a consortium of twenty-three municipalities that committed themselves to send waste to North Andover. In 1985, these towns signed a twenty-year contact with Wheelabrator, under which they are required to ship a guaranteed annual tonnage of waste to North Andover. In addition to NESWC's trash, the Wheelabrator plant receives about the same amount from spot-market haulers, who take truckloads to North Andover in hopes of finding space available. The spot price is usually about half that paid by the NESWC communities.

In 1985, NESWC's contract with Wheelabrator seemed like a good deal to the participating municipalities, but in the 1990s the reverse has proved to be true. These towns receive most of the revenue from the sale of electricity generated from trash-burning—but now the price of this power is uncompetitive, and the North Andover plant is unprofitable (Kennealy 1997a). Under the twenty-year contract, most of the payments on bonds for the plant's construction fall due at the end of the period, that is, in 2000–2005. The per-ton price of trash (or tipping fee) is currently about twice what NESWC municipalities would have to pay if they used other incinerators, and will rise further as the large bond repayments fall due. Some NESWC communities currently cannot fulfill their trash quotas, so they are paid through the open market to take additional waste from other communities or trash haulers, and then ship it to North Andover (Talbot 1997).

Thus the NESWC contract is a major issue for the participating towns—although in 1999 some of the financial burden was eased. By December 2000, all the incinerators in Massachusetts must be retrofitted to comply with the 1990 federal Clean Air Act amendments. After much wrangling, it was agreed that the retrofit at North Andover would cost $35.5 million, with the NESWC towns paying $17 million. In addition, the state will pay the twenty-three towns $3 million in fiscal year 2000 toward the retrofit, and an additional $18 million is expected from the state before the contract expires in 2005 (Macone 1999).

Arrangements for shipping waste to the Haverhill incinerator are less complex. About 20 percent of its waste comes from municipalities that have long-term contracts with Ogden-Martin, but are not in a consortium like NESWC. In addition, waste is sold to Haverhill on the spot market as it is in North Andover. A third waste input here is commercial trash, contracted for several months at a time with private haulers. Like North Andover's, this plant will have to be retrofitted, but at a lower cost (MVEC et al. 1997, 13; Perkins 1998; Quimby 1998).

Turning to the health and environmental effects of Greater Lawrence's incinerators, the air pollutants of greatest concern there are mercury and dioxins. When it enters the food chain, mercury (found in batteries and thermometers) leads to serious nervous-system disorders, and incinerators are the leading source of this metal in Massachusetts. Fish in Kenoza Lake, Haverhill, were found to have the highest mercury level of fish from any lake in Massachusetts (Rodriguez 1998), and the Merrimack Valley has been identified as a "mercury hot spot" by the North Eastern States for Coordinated Air Use Management, based on this agency's modeled study. Dioxins, which mainly result from the burning of materials containing chlorine, such as paper and plastics, are very potent and long-lasting carcinogens. At the North Andover incinerator, dioxin levels exceeded state maximums on four occasions between 1991 and 1996, and all the MSW plants in the region have been frequently cited for violations of state and federal health and environmental laws. Problems with asthma in Lawrence have already been mentioned. In a separate study by the local Community Health Network Agency, the particular zip code where the MSW incinerator was located was found to have an even higher asthma-related hospital in-patient rate than other parts of the city. All of these problems are aggravated by the Merrimack Valley's incinerators all being within five miles of each other, and very close to the Merrimack River (MVEC et al. 1997, 4–10).

In June 1999, the state approved Wheelabrator's emission control plan for the North Andover plant's retrofit (Shohl 1999). This is supposed to reduce mercury and dioxin emissions by 85 and 87 percent respectively. There will remain, however, high concentrations of toxins in incinerator ash.

The Merrimack Valley: The Main Political Players

The firms that operate the region's incinerators have already been mentioned. They are the three biggest waste-management companies in the world, with enormous political clout. Waste Management, the owner of Wheelabrator, has been guilty of particularly unethical and illegal behavior (Crooks 1993). In 1998 Waste Management merged with USA Waste, a smaller corporation that has expanded rapidly in recent years (Lazaroff 1998b).

In the public sector, it should be noted that local governments are important players in Massachusetts, because they jealously guard their autonomy, and because county government is virtually nonexistent. In the Merrimack Valley, most local officials, both elected and appointed, are ignorant about incinerators or supportive of them. They are mainly concerned about these plants' fiscal aspects. In addition, in NESWC towns, many officials also fear being sued by Wheelabrator if they push for closing the incinerator. In North Andover, which is to receive about $950,000 a year from Wheelabrator between 2000 and 2004 through a Host Community Agreement (Lazaroff 1999), town boards have been much more responsive to Wheelabrator officials than to incinerator critics.

In Lawrence and Haverhill—the other host communities in the region—the situation is rather different. In the former, elected officials are generally opposed to incinerators; in Haverhill, the mayor has mostly favored trash-burning, but most elected officials have been opposed. Unlike North Andover, Lawrence has no host-community agreement with incinerator operators; Haverhill has one, but it generates virtually no revenue.

The NESWC board has to be unanimous on all important decisions—a difficult task given the demographic diversity and geographic dispersion of the member towns. Still, the cohesion of the board has increased in the past few years, with Steven Rothstein as its executive director or consultant. He has supported the Wheelabrator incinerator, and runs a consulting firm, which is paid more, the greater the volume of trash shipped to North Andover—but it seems he is mainly interested in his own remuneration, regardless of its source.

At the state level, officials in the Department of Environmental Protection (DEP) have been accused of being too lax about enforcing regulations governing incinerators and of having improper relationships with incinerator companies (MVEC 1998). The top environmental official in Massachusetts, the Secretary for Environmental Affairs Trudy Coxe, in 1990–1998, and Robert Durand since 1999, have reputations as strong environmentalists, but they have not been willing to oppose DEP decisions actively.

State senators and representatives from NESWC communities have formed their own caucus, which has not been very cohesive and has pushed chiefly for increased funding for member towns. Outside the caucus, no state legislator has been a particularly effective opponent of incinerators.

The federal Environmental Protection Agency (EPA) has mainly left to the state DEP the enforcement of federal laws regulating incinerators, and EPA officials have been accused of not seriously engaging in dialogue with ordinary citizens. Still, former EPA Regional Director for New England John DeVillars was widely regarded as bold and energetic, but because he had to enforce the federal Clean Air Act amendments, he claimed he had no legal authority to close trash-burners in the Merrimack Valley or halt their retrofits. In early 2000 DeVillars resigned, and was replaced by Mindy Lubber, who previously worked for Massachusetts Public Interest Group, an environmental and consumer advocacy organization.

Another important political player is the media. Television stations in Boston have rarely covered incinerators in the Merrimack Valley, and they are rarely mentioned by local radio or cable-TV stations, or by the two major Boston newspapers (the *Globe* and the *Herald*). Coverage is better in local print media, however. The most widely read newspaper, the *Lawrence Eagle-Tribune*, which previously had never mentioned incinerator issues, shifted in 1997 to quite extensive and balanced coverage. The *Haverhill Gazette* has generally opposed incinerators. The *North Andover Citizen* provides detailed coverage and has editorialized against the town's incinerator ("A Boon for the NESWC Towns" 1997). The *Citizen* is part of a chain that also publishes local papers in several distant NESWC communities and reproduces articles on incinerator issues in those towns.

The Movement against Merrimack Valley Incinerators
A Brief History and Background

When plans surfaced to build incinerators in Lawrence and North Andover, there was virtually no organized opposition. When the siting of the Haverhill plant was proposed, however, some groups protested, and the same was true in the late 1980s when operators wanted to burn municipal sludge along with MSW in the region. In North Andover after a bitter fight, the state's attorney general ordered the Wheelabrator plant closed temporarily, and in addition, the incineration of sludge ended. New proposals by both Ogden-Martin and Wheelabrator to incinerate both sludge and MSW at various sites sparked new citizen opposition in 1994–1997, and, in the end, both companies pledged not to burn sludge. In 2000 a fresh proposal to build a plant in North Andover to convert sludge to fertilizer was rejected by the town.

After this struggle, anti-incinerator groups in Andover, Haverhill, Lawrence, and North Andover joined in 1996 to form the Merrimack Valley Environmental Coalition (MVEC), Anti-incinerator groups in these towns have a dedicated core of five to twelve leaders—mainly white professionals—however, by 2000 most members of the Lawrence anti-incinerator groups were Latinas and Latinos. Local groups meet regularly among themselves and with other MVEC leaders; they have become well informed and developed great mutual trust. A few other groups in the region now belong to MVEC, but outside these four communities, there is not much anti-incinerator activism at this writing (August 2000). Still, the persistence of NESWC's problems keeps the scattered activists in affected towns in touch with each other. In Lawrence, organizing among Latinos/Latinas and other low-income groups was begun in 1997, mainly by the Lawrence Environmental Justice Council (LEJC). The council is now taken seriously in the city, partly because it takes every opportunity to state that it expects significant resources to be committed to environmental justice.

I also consider as part of the movement opposed to Merrimack Valley incinerators several state-level environmental and environmental-justice organizations, with which MVEC and LEJC have developed close ties. These groups provide MVEC and LEJC with legal, scientific, and political expertise and help them secure funding. The local and state-level organizations have formed an informal group of about ten leaders, who are in close contact and divide specific responsibilities among themselves. Some members of this group have gained access to key state and federal officials on a fairly regular basis.

Movement Strategy: Building the Base, Forming Coalitions

The public in the Merrimack Valley is generally opposed to incinerators. In a 1995 referendum, sludge-burning in Haverhill was rejected by a three-to-one margin. In North Andover, the town meeting overwhelmingly voted in 1998 and 1999 to move toward closure of the Wheelabrator plant. A poll in North Andover found most citizens opposed to incinerators (Shannon 1997). But the challenge for the movement has been to translate this sentiment into active opposition.

MVEC and LEJC have made conscious efforts to educate the public and their active adherents through, for instance, letters to local newspapers, appearances at municipal hearings, and training speakers for such hearings. While they have worked extremely hard, the suburban groups could perhaps have done more outreach. LEJC has been hampered in building a base of sustained adherents by the lack of resources. It could do little about the practical problems that keep many poor Latinos from involvement, such as translating everything said at meetings into Spanish or holding meetings right in people's neighborhoods. Still, the problem of child care for young parents has been partially resolved through LEJC's providing stipends to those attending training workshops. In addition, while almost all the community organizing for LEJC has so far been done by volunteers, a two-year EPA grant to the council, in fall 1999, enabled it to hire a half-time, low-income Latina organizer. That move should facilitate more sustained popular involvement. In addition, in Lawrence there are twelve neighborhood associations, led largely by blue-collar whites; the MVEC member group in Lawrence has made presentations to about a third of these associations.

A key part of changing supporters into active adherents is building solid alliances with local citizen groups that are not primarily environmentalist. A very important case in point is the Merrimack Valley Project, which is itself a coalition of some forty-five groups, mainly churches and labor unions organized around four local chapters and concerned with saving jobs, public safety, and the like. The project has made stopping incinerators one of its top priorities and has helped MVEC bring hundreds of additional people out to important public events. Thus, the anti-incinerator movement avoided opposition from local organized labor, and could portray itself as more than an elite clique of environmentalists.

Another important local ally is Community Health Network Area 11, which is the state-mandated agency covering Andover, Lawrence, North Andover, and two other towns. Pediatric asthma has been chosen as one of the agency's top three priorities for detailed study, education, and advocacy, and because of this disease's connections to trash-burning, the agency has cosponsored a number of educational events on incinerators and public health. The agency does not officially oppose incinerators, but its work on the issue has provided important new ways for MVEC activists to collaborate with local health and human-service professionals.

Movement Strategies: Devising and Implementing Campaigns

The broad goal of MVEC, LEJC, and their state-level partners is to close the North Andover MSW incinerator and the Lawrence medical-waste burner. Activists admit the Haverhill plant will not be closed in the foreseeable future, but they advocate more stringent regulation of it. They oppose retrofitting the North Andover incinerator, on the grounds that this will allow continued unsafe emissions of toxins such as dioxins and mercury and also shift the site of the pollution from the air to the ash left after combustion. Environmental groups press for more government spending on recycling and waste reduction as ways to make incinerators unnecessary (MVEC et al. 1997).

It is hard for opponents of incinerators to determine who their main targets are, because in Massachusetts no one agency or legislative body has sole responsibility for incinerator policy. Accordingly, the movement's strategy is twofold. First, it must keep pushing on all fronts and constantly build up the base of support. Second, for the shorter term, it should target those players who, at the time, seem most likely to yield important victories and give supporters a sense of making headway. A key path to both goals, in the eyes of environmental leaders, is to demand tighter statewide health and environmental standards. Not only will this reduce ecological harm, but it will rebut criticism of the Merrimack Valley groups as being interested only in their own backyard and raise the cost of operating and retrofitting incinerators. Another important issue that environmentalists take up whenever possible is recycling and other forms of waste reduction. At times, environmentalists act as insiders—for instance, negotiating over regulations with DEP officials—and at other times, act as outsiders—for example, mobilizing hundreds of supporters to come to hearings.

The movement's campaigns directed toward local government have failed so far. As indicated, North Andover town boards have been unresponsive, even though the town meeting has expressed opposition to the Wheelabrator plant. An important example of this is that activists found in 1997 that the North Andover board of health never held public hearings about safety aspects of siting the plant before construction, as required by state law. When activists protested against this, the board responded by signing an agreement with Wheelabrator promising that the town would never legally challenge the company's operations.

In Lawrence in 1999, however, another campaign was launched that may make more headway. This was a petition drive to get the city to ban emissions of dioxins, mercury, and lead. Local environmentalists claimed incinerators violate the Civil Rights Act (Griffin 1999; Quimby 1998).

In 1998 the main campaign was at the state level, targeting DEP's regulations on emissions. This was urgent because the department had to show, by 2000, that it was complying with the 1990 amendments to the federal Clean Air Act. Environmentalists filed a legal document with the state's attorney general giving details of alleged lax oversight by DEP. Such documents are usually treated as confidential, but DEP leaked this one to Wheelabrator. After learning about the leak, environmentalists released their document to the press (MVEC 1998). DEP was furious at this, but became more responsive to citizen concerns during subsequent hearings. The draft emissions regulations were the topic of a public hearing in Haverhill on April 8, 1998, to which anti-incinerator activists brought about 175 members of the public. That summer, DEP maintained close contact with environmentalists in discussions that normally involved only incinerator operators. In the light of the final version of the regulations, the chances of closing the North Andover incinerator were not much greater. But the regulations would at least ensure that trash-burners operate more safely, for instance, through the tightest standards in the country for mercury and dioxins. In addition, the standards require consideration of cumulative effects, that is, the impacts of toxic exposures where two or more incinerators are very close to each other—as is the case in Greater Lawrence.

When these new regulations were applied the following year, during DEP's examination of the emissions control plan for the North Andover retrofit, environmentalists objected that the cumulative-effects standards were not properly applied. In 1999, however, the main focus of the movement shifted to the upcoming revision of the state's solid-waste master plan and ensuring that it contained stronger provisions for recycling and waste reduction.

Movement Strategies: Framing the Issues

The symbols and rhetoric most commonly used by incinerator foes concerns threats to health and the environment. A bumper sticker sold by MVEC reads "Incinerators Are Dioxin Factories." A flyer from the NESWC facility, headed "Where Does Your Trash Go?" has a picture of pollutants going into a human body from incinerators. It is easier to persuade the public about the link to health and environmental problems when the focus is on the positive issue of recycling rather than on the more negative problem of incinerators. A second frame used an economic one, stressing the fiscal losses the North Andover incinerator imposes on NESWC communities. Here, environmentalists face a dilemma: While they object to the taxpayer burden and the disincentive to recycling, they also recognize that if the cost to towns was reduced they would push harder for the retrofit rather than closure. This latter argument gained power in 1999, thanks to the fiscal relief provided to NESWC towns by the state and Wheelabrator. The third frame used quite often is local autonomy—and state and local officials' unresponsiveness. Often the Merrimack Valley is described by environmentalists as "the trash capital of Massachusetts," which evokes a ready response in the public. A final frame is social justice—the unfairness to poor communities having so many incinerators in their midst. This resonates well with Latinos/Latinas (even conservative ones) and low-income people, but not so much with middle-class audiences.

Environmentalists frame their messages in simple terms, which they say are readily understood and generally accepted. They also usually shy away from provocative symbols—though at a December 1998 hearing in North Andover, one activist held a dead fish and another dressed as a cow, to highlight the state's neglect of "motherhood, milk and mercury" (Lazaroff 1998c). In addition, while leaders of MVEC, LEJC, and their state-level partners try to ensure a steady flow of letters to newspapers and cultivate relationships with sympathetic journalists, they have not developed a broader media strategy.

Conclusions

In the Merrimack Valley there is an exceptional concentration of municipal and medical-waste incinerators that pose serious environmental and health problems, especially to the host communities. The approach adopted by state officials is to retrofit these facilities to comply with amendments to the federal Clean Air Act. The incinerator in North Andover has proved costly for the twenty-three towns in the NESWC consortium that are required to send large amounts of trash to that expensive facility. This burden will be reduced, however, through the relief provided in 1999 by the state and the operator. A protest movement has developed, demanding closure of the region's two most dangerous incinerators and tighter regulation of the third one. The movement consists of white professionals, mostly in the suburbs, of environmental-justice advocates working with poor Latinas/Latinos in Lawrence, and of statewide environmental partners who provide specialized assistance.

All social movements have to take into consideration the political opportunity structure affecting their issue. In the Merrimack Valley, this structure is generally unfavorable: the only potential allies that have significant power are a minority of local, state, and federal officials and the local newspapers. In addition, social movements have to make strategic decisions on three kinds of issues. On the first of these, mobilization and coalition building, Merrimack Valley environmentalists and their statewide partners have tried hard to recruit adherents, but have been hampered by various obstacles, such as the lack of staff to enable the Lawrence Environmental Justice Council to address mundane problems such as holding meetings in Hispanic neighborhoods. Local environmentalists, however, have built some solid alliances, for example, with an important coalition of churches and labor unions. Second, like all social movements, the one opposed to incinerators in the Merrimack Valley has to decide about targets and campaigns. Here, a major problem is the lack of a single incinerator-policy body in state government; to counter this, environmentalists target what they consider the most vulnerable agency and issue at the time, for example, the state's solid waste master plan, which was revised in 1999–2000. Third, social movements have to decide how to frame their issues so as to maximize public support and involvement. The frame most frequently used is the threat to the environment and health posed by incinerators in the Merrimack Valley, though other frames, such as social injustice in Lawrence, have been used.

What lessons does the anti-incinerator struggle in Greater Lawrence offer advocates of community sustainability? Closing existing facilities is harder than stopping construction of proposed trash-burners (Walsh et al. 1997), and the incinerator industry wields great political power. So the political opportunity structure is not promising for environmentalists—for example, in the Merrimack Valley—who want trash-burning stopped. Despite this, the movement I describe has been successful to the extent that it is now a player that local, state, and federal elites have to take seriously. It is also interesting that the movement in Greater Lawrence combines both suburbanites and inner-city residents, all of whom have demonstrated great tenacity, with activists in one community helping out those in another wherever possible. Still, there could perhaps be a more persistent and creative outreach effort in the suburbs.

Regarding targets and campaigns, the movement's flexibility in its choice of targets makes sense. Focusing on the minutiae of state regulations in 1998 was probably another smart choice: This engaged citizens to an unusual degree in detailed policymaking, but activists also insisted on including in the new regulations an exceptionally broad range of concerns such as cumulative effects. Provided these concerns can be translated from technical jargon into readily understandable language, such regulatory campaigns promise to be effective. In framing, the focus on readily understandable words and symbols makes sense. But both middle-class environmentalists and environmental-justice advocates could at least on occasion highlight broader issues, such as the political economy of the global waste industry, wider ecosystem threats, and the antidemocratic and racist nature of many industrial decisions (Field 1998; Sclove 1997). In addition, more large-scale or disruptive approaches to framing could be used at times, for instance by dramatizing stories of harm to particular families or neighborhoods, staging large statewide or regional demonstrations, or picketing corporate offices. Although these tactics alienate some supporters, they also gain media attention, inspire adherents, and forcefully challenge mainstream paradigms (Szasz 1994, 44; Tarrow 1994, 113–14).

How does this analysis compare with that of other essays in this part of the book, concerning the dynamics of moving toward a sustainable society? Although this chapter, unlike the others, focuses on the community as the unit of analysis, there are several aspects of the search for sustainability evident both here and elsewhere. First, the analysis of industrial waste by Geiser, Greiner, and others (part II herein) has much applicability to community-level problems, because many of the same industrial changes that could prevent pollution in manufacturing can also prevent pollution in consumer products, for example, through eliminating dioxin-producing chemicals in computer components. Second, in addition to promoting ecosystem preservation, those concerned with sustainability must link this value to other key ones in the U.S. culture, such as business viability and worker–citizen health. Third, advocates of sustainability must translate their broad concerns into specific policies designed for the workplaces or communities they are involved with. Finally, there is no clear-cut path to sustainability, so its advocates must continually adapt their strategies and organizational forms.

Finally, a few comments on the role of universities. There are several research centers at the University of Massachusetts Lowell that already, or potentially, provide services of value to local anti-incinerator activists. These include compiling health data in Lowell (some of it connected to trash-burning), assisting businesses that use as raw materials otherwise-incinerated plastics and other wastes, and helping hospitals reduce their waste. The university, however, has little expertise in community-oriented recycling or waste-reduction campaigns and has no resources currently available for pooling all the health and environmental data connected to the region's incinerators.

This raises three general points. First, issues of trash management and sustainable alternatives involve a large number of disciplines. No university can be expected to have faculty members or staff experts in all relevant fields. Whenever a university lacks its own expertise in one of these fields, it should serve as a clearing-house of information about where local activists can readily get assistance from other universities or agencies with the needed expertise. Second, regarding the form of university centers' partnerships with local agencies and citizen groups, some of them will be direct, through meetings and joint projects directly involving local sustainability advocates. But, in addition, some of the collaborations will essentially be indirect, through local activists using the centers' Web sites, training programs, and the like. Both forms of cooperation are useful, but it is important for academicians to remember that, even in this electronic age, low-income partners may need help in navigating the Internet. Another challenge for university centers is conducting constant follow-up on their contacts with activists (see Forrant, "Random Acts," in this volume). Third, it is important for universities to provide space, both political and physical, for open discussion of a wide range of analytical perspectives and practical options regarding complex and contentious issues like waste reduction and recycling. "Deviant" ideas and information usually receive a more detailed and respectful hearing in academic institutions than they do in other places such as government hearings. Thus universities can play a very useful role as institutional homes for important parts of the environmental-justice and environmental movements (cf. Tarrow 1994, 146).

NOTES

I am immensely grateful to Ed Meagher for enormous amounts of data and guidance. Helpful comments on earlier drafts were provided by Bob Forrant, Kathy Moyes, Jean Pyle, John Rumpler, Ann Wein, and Eric Weltman. Very useful information was provided by all the interviewees: In Merrimack Valley organizations: Brent Baeslack (Haverhill Environmental League), Joan Kulash (People for the Environment, North Andover), Jonathan Leavitt (Lawrence Environmental Justice Council), Ed Meagher (People for the Environment), Kathy Moyes (Lawrence Environmental Action Group), Claire Paradiso (Community Health Network Area 11), and Anne Wein (Concerned Andover Residents for the Environment). In statewide or non–Merrimack Valley organizations: Lee Ketelsen (Clean Water Action), John Rumpler (Alternatives for Community and the Environment), and Jill Stein (Lexington Solid Waste Action Team). Thanks are also due to Sarah LeBlanc, Michelle Morelli, Kimberly Rauscher, Mitch Schuldman, Kristine Vlahos, and Jennifer Zdon for clerical and computer help, and to Debbie Friedman, Ron Karr, Tom Sheehan, Craig Slatin, and Richard Slapsys for assistance in locating data sources.

1. Ash from the Haverhill incinerator is buried on site; ash from North Andover is trucked to a specialized landfill at Peabody, fifteen miles away.

2. The Lawrence medical waste incinerator was sold in 1999 to Stericycle Corporation, a major operator of medical-wasteincinerators.

REFERENCES

A boon for the NESWC towns. 1997. Editorial, *North Andover Citizen.* November 19, 5.

Crooks, Harold. 1993. *Giants of garbage.* Toronto: James Lorimer.

Declercq, Eugene. 1998. *The health of the Merrimack Valley.* Lawrence: Lawrence Prevention Center.

Edwards, Bob, 1995. With liberty and environmental justice for all. In *Ecological resistance movements.* ed. Bron R. Taylor. Albany: SUNY Press.

Einwohner, Rachel L., 1999. Practices, opportunity and protest effectiveness. *Social problems* 46 (2): 169–88

Field, Roger C. 1998. Risk and justice: Capitalist production and the environment. In *The struggle for ecological democracy,* ed. Daniel Faber. New York: Guilford Press.

Gamson, William A., 1990. *The strategy of social protest,* 2d ed. Belmont, Calif.: Wadsworth.

Gottlieb, Robert, 1993. *Forcing the Spring: The transformation of the American environmental movement.* Washington D.C.: Island Press.

Griffin, Rebecca, 1999. Blowing smoke a civil rights violation? *Lawrence Eagle-Tribune,* June 29, 16.

Kennealy, Rebecca. 1997. Why is NESWC so expensive? *North Andover Citizen,* June 4, 3.

Kirchofer, Tom. 1999. Landfill shortage is raising cost of trash disposal. *Quincy Patriot-Ledger,* April 5, 1.

Lazaroff, Tovah. 1998a. Trash incineration in Lawrence is shut down. *North Andover Citizen,* June 10, 1.

———. 1998b. Incinerator may get a new owner. *North Andover Citizen,* July 16, 4.

———. 1998c. State says the air is safe. *North Andover Citizen,* December 23, 1, 3.

———. 1999. Host agreement yields few dollars, *North Andover Citizen,* January 20, 1, 5.

LeBlanc, Steve. 1997. MassPIRG becomes latest foe of Wheelabrator incinerator. *Arlington Advocate,* September 11, 7.

Macone, John. 1999. $3 million in trash relief coming. *Lawrence Eagle-Tribune,* May 4, 5.

Massachusetts Department of Environmental Protection (DEP). 1997. *Massachusetts Solid Waste Master Plan, 1997 Update.* 2 vols. Boston: D.E.P.

McCarthy, John and Mayer Zald. 1977. Resource mobilization and social movements. *American Journal of Sociology* 82 (May): 1212–41.

Merrimack Valley Environmental Coalition (MVEC). 1998. White paper: the best available technology.

———, Clean Water Action, Toxics Action Center, and Alternatives for Community & Environment. 1997. The case for closure: phasing out incineration in the Merrimack Valley. Unpublished paper. Clean Water Action. Boston.

Perkins, Anita. 1998. Trash plant closes, 70 jobs lost. *Lawrence Eagle-Tribune,* June 5, 5.

Quimby, Beth. 1997. State eyes trash plants as local breast cancer leaps. *Lawrence Eagle-Tribune,* July 26, 1, 31.

———. 1998. Waste plant deserving of 'dirty dozen' charges? *Lawrence Eagle-Tribune,* December 3, 6.

Rodriguez, Nancy C. 1998. Council: do not eat city fish. *Lawrence Eagle-Tribune,* November 25, 4.

Sclove, Richard E. 1997. Design criteria and political strategies for democratizing technology. Unpublished paper. Loka Institute. Amherst, Mass.

Shannon, Christina M. 1997. Protests against incinerator continue. *North Andover Citizen,* October 29, 1, 5.

Shohl, Dan. 1999. DEP approves incinerator upgrade. *Westford Eagle,* July 1, 3.

Snow, David A., et al. 1986. Frame alignment and mobilization. *American Sociological Review* 51:464–81.

Szasz, Andrew. 1994. *EcoPopulism.* Minneapolis: University of Minnesota Press.

Talbot, David. 1997. Arlington taxpayers 'penalized' for recycling. *Boston Herald,* April 6, 17.

Tarrow, Sidney. 1994. *Power in movement.* New York: Cambridge University Press.

Walsh, Edward J. et al. 1997. *Don't burn it here: Grassroots challenges to trash incinerators.* University Park: Pennsylvania State University Press.

Sustainable Production

A Proposed Strategy for the Work Environment

Margaret M. Quinn, David Kriebel, Ken Geiser, and Rafael Moure-Eraso

THE restructuring of the economy, the globalization of competition, and the rapid pace of technological development are dramatically altering the structure of the business firm and the organization of the workplace. Although there are many opinions as to the outcomes of this change, it is clear that new market demands are motivating industries to evolve more flexible business models, to develop continuous-improvement management systems, and to use resources more efficiently. Some successful industries are beginning to design for environmental concerns as well as productivity in a far more integrated way than has occurred in the past (Fischer and Schot 1993; Huisingh et al. 1986; Schmidheiny 1992; Sullivan 1992). In recent years, the fields of pollution prevention, clean production, toxics-use reduction, and industrial ecology have developed to address concerns for ambient pollution that occurs as a result of production processes (Gottlieb 1995; Graedel and Allenby 1995; Jackson 1993; National Academy of Engineering 1994). Although there are differences among these models (Oldenberg and Geiser 1997), we will refer to them collectively as "pollution prevention" for the purposes of this discussion.

From the occupational health and safety perspective, it is significant that the pollution-prevention models do not include explicit terms for addressing human health, safety, and the work environment (Ashford et al. 1996). In the United States, this conceptual separation of the work and ambient environment is derived, in part, from regulatory structures with very little overlap (Ashford 1976; Froines et al. 1995). Traditionally, the control of workplace pollution has been seen as the responsibility of occupational health and safety, regulated by the Occupational Safety and Health Administration (OSHA), while the control of ambient pollution has been seen as environmental protection, regulated by the Environmental Protection Agency (EPA). Separate regulatory structures and professional cultures have divided environmental quality and occupational health and safety, creating a significant gap in our understanding of the full range of options for comprehensive pollution prevention (Mazurek et al. 1995). At the same

205

time, concerns for environment and health are often viewed as constraints on our concern for economic development.

Occupational and environmental health and safety practice frequently aims at the development of solutions to ambient and work environment problems only after there is significant evidence of adverse impacts. These are called "end-of-pipe" solutions, referring to the focus on the final product, emissions, or waste stream resulting from a process. Typically, the development of end-of-pipe solutions begins with the identification and quantification of the harm that is expected to be inherent in production processes, rather than from a review of the purpose and function of the production process itself (Gottlieb and Smith 1995).

In this chapter we maintain that improvements in environmental quality both inside and outside the workplace need not undermine the economic viability of a firm, when environmental, health, and safety considerations are incorporated into the fundamental design of production processes, rather than being viewed as secondary to production. This approach is called "sustainable production" to emphasize a focus on the fundamentals of long-term viability and productivity. After briefly presenting the sustainable-production framework, we address the question how occupational safety and health practice, education, and research might evolve in the directions suggested by the sustainable-production approach. This proposal is intended to promote a dialogue among occupational and environmental health and safety professionals as to how our roles could be expanded to include healthy, safe, and environmentally sound production-process redesign and what we would need to change in order to respond seriously to these new responsibilities.

Definition of Sustainable Production

In this chapter, the term "sustainable production" is used to define systems of production that integrate concerns for the long-term viability of the environment, worker health and safety, the community, and the economic life of a particular firm. Sustainable production is a system that unifies the conventionally fragmented components of environmental and occupational health and safety and uses their interdependence to the advantage of each of these areas of concern. Given the dangers inherent in coining new terminology, a brief explanation of the choice of words is advisable.

The most frequently cited definition of sustainability comes from the term "sustainable development," popularly identified in the 1987 report of the World Commission on Environment and Development (WCED), *Our Common Future* (Bruntland 1987). In that formulation, "sustainable development" means "development that meets the needs of the present without compromising the ability of future generations to meet their own needs." Since the publication of the WCED report, the term "sustainable" has been used broadly to refer to the conservation of natural resources in general and the promotion of economic development that does not jeopardize environmental quality. Lélé (1991) summarized the critical objectives of the WCED report, which are generally accepted as the main principles of sustainable development by international environmental agencies such as the United Nations Environment Program (UNEP), the World Wildlife Fund, and development agencies such as the World Bank, the U.S. Agency for International Development, and the Canadian and Swedish international development agencies. These objectives include: (1) reviving growth; (2) changing the quality of growth; (3) meeting essential needs for jobs, food, energy, water, and sanitation; (4) ensuring a sustainable level of population; (5) conserving and enhancing the resource base; (6) reorienting technology and managing risk; (7) merging environment and economics in decision making; (8) reorienting international economic relations; and (9) making development more participatory (Lélé 1991).

The framework of sustainable production presented here is facility specific. It focuses on specific production processes within a particular business enterprise, rather than on global development. The idea of sustainability as used here includes the health and safety of workers and the community. By workers we mean all those employed or who work at a facility. By community we mean those living around a particular production facility or those affected by the facility's occupational and environmental impact. Thus, terms such as "green manufacturing' or "green business," often used to mean environmentally sensitive manufacturing processes or businesses (Hopfenbeck 1992; Leff 1995; Sullivan 1992), here also includes respect for worker and community health and safety. The idea that environmentally sound business practices should include community concerns is already well-developed in the environmental-justice movement (Hofrichter 1993).

In this essay "production" refers to those activities intended to create and distribute goods and services for economic and/or social benefits. This broad definition includes such activities as care giving, food preparation, health care, office work, and cleaning services as well as agriculture, construction, manufacturing, and other industrial activities. Historically, many environmental advocates have viewed production activities as threats to environmental quality. As a result, they have chosen strategies that appeared both antibusiness and antilabor. For example, environmental groups have promoted strategies to reduce emissions or consumption without engaging in debates about how goods and services are produced. Such strategies are end-of-pipe solutions that take the final product and the way it is made as a given. Similarly, labor unions have often pursued end-of-pipe strategies that focus on job classifications, wages, and working conditions resulting from work processes that are accepted as givens. The framework proposed here begins with the premise that production is central to human life and that production is both the root of environmental pollution and occupational risk and the key to their elimination (Commoner 1990).

Criteria for Sustainable Production

Sustainable production means production systems that are nonpolluting, conserving of energy and natural resources, economically efficient, safe and healthful for workers, neighbors, and consumers, and socially and creatively rewarding for employees. Enterprises that seek to develop production that is sustainable should be guided by the following principles:

1. Products and packaging are designed to be safe and ecologically sound throughout their life cycle; analogously, services are designed to be safe and ecologically sound;

2. Wastes and ecologically incompatible by-products are reduced, eliminated, or recycled;

3. Energy and materials are conserved, and the forms of energy and materials used are most appropriate for the desired ends;

4. Chemical substances, physical agents, technologies, and work practices that present hazards to human health or the environment are eliminated;

5. Workplaces are designed to minimize or eliminate physical, chemical, biological, and ergonomic hazards;

6. Management is committed to an open, participatory process of continuous evaluation and improvement, focused on the long-term economic viability of the firm;

7. Work is organized to conserve and enhance the efficiency and creativity of employees;

8. The security and well-being of all employees is a priority, as is the continuous development of their talents and capacities; and

9. The communities around workplaces are respected and enhanced economically, socially, culturally, and physically; equity and fairness are promoted.

Clearly, these criteria define broad objectives for a long-term goal. Working toward the goal of sustainability is a complex process in which different concerns may be emphasized at different times. Yet it must be recognized that achievement of some, to the total exclusion of others, will not, in the long-run, meet the goal of sustainable production.

Current Limitations of Environmental and Occupational Health Methods

Environmental Pollution

For nearly three decades, the United States has made an intensive and very costly effort to reduce the burden of pollution, such as smog, acid rain, toxic emissions, contaminated effluents, and hazardous waste. Although there has been notable progress, the limits of these remedial efforts have taught us an important lesson: Pollution is best prevented at its source. For example, sources of sulfur dioxide and nitric oxide (major contributors to the formation of acid rain) originate in chemical reactions within the combustion systems of energy-generating power plants. Efforts to control these emissions with extended exhaust pipes and various flue gas-cleaning devices have not been highly effective, as evidenced by the continuing damage to fragile ecosystems in the northeastern United States. Such control devices are often inefficient because the technologies are costly, require continuous maintenance, and do not contribute directly to making a product or supplying a service (Commoner 1990). Far more effective have been programs that switch power plants to cleaner fuels, substitute more environmentally friendly sources of energy, or invest in energy-conservation technologies and practices.

Until recently, government regulatory agencies have been guided almost exclusively by the strategy of control in their regulation of environmental and occupational health. This has resulted in lengthy, acrimonious, and expensive debates over what constitutes "safe" levels of pollution or "acceptable" health risks. Today, some European environmental agencies and key domestic environmental agencies are recognizing that pollution-prevention strategies that go to the root of the problem by designing out the environmental hazard or pollutant may be more effective (Ashford 1994; Becker and Ashford 1995). Thus far, however, pollution-prevention research and methods development have been disconnected from work environment concerns. For example, process engineering design models used by engineers to design for environmental concerns or safe products, such as "Design for Environment" or "Design for Product Safety" typically do not include the workers and job requirements essential to the operation and maintenance of the production process being redesigned (National Academy of Engineering 1994). In a review of cleaner-production case studies promoted in the United Nations Environment Programme (UNEP) database, it was found that job requirements and workers were seldom mentioned. Some cases even presented new processes, technologies, and materials to protect the environment with no acknowledgment that they were likely to increase the hazards experienced by workers (Ashford et al. 1996). In addition, the cases did not specify proper work practices, especially those related to storage, maintenance, and disposal, to ensure that the newly proposed production technologies were managed in a safe manner.

Workplace Hazards

The hazards of the workplace, including toxic chemical exposure, physical and ergonomic risks, and psychological stress, also originate from the design of production processes. Industrial hygiene, one of the core disciplines of the occupational safety and health profession, has long recognized that disease prevention is far more effective when a hazardous process or chemical is eliminated altogether than when the emissions of one technology are controlled by means of another technology. Industrial hygienists or other occupational health and safety professionals, however, are seldom involved in the fundamental decisions about purchasing new production equipment or designing new production processes. Ventilation, for example, is often considered an appropriate response to airborne toxic agents, because it is high in the industrial hygiene hierarchy of controls for prevention of worker overexposure (Raterman 1996). Yet ventilation moves the pollutants from the inside air to the outside air, which must then be protected by installing pollution control devices at the emission stack. These control devices may reduce air pollution to some degree, but they only displace the problem. The control systems must be cleaned and the trapped toxic agents disposed of, often exposing workers and, ultimately, the environment to the same hazards.

In 1994–95 OSHA made a modest effort to explore the potential overlap between pollution prevention and occupational health and safety. Some of the main occupational health and safety and pollution-prevention activities were evaluated by their level of prevention, such as primary (complete hazard elimination), secondary (hazard control), or tertiary (management of the damage resulting from a hazard). OSHA concluded that, because of their primary prevention effect, substitution and process changes should be emphasized more than the other elements of the industrial hygiene hierarchy of controls (Moure-Eraso 1995). Unfortunately, OSHA's budgetary constraints have limited progress in this area.

As government resources for public programs are being cut, individual firms are called on to carry an increasing responsibility for health and the environment. Within firms, workers as well as managers have many resources to contribute to sustainable production. A worker, however, who does not have a voice in routine work decisions, or who is concerned about job loss, will not be a constructive contributor to change. Sustainable forms of production must not only protect workers from occupational hazards, but should also involve workers in the design of workplaces that are both productive and rewarding.

Economic Development

The focus on controlling pollution rather than preventing it is one of the reasons for the perceived conflict between environmental quality and economic development. The cost of a control device typically rises exponentially with its efficiency, so that a device designed to remove close to 100 percent of a pollutant becomes prohibitively expensive. Moreover, since the controls cannot reduce emissions to zero, as economic activity expands and production increases, total emissions rise and eventually cancel the cleaner environmental value of the devices (Commoner 1990). Such an approach to environmental protection can only be effective if it limits production and hence curtails economic development. This threatens the economic viability of the company as well as its jobs.

In contrast, sustainable systems of production can be economically constructive. Redesigning production to achieve higher levels of environmental performance can achieve higher levels of efficiency in the management of materials and energy and reduce wastes that would otherwise require costly regulatory compliance and expensive treatment and disposal. Nonetheless, sustainable production is not limited to cost avoidance. Manufacturing electric vehicles or photovoltaic cells, developing new nontoxic chemicals for industrial processing, or developing degradable biopolymers from agricultural products stimulate the development of new industries and new enterprises, generates jobs, and by conserving resources, can improve economic productivity. The transition to sustainable systems of production opens up new markets for new materials, technologies, and services; and the need to retool and transform existing production systems generates its own economic activity.

The Role of the Occupational Safety and Health Professional in Sustainable Production

The conventional end-of-pipe focus on hazard measurement for compliance and hazard control and remediation places occupational health and safety in a role that is ancillary to production. Redesigning for sustainable production involves setting health, safety, and the environment as production-design parameters, thus integrating occupational and environmental health and safety into the design and evaluation of production processes. For a company to move toward sustainable production, any new design or redesign effort should include an occupational health and safety professional trained in sustainable production as part of a team put together to oversee the design/build effort.

In our recent work with several different industries, we have identified or developed some practical tools that can be used by the occupational health and safety professional interested in moving toward sustainable production. Here we list several techniques that were developed with a company that was rebuilding a manufacturing facility from the ground up. We recognize that, much more frequently, occupational and environmental health professionals are faced with retrofitting old equipment and processes. Retrofitting is technically difficult and costly, because it attempts to modify an established process in a way that does not contribute directly to production. The methods and tools presented here, however, could be applied on a smaller scale as a company periodically replaces equipment or reconfigures existing space.

Hazard-Analysis Team Process

As a first step in the efforts of a design team, the work environment professional can facilitate environmental and occupational process hazard and job safety reviews with all of the team members. OSHA already strongly encourages companies to conduct similar activities, called Process Hazard Analyses (PHA) and Job Safety Analyses (JSA), to implement its Voluntary Protection Program (OSHA 1996). From the sustainable production perspective, it is important that the process and job analyses include an evaluation of environmental hazards as well as occupational hazards. Thus, the design team should include participants involved in all aspects of the process to be designed or modified, such as engineers, worker-operators, maintenance and repair personnel, production supervisors, and purchasing and marketing personnel. In the hazard and job analyses, each production process or unit is reviewed for its overall flow and product output, materials, equipment, jobs, standard operating procedures, maintenance and repair needs, and the physical space in which the process occurs. The hazards associated with each are identified and prioritized. If necessary, consensus methods can be used to prioritize hazards and solutions. Solutions to each problem then are researched or developed by the team. As with all team efforts, it is essential that the work of the redesign team have an avenue of access within the organizational structure so that recommendations will be communicated and implemented effectively. This work should follow the quality engineering model of "continuous improvement," with the team reviewing the new process and job requirements at regular intervals throughout their piloting and standard operations.

Work Environment Checklists for Purchasing Specifications

Another useful tool is a checklist for production, engineering, and purchasing personnel that encourages them to consider sustainable production criteria when ordering new equipment and materials. The checklist should include a review of all occupational and environmental chemical, physical, ergonomic, and safety hazards. The work environment professional should work closely with those finalizing the equipment and materials orders so that the checklist procedure is effective and can be implemented with as little disruption as possible. The checklist can be a practical and specific tool for creating awareness about health and safety throughout departments that typically have not been involved with such issues. It also can encourage equipment and materials suppliers to "go green" by specifying occupational and environmental considerations in the purchase orders. Our experience is that many equipment and technology specifications for improved occupational and environmental health and safety do not cost much more than standard features, if the specifications are made in the initial design. For example, it was not more costly to specify the height and design of a platform at the charging port of a tank with ergonomic and safety measures than to select the standard platform designed without work environment considerations.

Development of Standard Operating Procedures, Job Requirements, and Job Training

The work environment professional should work closely with team members to write standard operating procedures (SOPs), job descriptions, and job requirements for the new production process. For example, rather than having a separate chapter for health and safety practices in an operating or training manual, sound occupational and environmental health and safety practices should be incorporated in every step of the production process SOPs and job requirements. This strategy incorporates health and safety in production rather than making it ancillary. It is an extension of the idea already practiced in some industries of making production processes "inherently safe."

Development of Production Process Evaluation and Predictive Models

The work environment professional working toward sustainable production can use production records such as engineering, materials, product and job specifications to construct statistical or empirical models to predict the effect of production parameter changes on particular work environment hazards such as the airborne concentration of dust. Production data often are routinely collected by quality control, research and development, and/or production supervisors, but are seldom used in a systematic way by industrial hygienists. The models could be used to guide production process redesign by identifying the most significant production process determinants of workplace or environmental pollution or other hazards, which, then, could become the most significant targets for redesign. In addition, the models could be used to predict the effect of process changes on particular work environment outcomes before the changes are actually made so that disruptions are minimized. This approach uses methods that have already been developed for retrospective exposure assessment (Quinn

1992; Schneider et al. 1991), but shifts the emphasis from the past to the evaluation of current or future production processes.

Professional Training for Sustainable Production

Training occupational health and safety professionals with a focus on sustainable production will require an expanded curriculum. This new training should include the conventional identification, quantification, and prevention of chemical, ergonomic, physical, and other hazards to human health and the environment, including the prevention of psychosocial stressors of the workplace. But, in addition, a link with engineering education will be important so that the identification of production-related health and environmental problems is closely tied to the design of new ways to prevent them. Although no one individual will be expected to perform all of these tasks at a high level of proficiency, the effective work environment professional should be able to work on a team with design engineers, managers, shop-floor workers, and the marketing staff to find new design solutions that are inherently safe. Thus, education for the work environment professional should include the development of team-building skills, organizational management capabilities, and expertise in training and communication, as well as the traditional technical skills. Because the nature of workplace hazards will evolve with the technologies, the training should emphasize problem solving rather than rote learning.

Academic Research on Sustainable Production

The concepts needed to integrate environment, health, and safety into production are just now evolving. Much work needs to be done on elaborating and evaluating the framework of sustainable production, and much of this requires research and testing such as:

1. Research is needed to refine and elaborate the criteria for creating sustainable production processes. This essay can only present a brief introduction to this new approach; more detailed methods are needed to assist companies and work environment professionals who are considering a technological change to move toward a cleaner and safer environment. Rosenberg (1995) has proposed a Work Environment Impact Assessment for this purpose.

2. Research is needed to develop indicators of sustainable production that can be applied both within firms wishing to evaluate their progress and by communities planning their own sustainable economic development.

3. Curricula for sustainable production must be developed that are directed to the fields of industrial hygiene, ergonomics, safety, epidemiology,

toxics-use reduction/pollution prevention, process engineering, and occupational health policy.

4. Research is needed to adapt quantitative modeling methods from other disciplines to predict the effects of design changes on the work environment. Closely related fields such as epidemiology, retrospective exposure assessment, and risk assessment use analytic methods that might be used to assist production process evaluation and redesign efforts.

5. Technology designed to reduce repetitive strain and other injuries should be evaluated for its impact on job quality and work organization. Robotization may lead to reduced musculoskeletal strain on the one hand, but it also may reduce employment, and the remaining jobs may become more routinized. Research is needed to understand how to translate economic efficiencies derived from mechanization into opportunities to expand environmentally sound production and to use this expansion to create more fulfilling, healthy jobs.

6. The human health effects of processes and materials that are viewed as "clean production" need systematic evaluation. Processes viewed as environmentally benign may not necessarily improve worker health and may even present new occupational hazards. For example, water pollution control laws have created strong incentives for industries to reuse cleaning and cooling water, but, warm, organically contaminated water is an ideal growth medium for bacteria and fungi, which may pose a variety of human health hazards. Re-aerosolization can cause particularly severe health effects, including hypersensitivity pneumonitis, asthma, and other obstructive lung conditions (Milton et al. 1996).

7. The health implications of worker exposure to enzymes, bacteria, biodegradable compounds, and bioengineered agricultural and manufacturing materials should be evaluated. Pollution-prevention efforts have led to increasing use of these biological agents in place of environmentally damaging chemicals. Some of the materials, like enzymes, can have irritant effects on mucous membranes and cause allergic responses that affect the skin and respiratory system (International Labour Office 1983).

Conclusion

In this essay we offer an initial presentation of the idea of sustainable production rather than a comprehensive analysis of the effects that these ideas may have on a firm, or even within the field of occupational health and safety. More experience in applying the framework is needed. Indeed, the prospects of exploring this new approach should not be considered as an invitation to replace current practices. Government, labor, and occupational health and safety professionals should not abandon their conventional activities related to the enforcement of occupational and environmental regulations.

These activities continue to provide the major motivation for industries to reduce pollution and health and safety hazards. Sustainable production involves expanding upon, not replacing, these traditional functions.

We are in a period of significant change in the global economy and the structure of work. This rapid change is requiring business enterprises to provide more flexible production processes that can respond quickly to fluctuating market demands. Such changes require a new, dynamic approach to the design of the workplace. This shift presents a challenge to conventional occupational health and safety practice, which has developed professional strategies primarily based on the identification and control of hazards from technologies whose fundamental design has been viewed as beyond the scope of the field. There is in these global, economic dynamics a significant opportunity to reconsider and redesign the practice of occupational safety and health protection. An approach that anticipates and motivates these changes could place the future work environment professional in a more strategic position to avoid and prevent the risks of workplace production. In response to this opportunity, we are proposing a new framework called sustainable production that expands the role of the environmental and health and safety professional to become a more active and effective participant in the design of the production processes of the future.

NOTE

The study was previously published in the *American Journal of Industrial Medicine* 34 (1998): 297–304. Copyright © 1998, Wiley-Liss, Inc., a subsidiary of John Wiley & Sons, Inc. We gratefully acknowledge Dr. Barry Commoner, director of the Center for the Biology of Natural Systems, Queens College, New York, for valuable discussions related to the development of this work. We thank Robert Forrant, Charles Levenstein, Craig Slatin, Cathy Crumbley, Janet Clark, Beverly Johnson, and Jack Luskin of the Lowell Center for Sustainable Production for their thoughtful comments on this proposal. This work was supported in part by grants from the Committee on Industrial Theory and Assessment (CITA), University of Massachusetts Lowell, and by the Jesse B. Cox Charitable Trust, Boston.

REFERENCES

Ashford, N. A. 1976. *Crisis in the workplace.* Cambridge: MIT Press.
———. 1994. Government strategies and policies for cleaner production. Paris: United Nations Environment Programme, Industry and Environment.
Ashford, N. A., I. Banoutsos, K. Christiansen, B. Hummelmose, and D. Stratikopoulos. 1996. Evaluation of the relevance for worker health and safety of existing environmental technology databases for cleaner and inherently safe technologies. Report to the Commission of the European Union (SOC 94 202215 05F02). Athens: Ergonomia, Ltd.

Becker, M. M., and N. A. Ashford. 1995. Recent experience in encouraging the use of pollution prevention in enforcement settlements. NTIS No. PB95–232781. Report No. EPA/300/R-95/006. Washington, D.C.: U.S. Environmental Protection Agency.

Bruntland, G. 1987. *"Our common future": Report to the World Commission on Environment and Development.* Oxford: Oxford University Press.

Commoner, B. 1990. *Making Peace with the Planet.* New York: Pantheon.

Fischer, K., and J. Schot. 1993. *Environmental strategies for industry: International perspectives on research needs and policy implications.* Washington, D.C.: Island Press.

Froines, J., R. Gottlieb, M. Smith, and P. Yates. 1995. Disassociating toxics policies: Occupational risk and product hazards. In *Reducing toxics: A new approach to policy and industrial decisionmaking,* ed. R. Gottlieb. Washington, D.C.: Island Press.

Gottlieb, R., ed. 1995. *Reducing toxics: A new approach to policy and industrial decisionmaking.*Washington, D.C.: Island Press.

Gottlieb, R., and M. Smith. 1995. The pollution control system: Themes and frameworks. In *Reducing toxics: A new approach to policy and industrial decisionmaking,* ed. R. Gottlieb. Washington, D.C.: Island Press.

Graedel, T. E., and B. R. Allenby. 1995. *Industrial ecology.* Englewood Cliffs, N.J.: Prentice-Hall.

Hofrichter, R., ed. 1993. *Toxic struggles: The theory and practice of environmental justice.* New York: New Society Publishers.

Hopfenbeck, W. 1992. *The green management revolution: Lessons in environmental excellence.* New York: Prentice-Hall.

Huisingh, D., L. Martin, H. Hilger, and N. Feldman. 1986. *Proven profits from pollution prevention: Case studies in resource conservation and waste reduction.* Washington, D.C.: Institute for Local Self-Reliance.

International Labour Office. 1983. Enzymes in industry. In *Encyclopedia of occupational health and safety,* 1:766–68. Geneva: International Labour Office.

Jackson, T., ed. 1993. *Clean production strategies: Developing preventive environmental management in the industrial economy.* London: Lewis.

Leff, E. 1995. *Green production: toward an environmental rationality.* New York: Guilford Press.

Lélé, S. M. 1991. Sustainable development: A critical review. *World Development* 9(6): 607–21.

Mazurek, J., R. Gottlieb, and J. Roque. 1995. Shifting to prevention: The limits of current policy. In *Reducing toxics: A new approach to policy and industrial decisionmaking,* ed. R. Gottlieb. Washington, D.C.: Island Press.

Milton, D. K., D. Wypij, D. Kriebel, M. D. Walters, S. K. Hammond, and J. S. Evans. 1996. Endotoxin exposure-response in a fiberglass facility. *American Journal of Industrial Medicine* 29:3–13.

Moure-Eraso, R. 1995. Pollution prevention/source reduction and occupational safety and health: Final report. Washington, D.C.: Occupational Safety and Health Administration.

National Academy of Engineering. 1994. *The greening of industrial ecosystems.* Washington, D.C.: National Academy Press.

Occupational Safety and Health Administration (OSHA). 1996. Revised voluntary protection programs (VPP): Policies and procedures. Washington, D.C.: OSHA.

Oldenberg, K., and Geiser, K. 1997. Pollution prevention and . . . or industrial ecology. *Journal of Cleaner Production* 4(4) (Spring).

Quinn, M. M. 1992. A biologically-based quantitative method for characterizing airborne fiber exposure. Doctoral dissertation. Ann Arbor. UMI Dissertation Information Service.

Raterman, S. M. 1996. Methods of control. In *Fundamentals of industrial hygiene,* ed. B. A. Plog, J. Niland, and P. J. Quinlan. Chicago: National Safety Council.

Rosenberg, B. J. 1995. The best laid bans: Impact of pesticide bans on workers. Doctoral dissertation. University of Massachusetts Lowell, Department of Work Environment.

Schmidheiny, S. 1992. *Changing course: A global business perspective on development and the environment.* Cambridge: MIT Press.

Schneider, T., L. Olsen, O. Jorgensen, and B. Lauersen. 1991. Evaluation of exposure information. *Applied Occupational and Environmental Hygiene* 6(6): 475–81.

Sullivan, T. F. 1992. The greening of American business: Making bottom-line sense of environmental responsibility. Rockville, Md.: Government Institutes, Inc.

Innovation and Adoption of Cleaner Production Technologies

Ken Geiser and Tim Greiner

TECHNOLOGY is a critical element of sustainability. Many of the technologies that we employ today to transport us, to provide our energy, and to manufacture the products that maintain and enrich our lives are not sustainable. They are resource depleting, waste generating, energy intensive, and polluting. Decades of invention and innovation have sought to optimize their functional performance and cost characteristics, but because environmental and public health concerns have not been factored into their design, their consequences in these areas are less than satisfactory.

Recent attention to the health and environmental factors of production technologies, driven by government regulations, liability concerns, and a more sensitive market, have generated a host of new technologies that are more energy efficient, less polluting, and safer to use because they avoid the use of highly dangerous chemicals. Many of these technologies are earnestly marketed as "good for the environment" and as good investments. After relatively short payback periods, some of these technologies have demonstrated remarkable economic savings in terms of more efficient energy and material use, lower waste-management costs, lower regulatory burden, and lower occupational health costs (Jackson 1993).

In view of these various benefits, it is surprising to hear the multitude of stories about how difficult it has been to get current manufacturing firms to adopt these more sustainable technologies (Bierma and Waterstraat 1995). It is not the economics that acts to inhibit these sales—the numbers often add up to a good investment. Rather, there appears to be something about environment and health as factors of technology that results in slow adoption.

This study examines the problems that new, cleaner production technologies confront in achieving adoption by way of case studies of particular efforts to promote manufacturing innovation through the adoption of cleaner production technologies. Over the past decade the concept of cleaner production (here in the United States, the term "pollution prevention" is used alternatively) has emerged internationally as an industrial

219

program for helping businesses move towards principles of sustainability. Cleaner production is defined by the United Nations Environment Programme as the continuous use of industrial processes and products to prevent the pollution of air, water, and land, reduce waste at source, and minimize risks to the human population and the environment. (United Nations 1994). While the study of the demand and need for cleaner production has progressed steadily, comparatively little research has been done on the suppliers of cleaner technologies.[1]

This study examines the innovation process and how pollution-prevention and cleaner production innovations occur and how such technologies are adopted in the market. We begin with an investigation of the process by which technologies are adopted in the market with particular attention to obstacles faced by cleaner production technologies when competing with established technologies. We present three cleaner production-process-equipment innovation cases. In one case, the technology was developed by the owner of a small plating firm; in the other two, the technology was developed by external specialists—a consultant in one instance and an academic in the other—who went on to try to market and sell their inventions. The studies reveal several important aspects that must be considered and overcome in promoting technological solutions more compatible with the principles of sustainability. We sum up our findings and the directions for further research they suggest.

Conceptualizing Technology Innovation and Adoption

Significant obstacles lie in the path of any new technology—be it "green" or not. New production technologies must fit into the manufacturing systems of prospective clients and, at the same time, satisfy numerous other requirements (quality, user friendliness, cost, throughput, etc.). If the new technology requires significant modification of the user's existing production system, the spread of new technology may proceed slowly (Matalin 1996).

In addition to the need for the innovative technology to fit into the organizations' existing production system, a new technology must compete with established technologies in two other areas. First, established technologies benefit from the effects of scale since manufacturing costs per unit decline as production volume increases (in economist terms, the marginal cost decreases as production volume increases). Second, existing technologies are farther down their "learning curve," meaning that the manufacturer has accrued experience in making the product over time. In general, the more experience a firm has manufacturing a product, the greater its reliability and the lower its cost (Utterback 1996).

Cleaner production innovations are at a significant disadvantage in the early stages of their development. Their costs are uncertain and there are few (if any) demonstrated installations. Cleaner production technologies may also face more significant obstacles to commercialization since they may require employee training or a temporary halt in production operations while the new process is installed and characterized. Potential users have little incentive to be the first user of the technology since there are often many early problems with the technology that will not be eradicated until multiple units have been installed. Conventional environmental technologies that control pollutants at the ends of discharge pipes—so-called end-of-pipe technologies—on the other hand do not face these obstacles. First, product costs are typically well known. Second, production operations do not usually have to be halted for testing and installation. Third, end-of-pipe technologies rarely require changing the management of production or the job duties of line operators.

Not all innovations face these barriers to adoption to the same extent. Everett Rogers has found that certain characteristics of an innovation, as perceived by potential adopters, help to explain the rate at which innovations are adopted (Rogers 1995).

Relative advantage is the degree to which an innovation is perceived as being better than the technology it replaces. Relative advantage can be measured in economic terms as well as by quality and throughput. The greater the relative advantage, the higher the rate of technology adoption.

Compatibility is the extent to which the innovation is perceived as being consistent with the core values, experiences, and needs of the potential adopter. The more compatible an innovation is in terms of these factors, the higher the rate of technology adoption.

Complexity is the extent to which the innovation is difficult to understand and use. The more readily an innovation is understood and easily used, the higher the rate of technology adoption.

Trialability is the degree to which an innovation can be experimented with on a limited basis. The more readily an innovation can be tested in a production or near-production setting, the higher the rate of technology adoption.

Observability is the extent to which the results of an innovation are visible to others. The more easily an innovation's results are observed (and the less they are shielded by confidential business information concerns), the higher the rate of technology adoption.

Technology diffusion research shows that adopters depend on not only scientific studies, but also subjective evaluations conveyed by individuals who have previously adopted the innovation. Thus technology diffusion is essentially a social process carried out among potential adopters. The more alike the potential adopters are, the more effective their communication.

"When [potential adopters] share common meanings, a mutual subcultural language, and are alike in personal and social characteristics, the communication of new ideas is likely to have greater effects in terms of knowledge gain, attitude formation, and change, and overt behavior change" (Rogers 1995, 19).

One of the problems with technology diffusion processes is that the potential adopter and the innovator are typically rather unalike. The innovator is often more technically competent than potential adopters. Furthermore, if the innovation is truly radical, the innovator may have a background completely incongruent with the potential adopter (for example, in one of our cases, the innovation required knowledge of physics while the adopter community was predominantly wet chemists).

Innovation, itself, is colored by the social relations of the innovator. Eric von Hippel (1988) finds that the functional relationship by which innovators derive benefits from the innovation affects the way in which the innovation emerges and is subsequently adopted. For instance, users of a technology may innovate by adapting current technologies, while suppliers of the technology may start with a totally new concept.

Over the past several decades, various studies of the diffusion of innovations show that innovations undergo a period of slow initial growth, followed by rapid exponential adoption, transitioning into slow adoption as the innovation reaches its natural limit (Abernathy and Utterback 1978; see figure 1). Rogers developed a classification of adopters that in part de-

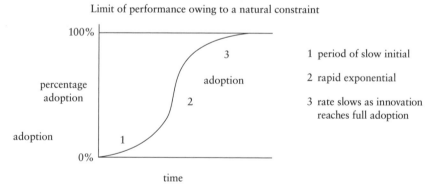

Figure 1. S-curve of technology adoption

scribes this phenomenon. The system classifies firms according to their order of technology adoption: early adopters, early majority, majority, late majority, and laggards. (Rogers 1995, 246). Early adopters are firms that tend to take greater technology risks than their competitors and are more likely to be open to cleaner production innovations than other firms. Firms that are among the last to adopt, termed "laggards," are risk- and change-adverse. Innovators tend to focus their resources identifying and working with early adopters—hoping that adoptions by these lead users will spur "copycat" adoptions by more risk-adverse firms.

There is, however, no predetermined rate at which a particular technology diffuses. Nor does a technology necessarily achieve 100 percent adoption (see figure 2).

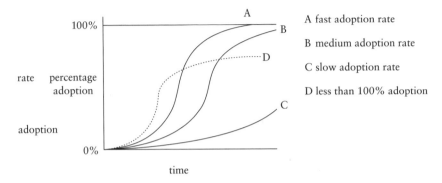

Figure 2. Technology adoption rates

These general findings concerning the adoption of new technologies offer a framework for considering the specifics of adoptions of cleaner production technologies. Thus, whether and how quickly a specific cleaner production innovation is adopted depends upon the specific characteristics of the innovation and the social dynamics among the potential adopters. In addition, the adoption of cleaner production innovations, like other environmental technologies, is directly affected by current environmental regulations and the prospect of new regulations. Nicholas Ashford finds that certain factors of environmental regulation spur technological innovation, including the type of regulation (on products or on processes), the mode of the regulation (design specification or performance), the time limit for compliance, the stringency of the regulations, and the existence of other economic incentives. Of these, Ashford finds that regulatory stringency may be the most powerful in determining the frequency of innovation and the rapidity of adoption (Ashford 1993, 277–310).

Our knowledge of how innovations are adopted by industry points to gaps in the commercialization and adoption of cleaner production innovations (Bierma and Waterstraat 1995; Lindsey 1998). These gaps include: a lack of information regarding the innovation's performance for consumption by potential customers—for example, information regarding the technology's operating costs, technical capability, reliability, throughput, and the like; a reluctance among users to be the first to adopt the innovation because of the risks associated with the new technology; and difficulties between innovators and the adopter community in terms of values, experience, and culture.

Case Studies in Cleaner Production

To examine cleaner production equipment innovation and adoption in greater detail, we researched three cleaner production innovations. The research involved compiling a list of innovations in several targeted industrial sectors. Three examples were selected for investigation in case study form. In each case, the research began with the first firm to develop and commercialize the innovation. After establishing the innovating firm, we interviewed employees at the firm who claimed substantial, firsthand knowledge of the innovation work. Each interviewee was asked to provide the names of others he or she felt might have important information to contribute. Those persons were contacted and interviewed where possible. Following the interviews, information from the various sources was assembled and analyzed. Table 1 introduces the three case studies.

Each of the following case studies examines the innovation process from its earliest stages through early commercialization. The environmental benefits of the innovation are discussed as are the roles various private and public entities played in the development of the technology.

Case 1: Zero Discharge, Inc.

In 1983, Eddie Ondrick, owner of a job-shop nickel plating firm in Chicopee, Massachusetts, was faced with meeting stringent metal limits on his wastewater discharge. The Environmental Protection Agency (EPA) had forced the sewer authority in Ondrick's area to tighten up pollution limits on firms discharging to the sewer. At the time, Ondrick's firm was discharging 80,000 gallons of wastewater per day. According to Ondrick, he was told that, "If you don't have a discharge . . . if your discharge equals zero, then you can get out of the permit." Ondrick was also told that there was "no way you can go to zero discharge."

With stricter permit limits looming, Ondrick began experimenting with custom-built separation and recovery equipment to close the loop on the water use in his plating shop. Attempts to utilize turn-key equipment proved unsuccessful. Ondrick then began to modify and develop equipment himself, in the course of which he generated patentable ideas. Over a ten-year period, he not only closed the loop on his plating shop, but also developed four different separation and recovery equipment systems, received two patents, and founded a company called Zero Discharge, Inc. to

Table 1. The case studies

Zero Discharge, Inc.	Metal finishing recycling and recovery equipment developed by a small metal-finishing company to comply with discharge regulations
Legacy Systems, Inc.	Semiconductor cleaning process using chilled ozone and water process developed by an equipment manufacturer
The Radiance Process	Chemical-free laser and inert gas cleaning process discovered by a graduate student and a professional colleague

commercialize and sell his process equipment innovations. Two of these process innovations are described below.

Acid Recovery System

Waste acids typically contain a high concentration of heavy metal ions—a result of the chemical reaction between the acid and the metal substrates that are processed in metal-finishing operations. The traditional procedure for treating tanks of waste acid is to slowly "bleed" the acid into the company's wastewater treatment process. Many wastewater treatment processes, however, are unable to handle concentrated acids—resulting in wastewater treatment upsets. Such upsets often cause discharges over permitted levels. Waste acid recovery allows a firm to not only reuse its acids (and therefore save money), but also to avoid the compliance problems associated with wastewater treatment.

In 1987 Ondrick realized the need for recovery equipment that would allow for reuse of process chemicals (such as acids and plating bath metals). His research into existing systems for acid recovery had concluded that they suffered from slow recycling rates due to air entrapment in the system's vertical membranes. After three years of trial and error, Ondrick developed a reliable acid recovery system using a horizontal stack of membranes and a pumping configuration that eliminated the air-entrapment problem. In the first year of the system's operation, it reduced the company's nitric and hydrochloric acid use from 100,000 pounds per year to 1,000 pounds.

Early in the development of the system, Ondrick considered licensing the technology to a metal-finishing equipment manufacturer. Ondrick reconsidered when he was unable to negotiate a license that would afford him a satisfactory profit. According to Ondrick, "We thought we would be stupid to sign such an agreement, so we decided to try to market the systems ourselves."

In 1988, Ondrick set up a booth at a metal-finishing industry trade association conference to market his acid recovery system. At the time of the trade show, his only acid recovery unit was the one working on his shop floor. Several firms expressed interest in purchasing or renting a system at the trade show. At the same time, staff from the regional EPA office visited Ondrick's shop to see how he had eliminated his discharge. According to Ondrick, EPA told him, "You've got to sell this stuff. We know there are lots of companies with similar discharge problems that could use equipment like this."

Following the trade show, Pitney-Bowes, a Connecticut manufacturer, sent samples to Ondrick for testing. In 1989, after a comprehensive evaluation that lasted nearly a year, Pitney-Bowes purchased two acid recovery systems. Encouraged by his first sale and the endorsement of the EPA representative, Ondrick established Zero Discharge, Inc. in 1990 to commercialize his innovations. In 1990 and 1991, Ondrick applied for and received two patents for this acid recovery system, and the technology appeared to be on its way toward adoption.

Ondrick's next sale came from a 1991 trade association conference, where he met representatives from a Milwaukee metal finisher who was having difficulty with permit excursions when the company dumped and treated its spent acids. Like Pitney-Bowes, the Milwaukee firm purchased the tool after Ondrick successfully processed several waste acid test samples. At the same conference, Ondrick overheard a presentation on the waste acid problems at Lawrence Livermore National Laboratory. He contacted a laboratory representative about the talk. Six months later, Ondrick sold a unit to Lawrence Livermore. Shortly thereafter, he sold units to Sandia National Laboratory and Los Alamos National Laboratory as well.

By the end of 1991, Ondrick had sold six units, and the unit appeared to be on its way toward further adoption. In 1992, however, Ondrick found that a former employee was selling acid recovery units using Ondrick's technology. Ondrick sued and spent three years and $200,000 before winning a court settlement. Under the settlement, Ondrick's former employee was required to pay royalties on all prior and future acid recovery system sales. According to Ondrick, the suit not only was expensive, but it siphoned time and money away from improving and marketing his acid recovery system.

Following the litigation, sales of the acid recovery unit to industry increased. The military became interested in the system and began evaluating it for use in Navy and Air Force operations. In 1994, Lawrence Livermore National Laboratory published a report documenting the positive financial and environmental performance of the acid recovery system. That report and a track record of satisfied customers were instrumental in further sales. By 1997, roughly forty acid recovery systems had been sold. By that time, Ondrick had gradually handed marketing, sales, and technical support for the system over to an associate, preferring to devote his time to another cleaner production innovation—an electroless nickel purification system.

Electroless Nickel

In 1988, having developed an acid recovery system, Ondrick turned to developing equipment to recover nickel from his electroless nickel-plating process. Ondrick developed a system using electrodialysis to selectively remove contaminants (orthophosphite and sodium sulfate) that require electroless nickel solutions to be discarded. The system removes the contaminants and returns 80 percent of the plating solution to the process. The new system, which has an approximate capital cost of $20,000, is a quantum improvement over existing systems, increasing throughput by a factor of five.

Ondrick recognized that few metal finishers truly understood the complex chemistries involved in electroless nickel-plating baths. Although most metal finishers are capable of making simple adjustments to electroless nickel baths, they rely on their chemical suppliers for bath trouble shooting and maintenance. Over a two-year period (1991–1993) Ondrick discussed licensing agreements with the largest electroless nickel chemical suppliers. Each supplier sent samples for processing. Even though all of them concluded that the recovery process was beneficial and economical, none seriously entertained licensing talks with Ondrick. The suppliers had a natural ambivalence about the technology, because it would have drastically reduced electroless nickel sales. Although open to further licensing discussions, Ondrick decided to sell the units directly to electroless nickel metal finishers.

Ondrick's first sale was to Lawrence Livermore National Laboratory in 1993. Ondrick had already sold the lab a waste acid recycling unit and had discussed his electroless nickel recovery unit with an engineer at the lab. According to Ondrick, "One day I get a call from [the engineer] and he says he just got out of a meeting where they were talking about a $1,000,000 electroless nickel purification project. I told them that I knew a guy who already had a system that worked."

In 1994, Lawrence Livermore prepared a report touting the financial and environmental benefits of Ondrick's electroless nickel recovery system. The report helped persuade two other national labs to purchase systems in 1994 and 1995. Despite the national labs sale, Ondrick was unable to sell to industry. Although sales were slow, Ondrick continued tinkering with the system, reducing the unit's size by 50 percent while simultaneously doubling its throughput.

In 1995, one of the largest electroless nickel chemical suppliers, Fidelity Chemical, asked Ondrick to consider a technology licensing agreement. According to Ondrick, Fidelity said, "We know this technology is the future of electroless nickel, and we want to be the company to bring it to the market . . . we want exclusive rights [to the technology]." Ondrick and Fidelity agreed on a deal by which Fidelity would be the only chemical supplier licensed to sell the technology, but Ondrick would retain the right to sell the system directly to metal-finishing firms. Fidelity saw the system as a means to increase their revenue base by winning over metal finishers using other electroless nickel chemical suppliers.

Ondrick was pleased with the partnership, saying, "I don't want to be a chemical company and they don't want to be an equipment company." Ondrick and Fidelity agreed to underwrite the cost of two systems that could be transported to potential customer sites. The systems would be run on the customer's electroless nickel bath. If the evaluation was positive, the company could choose to purchase the system. Fidelity and Ondrick set targets of selling two units in the first year, three in the second, and five to ten in the third year. Fidelity targeted its sales efforts at its largest electroless nickel users. By 1997, the second year of their agreement, Fidelity had sold two systems. Interest in the system was high. The two test systems, which remained on a customer's site for two to five months, were booked more than two years in advance. As with the acid recovery unit, Ondrick continued to improve the system by making improvements based on the suggestions of equipment users.

Ondrick's process innovations arose out of his need for a solution to a wastewater discharge problem. Unable to find a solution in the market, Ondrick developed equipment for his own use. Only after installing the equipment in his own shop, did Ondrick discover the potential commercial value of his innovations.

Analysis

The five characteristics of an innovation discussed earlier make a useful lens through which to examine the acid recovery and electroless nickel adoption process. The chief obstacle to the widespread adoption of the acid recovery system was its lack of relative advantage over standard end-of-pipe systems. Unless firms are faced with exceedingly stringent discharge limits or are unable to handle concentrated acids in their existing treatment system, most of them will choose simply to use end-of-pipe treatment over acid recovery. Although the recovery systems cost from $7,000 to $20,000, acid raw material prices run in the pennies per pound. Furthermore, end-of-pipe treatment has the advantage of being a well-tested technology whose unit price has decreased as the technology has matured. Other adoption factors, such as system compatibility, complexity, trialability, and observability appear not to have hampered the technology's acceptance in the market (see table 2).

The electroless nickel system differs from the acid recovery system in that it has a huge cost advantage over end-of-pipe treatment (owing to the raw material cost of the electroless nickel-plating solution). The system, however, is largely incompatible with the experiences and expertise of most

Table 2. Zero disadvantages of technology adoption charcacteristics

Characteristic	Acid recovery	Electroless nickel
Relative advantage	Cost advantage over end-of-pipe in certain situations	Large cost advantage over end-of-pipe
Compatibility	Highly compatible with core values, experiences, and needs of adopter	Low, because few electroless nickel finishers understand their chemistry
Complexity	Fairly low	Fairly low
Trialability	High—samples can easily be processed in a lab setting	Low—requires testing customer's electroless in bath on site
Observability	High—laboratory test results easily verifiable	Medium—efficacy of a recycled plating bath difficult to prove to skeptics

metal finishers. Few metal finishers have the grasp of chemistry to understand how plating solutions operate. Trialability may also slow adoption since equipment to perform trials are scheduled years in advance (see table 2).

Although the adoption rate is difficult to predict a priori, based upon this analysis, we would expect that the acid recovery process would see its adoption rate climb as more stringent wastewater effluent limits make it costly for firms to treat spent acid baths. We would expect the adoption of the electroless nickel process to proceed more slowly given the complexity of the process in the eyes of potential adopters—despite the relative cost advantage of the process.

Ondrick's experience running a metal-finishing business also plays a role in the adoption process. In many ways, he shares the culture, skill, and characteristics of other metal finishers—meaning that he is likely to have an effect on the attitude, knowledge, and behaviors of potential adopters. As Zero Discharge's own marketing materials put it, "Made for metal finishers, by metal finishers."

Lastly, Ondrick made numerous key decisions in commercializing his innovations. Whether to patent his innovation and to whom to license the technology were among the more important decisions. Such decisions also play an important role in our other case studies.

Case 2: Chilled Ozone Cleaning

For roughly thirty years, the semiconductor industry has used a combination of concentrated sulfuric acid and hydrogen peroxide to clean photo resist off semiconductor wafers. This process, known as the piranha process, consumes roughly 4,200,000 gallons of hazardous chemicals across the nation each year. Robert R. Matthews, founder and president of Legacy

Systems, Inc. developed a substitute process that uses 100 percent oxygen and water as the sole raw materials to perform the same function.

Matthews discovered the chilled ozone process by accident. In December 1992, working at his warehouse facility, Matthews was testing wafers for a client. Matthews wanted to measure the displacement when ozone gas (O_3) was bubbled into a tank of deionized water. Since Matthews was conducting the test during the Christmas and New Year holidays, the warehouse heating system had been turned off. Matthews figured that running the experiment with water near freezing temperature would not have an effect on water displacement in the tank. He inserted the wafers in the water-filled tank. Upon pulling the wafers out, he saw that all the photo resist on the wafer had been removed, although the bath contained no sulfuric acid. After repeating the test, Matthews realized he had stumbled on an incredible breakthrough in semiconductor processing. The lower water temperature had significantly increased the concentration of ozone in the tank, dramatically increasing the photo resist etch-rate. As a process engineer [with more than seventeen years of experience working for semiconductor manufacturers on piranha and other wet chemistry operations,] Matthews realized the commercial potential of his innovation.

High-temperature piranha solutions generate sulfuric-peroxide fumes that are hazardous for workers and extremely aggressive on expensive quartz-heated tanks and fume scrubbers. The deionized chilled ozone process eliminates not only the use of chemicals, but also introduces few microcontaminates into the process. Furthermore, its throughput rate can be greater in comparison with the piranha process. Table 3 compares the chilled ozone and piranha processes.

In 1993 Matthews applied for a patent on the process, which he received in November 1995. Throughout the spring, summer, and fall of 1995, Matthews began putting together a strategy to commercialize his innovation. A key part of the strategy was to retain operating control in his current company, Legacy Systems, and avoid venture capital financing. Matthews wanted to build a company that could take advantage of his family's significant industry skill base and offer employment–ownership options for his grandchildren. Throughout 1995, Matthews developed the firm's strategy with his three brothers who had experience working in project engineering, manufacturing management, and accounting. Matthews also wanted to take advantage of his company's existing strengths—selling gas diffusion plates to wet chemistry equipment manufacturers for use in semiconductor processes such as the piranha.

The market for Matthews' innovation fell into two main segments: The first was retrofit: This segment involved retrofitting existing wet etch stations with the chilled ozone process. Each of the thousand semiconductor fabrication facilities (fabs) worldwide had roughly three stations applicable

Table 3. Chilled ozone vs. piranha in annual amounts

Process Attribute	Piranha	Chilled Ozone Process
Sulfuric acid use (98% @ 1300°C)	6,300 gallons per year (gpy)	none
Peroxide use (30% @ 1300°C)	2,100 gpy	none
Water use	3,335,800 gpy	3,500 gpy
Workplace safety	Acid fumes	No acid fumes, although ozone exposure possible
Microcontamination	Metallic impurities	No metallic impurities
Maintenance and repair	Corrosive environment shortens equipment life	Noncorrosive environment

to the chilled ozone process. A typical retrofit cost around $350,000. Selling to the retrofit market took advantage of the numerous contacts Legacy had in the industry—many existing wet etch stations already used Legacy components such as gas diffusion plates.

The other market segment, new installations, involved manufacturing new wet stations that would employ a variety of wet chemistry operations, one of which would be the chilled-ozone process. The cost of newly installed wet stations runs from $700,000 to $3.5 million (depending upon their level of automation and process complexity). Selling new installations would be difficult since it would require learning about automation systems. It would also require significant working capital to manufacture systems costing over $3 million.

Legacy decided that the best strategy was to license its technology to a major wet station manufacturer to sell new installations—thus providing Legacy with a royalty stream to fund the retrofit market. Such a strategy avoided venture capital financing and fit well with Legacy's existing strength in selling gas diffusion plates and other hardware to wet station equipment manufacturers. The partnership would also provide Legacy with a bona fide clean laboratory in which to perform customer testing and host customer visits. Finally, such an agreement could provide Legacy with access to the retrofit market by leveraging the partner's installed base of wet cleaning stations.

In July 1996, Legacy signed a licensing agreement with Sub Micon Systems (SMS), a large wet chemistry semiconductor equipment manufacturer. The agreement permitted Legacy to sell new installations in the future, but prohibited Legacy from signing licensing agreements with other manufacturers.

Within months of signing its agreement with SMS, SMS and Legacy ran an ad in *Semiconductor International Magazine* and *Micro Magazine*. SMS's size lent itself to marketing efforts beyond those Legacy could have afforded alone. SMS set up demonstration equipment at semiconductor conferences, introduced the system during a worldwide marketing tour, and ran numerous advertisements in industry trade magazines. These ads marketed the process on its quality, throughput, cost of ownership, and chemical reduction/environmental attributes. Soon thereafter, Legacy and SMS began performing tests at SMS's facility for potential customers. Such tests included removing photo resist from test and production substrates and evaluating the cleanliness of the chilled-ozone stripping process. In 1996, Legacy applied for and won a "Green Chemistry Award" from the Office of the President, which evoked a large influx of customer inquiries.

One of Legacy's biggest challenges was to find a firm willing to be the first adopter. Legacy's strategy for finding a retrofit beta site was to begin discussions with contacts in the industry—and to find a company that would be receptive to the new technology and be willing to share its evaluation data. Legacy identified a major semiconductor manufacturer whose interest in a retrofit was due to its operating-cost savings. The firm purchased the unit in June 1995 and moved it from its research facility to a manufacturing site in June 1997. Tests on the system were successful; however, the manufacturer has been unwilling up until this point to release the data.

By September 1997, SMS had sold new installations to two Korean semiconductor and flat panel display manufacturing firms. One unit had been installed and was going through acceptance testing; the other units were in the process of being installed. Adoption in the United States has been slow. Several U.S. manufacturers signed agreements to purchase new installations and to produce user reports in exchange for reductions in the purchase price.

Although Asian firms were quicker to adopt the new technology than their U.S. counterparts, they were less willing to publicize their test results or allow their names to be revealed. Unlike their U.S. counterparts, the Asian manufacturers were unwilling to permit SMS or Legacy to publicize their identity. According to Carl Muti of Legacy, Asian manufacturers prefer to reveal as little information about their processes as possible. He believes that the Asians adopted the process faster because they have more severe water and chemical problems and because they produce higher volumes of semiconductor chip—and therefore realize greater financial savings from the process.

Analysis

Legacy's position in the semiconductor industry gave the company an excellent position to capitalize on Matthews' process innovation, with Matthews' seventeen-plus years of experience in semiconductor wet chemistry and Legacy's numerous contacts among equipment manufacturers and equipment users.

Examining again the five characteristics of an innovation introduced earlier indicates that, once Legacy and SMS can publicize the results of several successful installations, adoption should proceed at a rapid rate. First, the process has a significant relative advantage over the standard chemical- and water-intensive process in terms of cost, contamination, and throughput. Second, the process is highly compatible with the experiences and needs of potential adopters. Third, the process is simple from the standpoint of the end user. It involves a chiller (to keep the deionized water cold) and an ozone generator. Furthermore, the process has low maintenance needs and is safer to operate than the standard piranha process. Fourth, the process's trialability is high since samples can easily be processed and the results measured using well-established analytical techniques.

In addition to these advantages, both Legacy and SMS have significant experience in wet chemistry semiconductor and flat panel display processing. As a result, Legacy and SMS (innovators) share common language, norms, and experience with potential adopters (engineers working for semiconductor manufacturers).

The only significant barrier to faster adoption has been the lack of "observability" of successful installations. Early in the commercialization process, Legacy approached SEMATECH, a U.S. government-funded consortium of U.S. semiconductor companies, to ask SEMATECH to evaluate the process. However, SEMATECH's reception was cool. Rather than investing significant time convincing SEMATECH that the process was viable, Legacy chose to invest its time in meeting with potential customers and processing their test wafers. In 1997, SEMATECH changed course and selected the chilled-ozone process as one of a handful of processes to evaluate. Legacy's process is one of the planned programs for 1998.

As the next case study indicates, laboratory test data alone are insufficient for potential adopters. Even early adopters don't want to be the first firm to adopt a new technology. As one manager stated, "We are involved in the great race to be second."

Case 3: The Radiance Process

Surfaces free of contaminants are important in manufacturing environments such as optics, semiconductors, and precision tooling. Most traditional cleaning methods use toxic cleaning chemicals and large volumes of water to remove the contaminants—at significant cost to business and the environment.

A dry-cleaning method using neither toxic materials nor water was discovered by accident in 1987 by Audrey Engelsberg, then a doctoral candidate at Rensselaer Polytechnic Institute working on laser-assisted chemical vapor deposition of aluminum. During a failed experiment, she noticed that the laser treatment partially removed carbon contamination from the surface. When, at the urging of a colleague, Joe Dehais, she repeated the experiment with the same result, they realized that Engelsberg had stumbled upon a new dry-cleaning technique. This new technique, later named the Radiance Process (RP),[2] proved to be a general procedure for removing unwanted films and particles from delicate surfaces. There are two working components to the process: a light source and a flowing inert gas. The photon flux from a low-intensity laser light source lifts the particles from the surface. Particles are carried away from the surface with the aid of flowing nitrogen or argon gas. The process breaks the bonds of particulate and thin-film contamination without burning or melting the substrate.

According to Dehais, Engelsberg and he immediately realized the new process might have the potential to remove microscopic particles better than the standard process used in the semiconductor industry. Within six to eight months, the pair determined that the process was cheaper than the existing process—in some applications, one-seventh the cost of standard methods. Since the cost advantage of the process stemmed from the fact that costly deionized water and hazardous chemicals could be eliminated, Engelsberg and Dehais also realized that the process was "green." According to Dehais, recognizing the green aspects of the process was easy. "I had an engineering professor," said Dehais, "who said 'by-products are the result of stupid design.'"

The most striking application of the Radiance Process occurs in semiconductor industry cleaning. The standard industry process, known as the RCA wet cleaning process, uses large volumes of deionized water, hydrochloric acid, hydrogen peroxide, ammonia hydroxide, and isopropyl alcohol. Since the Radiance Process requires no water, acids, or other chemicals, it has the potential to cut semiconductor company capital outlays for new plants by roughly 10 percent and operating expenses by 10 to 20 percent. Other immediate advantages of RP over the traditional RCA cleaning process are:

Reduction in liquid chemical use
Reduction in toxic gas use
Reduction in deionized water use
Reduction in hazardous waste-
   disposal costs
Reduction in costly floor space

Reduction in employee exposure to
   toxins
Increased throughput from shorter
   processing time
Improved removal of contaminants
Reduction in costly core space for
   stand-alone support equipment

Moreover, there are the potential benefits of RP's further applications. Yield is a continuing concern for semiconductor manufacturers. Many in the industry believe that wet cleaning will not be effective for ultra-high-density semiconductor chip designs with minimum feature sizes below .18

or possibly .13 microns. Radiance cleaning has been shown to be capable of removing parts of molecules. Radiance's potential to clean beyond existing wet cleaning and improve yields may be its ultimate adoption-driver. Tests on other substrates indicate the process's cleaning and environmental benefits may also be applicable in numerous other sectors. Such benefits have been demonstrated in lab tests on substrates such as hard discs, photomasks, flat panel displays, tire molds, brass knobs, and optical equipment.

Engelsberg and Dehais filed for a broad patent on the process in 1988. After the patent was approved in 1991, they formed a company (Radiance Services) to commercialize and market the cleaning technique. Radiance is a wholly-owned subsidiary of Cauldron Company, the managing general partner of Cauldron Limited Partnership (Cauldron LP), which owns the Radiance Process patents. Radiance chose to license the process to equipment manufacturers (in semiconductor and other industries) in exchange for royalties. Given the equipment manufacturers' existing expertise and customer base, Radiance felt such firms were better equipped to perform research, development, and marketing, sales, and sales support for the various industrial applications the process is amenable to. In addition to the small licensing fee paid by equipment manufacturers that incorporate the process into their tools, end users must pay a separate royalty fee directly to Radiance's parent company.

Shortly after receiving the patents, Radiance introduced the process to semiconductor engineers and executives. The reception was encouraging, but industry needed more empirical data to be convinced of the process's benefits. To gather empirical data, Radiance began looking for potential partners to underwrite testing. Such testing is costly not only because of the tooling costs ($500,000), but also because any testing had to take place in an ultraclean environment. For credibility purposes, the test facility needed to be a recognized microelectronics cleaning research laboratory.

A series of contacts led Radiance to the director of the Department of Defense Microelectronics Research Laboratory (MRL). MRL was interested in Radiance because of the potential savings that would accrue to the U.S. government's semiconductor manufacturing activities. Through a Cooperative Research and Development Agreement (CRADA) in 1992, MRL became the first organization to receive a license to use the Radiance Process. Tests at MRL demonstrated that Radiance could remove significant levels of organic contamination from semiconductor wafers. Further tests in 1993 at the federal Potomac Photonics Laboratory using an excimer waveguide laser demonstrated that the process could remove contamination without altering the material being cleaned.

Interest in the process increased with the release of the MRL test data. Radiance elected to accommodate these requests by contracting additional laboratory facilities at Diffraction, Ltd. in Vermont, and by licensing the technology to two equipment manufacturers with a proven track record in laser equipment design and fabrication. Exitech Limited of the United Kingdom and Neuman MicroTechnologies, Inc., of New Hampshire were licensed to build and market Radiance Process equipment in Europe and the United States, respectively.

In 1994, IMEC (a Belgium semiconductor research lab) invited Radiance to research and develop the process at its laboratories. IMEC evaluates microelectronics technology and has an Ultra Clean Processing Program with European, Asian, and U.S. affiliates. The first production prototype radiance tool, built by Exitech Limited, was installed in IMEC in 1995. Work there led to major improvements in tool design. Exitech installed an improved processing chamber, and IMEC completed its tests in December 1996. Their report found significant cleaning on silicon wafers and recommended further testing.

The tool was then transferred to Rutherford Appleton Laboratory in England. In 1995, a Radiance Process demonstration project was chosen for an EPA Environmental Technology Initiative grant. Grant partners included MRL and EPA's National Risk Management Laboratory. Under these auspices, a new tool was constructed and installed at Motorola's Phoenix Corporate Research Laboratory. In addition to testing at these research facilities, Radiance opened its own research facility in Vermont in 1996.

Radiance sought a substantial presence in the news media and the technology press. Radiance has received awards and recognition, including its selection as an Industry Week Technology of the Year (a status shared with four other technologies including Sun Microsystem's Java, Sandia National Laboratory's micromachines, and American Superconductor's high-temperature superconducting wire). In 1997, Radiance was awarded the National Pollution Prevention Roundtable's Most Valuable Pollution Prevention Award and the Photonic Spectra Circle of Excellence. The early Radiance marketing strategy was to pursue the semiconductor market and sell the process to semiconductor executives based on its cost-saving attributes. Radiance purposefully avoided marketing the technology on its environmental attributes. According to Dehais, environmental marketing was a turnoff for customers: "Customers' first response when you told them it was a green product was to protect their wallets. We wanted to portray ourselves as truly professional business folks with a technology that has money in it for everybody." Dehais found green marketing more work than cost-effectiveness marketing because customers assumed green products would cost them more money. Radiance, however, got so many inquiries from environmental groups, regulatory agencies, state pollution-prevention technical assistance providers, and the like that the company decided to hire a marketing person with an environmental background who could speak to that community.

Although the company has pursued many potential clients, sales have been slow. Radiance has produced ample test data, but industry tends to trust information from its competitors more than information generated by independent labs. And although Radiance initially marketed its system on its economics, some of its most receptive prospects have been high-level corporate environmental officers who report to their chief executive officers. Such a reporting level gives Radiance more direct access to key decision makers than it might otherwise have when dealing with research and development engineers, scientists, and managers.

Radiance chose not to manufacture equipment using its process, but to focus its resources on laboratory testing and licensing. Radiance Services thought the proof of concept tests, if successful, would make commercializing the technology straightforward. Even though the technological success of the Radiance Process had been demonstrated in laboratory settings, commercialization proceeded slowly. The lag in adoption is in part owing to reluctance on the part of major semiconductor firms to invest in the development of semiconductor process equipment. In the 1980s, several major semiconductor firms (such as IBM) lost billions of dollars underwriting the development of process equipment.

Other factors also contributed to the slow adoption rate of the Radiance Process. Unlike the innovations commercialized by Zero Discharge, Inc. and Legacy Systems, Inc., the Radiance Process requires expertise in lasers, laminar gas flow, and physics, whereas the traditional semiconductor cleaning process involves wet chemistry acids, and deionized water.

Although the process enjoys a significant quality and cost advantage over the standard semiconductor cleaning process, it is complex from an end user standpoint. In fact, scientists have yet to fully understand the mechanism by which the process works. Lastly, although samples can easily be processed in a lab setting, the process's negative impacts on other parts of the chip manufacturing process may be difficult to measure. Thus, most semiconductor manufacturers are reluctant to adopt such a process until it has been extensively tested. Table 4 outlines the five factors affecting the technology's adoption rate.

Radiance's choice of licensing strategy would also be expected to slow the adoption rate. Radiance is neither an equipment manufacturer nor a semiconductor manufacturer. Although Radiance has contracted with equipment manufacturers, its use of a royalty arrangement whereby end users pay royalties is foreign to most engineers who are the technology's adopter community. The practice is common, however, and is familiar to corporate executives.

Table 4. Radiance's technology adoption characteristics

Characteristic	Radiance Process comparison with standard semiconductor process
Relative advantage	Quality and cost advantages over standard process
Compatibility	RP not compatible with expertise of adopters as it requires expertise in physics, not wet chemistry
Complexity	Complex from an end-user standpoint—how the process works is little understood
Trialability	Low—although samples can easily be processed in a lab, the process may be difficult to measure and even have unanticipated impacts on other parts of the chip-manufacturing process
Observability	Expected to be limited because semiconductor firms tend to be reluctant to publicize test data

Conclusions and Further Research

Our review of the innovation and the technology adoption process shows diffusion of cleaner production innovations is rather complicated. First, innovations in cleaner production technologies can arise from different functional actors. The acid recovery and electroless nickel systems were developed by a technology user; the chilled ozone process was developed by an equipment supplier; and the Radiance Process arose from a university laboratory. Second, cleaner production innovations face competition from well-entrenched end-of-pipe technology. End-of-pipe technology tends to be mature and therefore far down on its learning curve. And unlike many cleaner production innovations, these conventional technologies tend to fit the existing competencies, culture, and organizational systems of potential adopters.

Other factors contribute to the initial disadvantage faced by cleaner production innovations. Few firms want to be among the first adopters. Only the most trusting firms take innovator-derived test data at face value. Many potential adopters prefer to take their technology signals from competitors—figuring that if their competitor has adopted the technology, it must have proven itself. Ironically, many early adopters (particularly in high-tech industry) trying to protect confidential business information are unwilling to allow their names to be publicized.

Certain characteristics of the innovation—relative advantage, compatibility, complexity, trialability, and observability are thought to affect the adoption rate of an innovation. As we saw in the Radiance case study, even in cases where relative advantage is great, adoption may proceed slowly owing to other factors. For Legacy Systems, the lack of observability has served to retard rapid adoption of the technology.

Although environmental regulation obviously plays a role in the technology innovation and diffusion process (chiefly by affecting the relative advantage of one technology over another), many of the factors that speed or inhibit adoption are exogenous to such regulation. Even though no innovator can entirely sidestep many of these difficult diffusion issues, it seems that there may be an important role for the public sector in lowering some of the barriers inhibiting cleaner production innovators. Opportunities lie in four specific areas:

Support innovation research and development. The case studies suggest that new technology concepts can spring from users, suppliers, or research labs. University- or national agency-based labs can incubate innovations that can be spun off to private entrepreneurs (through licensing agreements) or form partnerships with private firms to conduct joint research programs.

Promote demonstration sites. One of an innovator's main challenges is to find a partner willing to adopt first. In two of the cases, national laboratory interest was an important aid. Tax credits and other financial and nonfinancial incentives could be provided to firms willing to take the risk of testing out new technology. Innovators could use assistance in setting up beta test sites to evaluate the equipment fully and willing to make public the test and related data.

Make demonstration tests results public. Potential clean production technology adopters lack information regarding an innovation's performance—for example, information regarding the technology's operating costs, technical capability, reliability, throughput, and the like. Government assistance could be useful in generating test data and disseminating that data through conferences, seminars, reports, newsletters, or Internet communications.

Assist with licensing agreements. Licensing can link innovators to those with commercialization skills and resources. All three firms in this study signed licensing agreements of one kind or another. Innovators could use assistance in finding licensing partners and in attracting agreements that promote the environmental benefits.

Government regulations can also encourage cleaner production innovations. Historically, environmental regulations spurred the development of end-of-pipe pollution-control technologies. Carefully redrafting these regulations could tilt the advantage to new process technologies and development of safer materials.

Our study was meant to shed light on the cleaner production innovation process and the earliest stages of the diffusion process. In our research, however, we were unable to talk directly to adopting firms to find out about their technology-adoption process. Such interviews would shed greater light on the important communication channels and decision-making processes that surround cleaner production technology choices.

To our knowledge, relatively little study has been performed on licensing agreements associated with cleaner production innovations. More research is needed on the types of licensing agreements likely to spur rapid technology diffusion. Further work should be done on developing government roles for providing guidance to innovators—for many of whom the licensing process is uncharted territory.

Finally, our study focuses on the earliest stages of technology adoption. The communication and diffusion processes in firms that are among the last to adopt cleaner production innovations are also of interest. What characteristics of these firms tend to put them among the last in an adopter community to embrace a particular technology? Can the speed of technology diffusion be increased by focusing on midterm adopters?

Such questions and others concerning the characteristics of adopting firms offer a rich array of potential research opportunities that should be further explored if we are to understand more fully and evaluate more acutely the potential for innovation and adoption of more sustainable technologies.

NOTES

1. Cleaner production can be achieved by simple changes in management procedures, by the redesign of products, by changes in operations and maintenance, or by alterations to the production process. Often the changes in production processes require changes in the production technologies—the equipment and materials used to manufacture products. These changes can involve process innovation and the adoption of new equipment and new production practices. (See United Nations 1994.)

2. "Radiance Process" is a registered trademark of Radiance Services Company.

REFERENCES

Abernathy, William J., and James M. Utterback. 1978. Patterns of industrial innovations. *Technology Review* 80 (7): 40–47.

Ashford, N. 1993. Understanding technological responses of industrial firms to environmental problems: Implications for government policy. In *Environmental strategies for industry: International perspectives on research needs and policy implications,* ed. K. Fischer and J. Schot. Washington, D.C.: Island Press, 1993.

Bierma, T. J., and F. L. Waterstraat. 1995. Marketing pollution prevention. *Pollution Prevention Review* 5 (2): 63–72.

Jackson, Tim, ed. 1993. *Clean production strategies: Developing preventive environmental management in the industrial economy.* Boca Raton, Fla.: Lewis Publishers.

Lindsey, T. 1998. Diffusion of P2 innovations. *Pollution Prevention Review* 8 (1): 1–14.

Matalan, R. 1996. Technological innovation for sustainable development: Generation and diffusion of industrial cleaner technologies. *Nota Di Lavoro* (November): 66–96.

Rogers, E. M. 1995. *Diffusion of innovation,* 4th ed. New York: Free Press.

United Nations Environment Programme. 1994. *Government strategies and policies for cleaner production.* Paris: United Nations.

Utterback, James M. 1996. *Mastering the dynamics of innovation.* Boston: Harvard Business School Press.

von Hippel, E. 1988. *The sources of innovation.* New York: Oxford University Press.

Indicators of Sustainable Production

Issues in Measuring and Promoting Progress

Vesela Veleva and Cathy Crumbley

HUMANITY is at a critical point. Environmental, economic, and social problems are spilling over political borders to produce depleted farmlands, fisheries, and forests, growing income disparities, global warming, choking urban pollution, infectious diseases, and refugees. By continuing this model of development, we are destroying the ability of future generations to meet their needs. Thus, the challenge of the twenty-first century is to develop sustainable societies: economic and social systems that the planet is capable of supporting indefinitely. As a crucial component of a more sustainable world, global production systems must begin to operate more in concert with society and the environment. It has been argued that corporations have "the resources, the technology, the global reach and ultimately, the motivation to achieve sustainability." Moreover, sustainable development is believed to constitute "one of the biggest opportunities in the history of commerce" (S. Hart 1997, 67). Yet this "opportunity" means that businesses need to make enormous changes that will transform their practices into more sustainable ones.

We believe that indicators of sustainable production can be helpful in developing the needed new paradigms and systems of production. Sustainability indicators should go far beyond measuring environmental impacts such as materials and energy used or toxics generated. Sustainability implies a much broader set of concerns, including equity within and among generations. There is a saying that "you are what you measure." The act of deciding what to measure and what to do with the measures or indicators causes us to think more deliberately about our activities and help communicate that information to others. They are thus a tool to help establish accountability.

Using indicators as a way to incorporate sustainable development into business decision making and strategies is gaining momentum worldwide. International bodies, such as the United Nations, the World Bank, and the Organisation for Economic Cooperation and Development (OECD), are promoting the use of sustainable development indicators at the national level. At the same time, governments, customers, and citizens are demand-

242

ing that environmental and social criteria be integrated into corporate decisions about products and materials, in addition to traditional economic criteria. Companies have started to search for tools to benchmark their long-term sustainability strategies. Various initiatives have been taken to promote environmental and other aspects of sustainable company management.[1]

Even as interest in sustainability indicators is growing, they are still poorly understood, and much research remains to be done. Some current issues in the development of different types of indicators of sustainability are presented here, with a focus on their use at the company level. First, the concept of indicators will be outlined, including their desirable qualities, scope, and interpretation. Next, an overview of environmental and sustainability indicators found in the literature focuses on company-level benchmarking and, in particular, the new International Organization for Standardization (ISO) indicators 14000. These indicators are further evaluated for their applicability at the level of an individual firm for guiding sustainable production practices. We then summarize difficulties with the present sets of environmental and sustainability indicators and provide recommendations for future research.

Sustainable Production

The concept of "sustainable production" follows from the notion of "sustainable development." The 1987 report of the World Commission on Environment and Development (WCED) defined sustainable development as development that "meets the needs of the present without compromising the ability of future generations to meet their own needs" (WCED 1987, 51). Sustainability is an all-encompassing concept. But at the same time all sustainability is local, according to McDonough (1998), since it aims to involve everyone and is a continuous process of improvement that never ends.

The challenge to promote sustainability in productive enterprises has led to the establishment of a new center at the University of Massachusetts Lowell. The Lowell Center for Sustainable Production (LCSP) was started in 1996 to promote new forms of industrial production that are safe, healthy, environmentally sound, socially accountable, and economically viable over the long term. The center is composed of a core of faculty and staff from several academic departments and research centers that work directly with industrial firms, nonprofit organizations, and government agencies to promote sustainable production.

LCSP defines sustainable production as "the creation of goods and services using processes and systems that are non-polluting; conserving of energy and natural resources; economically viable; safe and healthful for employees, communities, and consumers; and socially and creatively rewarding for all working people" (LCSP 1998, 2). This definition emphasizes not only environmental considerations but also the quality of life for workers and community as key elements of any production system. If firms are to develop sustainability, their operations should be organized "to enhance the environment, and benefit employees and communities, while at the same time leading, always in the long term and often in the short term, to more economically viable and productive enterprises" (LCSP 1998, 2).

With this unified concept, sustainable production aims at meeting objectives that have often been promoted independently, even antagonistically. For example, in order to protect the environment a company may use emission filters or waste recycling. Both practices, however, increase the risk of worker exposure owing to increased maintenance and handling of hazardous substances. On the other hand, in order to reduce worker exposure many facilities use local exhaust ventilation, which transfers air pollutants from the occupational setting to the external environment. Sustainable production assumes that worker health and safety as well as environmental protection are equally important elements of sustainable production.

Indicators of Sustainable Production

Sustainable production indicators help companies measure their progress toward sustainability goals. Indicators are different from primary data or statistics because they provide a meaning beyond the attributes directly associated with them. In this sense, they are a bridge between detailed data and interpreted information.

Indicators, both quantitative and qualitative, enable efficient monitoring and encourage ongoing revision and evaluation. They are signposts that can point the way to sustainable development. They are useful because they show trends and relationships in a concise way. They provide companies with key information about milestones achieved or about failures that frustrate progress.

Most efforts (both at the national and company level) until now have focused on the development of environmental indicators. Examples include energy use, air emissions, amount of water consumed, tons of hazardous waste discharged, and the like. Environmental indicators can be considered the first step in the development of sustainability indicators. The next steps will need to involve development of social and economic indicators to explore and evaluate these two crucial dimensions of sustainability.[2] It is clear, however, that these indicators by themselves cannot promote sustainability. They are only a tool for measuring results. Aggressive national and international policies as well as strong top-management support are among the factors moving companies toward sustainable production.

It is crucial that firms become more sustainable, and sustainable production indicators can be used to guide such a change. Indicators have important roles both internal and external to a company. Within a company, they can be used to monitor company performance. Externally, investors, consumers, other companies within the same industrial sector, and the broader public can use them to assess company performance. Although many firms perceive that it is not beneficial to use such indicators, others will choose to do so for reasons including the following:

Competitive Advantage

With globalization of the economy and increasing competition, it is becoming more difficult for many companies to maintain their market shares and stay in business. At the same time, as a result of increasing environmental awareness, a new "green" market has been developing worldwide. Firms that pursue sustainability as a serious source of competitive advantage may increase their chances of achieving long-term economic viability.

Financial Savings

There is evidence that improvements in environmental protection and occupational health and safety bring about significant savings for the companies. More efficient resource use, reduced medical insurance costs, or decreased liability exposure contribute to such savings. Indicators of sustainable production can help management identify such opportunities.[3]

Increasing Private-Sector Pressure for Voluntary Programs

It has been widely acknowledged that the "command-and-control" approach to environmental and occupational health and safety problems is becoming less efficient. Although a regulatory framework is still necessary, regulations can be expensive and often (especially for developing countries and countries in transition) difficult to enforce. Worldwide, there is corporate movement toward voluntary initiatives for environmentally responsible and even sustainable corporate management. ISO 14000, CERES, Responsible Care, The Natural Step, and the World Business Council on Sustainable Development are examples of this movement. Such organizations recognize that management programs reflecting a long-term vision of sustainability are an important tool for guiding business practices. Companies that join such initiatives may require that their business partners meet certain environmental standards, thus providing incentives for additional companies to join these efforts. In addition, some investors are taking an increasing interest in tracking and comparing environmental and even sustainability performance among companies. At the same time, it must be acknowledged that many of these changes are the result of stricter environmental policies, such as packaging and recycling regulations.[4]

Developing sustainable production indicators is not a simple, single-stage process. We suggest that it include the following steps:

Establish sustainability principles, policies, and/or goals.

Construct a set of indicators to reflect the above principles, policies, and/or goals. The set might consist of a core subset of indicators that is standard for all companies and a production-specific subset for the particular firm.

Implement the indicators ("feed" them with facility/company data).

Interpret and evaluate the indicators in accordance with sustainability policies, principles, and goals.

Amend and refine the indicators and/or sustainability policies and goals.

Constructing an empirical measure for a sustainable production indicator is only half the battle—the other half is interpreting and evaluating the findings. The latter may not be a simple task (Veleva 1994). Most companies have environmental, health and safety (H&S) policies and goals in place that can assist indicator evaluation. For example, if a company's goal is to reduce work-related accidents by 30 percent within a three-year period, a possible sustainable production indicator could be "the number of work-related accidents per year." At the end of each year, the number of accidents is compared to the baseline year and the percentage of change compared to the goal.[5]

In some instances, companies may not have such goals to use when evaluating sustainable production indicators. Here, the use of national and international goals and targets (e.g., CO_2 emission reduction, chlorofluorocarbons (CFCs) phaseout, and biodiversity enhancement) is highly recommended. Such sustainable-development targets help avoid subjectivity and the omission of key aspects of sustainability. For example, the use of an indicator, such as "tons of hazardous waste per unit of production" can be evaluated in light of the "zero waste" sustainability goal. A reduction in the hazardous waste would be considered as moving the company toward sustainability.

Before beginning to develop sustainable production indicators, it is important to identify their desirable qualities. Appropriateness to the task of assessing sustainable production is the first condition for a useful indicator. For example, an indicator such as "average annual background level of SO_2," is not appropriate for use as a sustainable production indicator, because it does not simply relate resulting environmental pollution to the particular firm's environmental performance. There are probably other sources of SO_2 emissions in the area. A better indicator in this case would be "SO_2 facility emissions per year," since it directly links environmental pollution to the firm's performance.

As much as possible, indicators should help to pinpoint the root problems and thereby create solutions. For example, if the indicators demonstrate an increase in worker health problems, regardless of the excellent environmental performance of a firm, they should also help to determine the underlying reasons for such a trend.

In general, it is better to have a set of indicators rather than a single indicator. A single measure will miss valuable information and lead to less reliable results. Indicators should address all aspects of sustainable production—environmental, social, economic, and occupational. If one or more of these fundamental aspects is missing, then the set is incomplete.

A major challenge is to balance the simplicity and the meaningfulness of indicators. Two forces pull in opposite directions: first, if an indicator is to be meaningful, it may need to be relatively complex and information demanding; but, then, if it is to be operational, it needs to be simple, easily understood, and based on relatively accessible data (Veleva 1994).

The set of indicators should consist of a manageable number. Too many indicators will make it difficult to use the method, yet too few indicators may miss important information. Furthermore, the indicators should be relatively easy to apply and evaluate; otherwise the complexity will deter companies from using them.

It is desirable to have measurable or quantifiable indicators, since these avoid subjectivity, yet this may not always be possible. Community quality of life, corporate image, or worker job satisfaction may not be directly quantifiable. Therefore, both quantitative and qualitative indicators should be used to encompass all issues deemed important.

Sustainable production indicators enable companies to benchmark their progress and better focus resources in pursuit of sustainability. It is also crucial to be able to compare performance across companies. The worldwide movement for standardization and comparability in the field of environmental indicators is gaining momentum, underlining the need for standardized indicators. Because it is impossible to have a single, standard set of sustainable production indicators applicable across all industrial sectors, the ideal system would consist of a core set of sustainable production indicators with each firm encouraged to create its additional production-specific indicators.

Achieving local sustainability is only possible if it is related to global sustainability goals and targets. Therefore, consistency is needed across national sustainable development indicators and sustainable production indicators.

Another important issue to consider is the scope of the indicators—that is, the main areas of production to be addressed. Building on the definition of sustainable production, the important areas are discussed below.

Conserving energy and material use is a central topic of sustainability. This can be achieved through improved process efficiency, housekeeping, reuse, and recycling. Developed indicators should reflect not only the quantity of materials and energy used but also their type (renewable vs. nonrenewable materials and energy sources, toxic vs. nontoxic chemicals). Furthermore, they should reflect the priorities in conserving energy and materials. For example, improved process efficiency is better than material reuse but reuse is better than recycling. In regard to transportation and the use of fossil fuels, indicators should attempt to measure reorienting technology away from high-energy usage or from individual transportation to mass transport.

Another component of sustainability, closely related to the quality of life, is the protection of the natural environment. This includes air and water quality, biodiversity, soil fertility, waste production and disposal, and the like. Indicators of sustainable production should reflect each of those issues, in particular the ones targeted under international agreements (global climate change, biodiversity, ozone depletion, etc.).

Maintaining the economic viability of a production system is a key element of sustainability, since it provides jobs and resources for community development. It is clear that long-term viability cannot be assessed by standard financial terms only, such as profit, stock price, and market share. Alternative measures of economic viability have been discussed at the global and national levels, but there is little understanding of how these translate at the level of the firm. Much more research needs to be done in this area, but potential indicators might reflect such factors as investment in research and development, innovation potential, corporate image, environmental liabilities, and environmental and social investment.

Community development and social justice are key elements of sustainability, but they have rarely been addressed in a consistent way beyond sustainable community indicators. Some companies now are going "beyond compliance" and engaging in "public–private partnerships." Examples include Marriott International, BankBoston, and Bell Atlantic (Kanter 1999). Key aspects of such partnerships include company openness, job creation, quality of life, community spending, and the involvement of citizens and workers in partnerships and decision making. Indicators, both quantitative and qualitative, need to be developed that address these issues and assist companies in measuring their attempts to promote community development and social justice. Recent surveys demonstrate that issues related to community and social equity receive the least attention in company benchmarking. Rather, there is a strong emphasis on environmental protection and much less on social equity (Krut and Munis 1998). This is likely to reflect the strength of the environmental movement in the last three decades and the related strengthening of environmental policies and legislation. This, however, is not the case for other social movements yet. As Kanter (1999, 123) points out, "Traditionally business viewed the social sector as a dumping ground for spare cash, obsolete equipment, and tired executives." Until this paradigm is changed, significant improvement in the social activities of companies cannot be expected. Indicators of sustainable production can only assist firms in measuring their achievements in the social arena, but cannot alone promote such a change.

Workers are a key element of any production system. Many sets of sustainability indicators or benchmarks, however, have ignored worker issues. For example, the use of toxic chemicals and the production of large amounts of waste and emissions affect both the workers and the environment. Yet in many cases the focus is on protecting the local environment and population and not the workers in the plant who are exposed to significantly higher levels of toxic chemicals and other harmful effects. To avoid shifting the burden of pollution from the external environment to the workers, an appropriate subset of indicators would reflect issues such as total lost workdays per year, number of work-related injuries and diseases, levels of hazardous exposure, training, job satisfaction, or worker participation. The difficulties of measuring worker health and safety should not be overlooked. Underreporting is a serious concern. Even such a drastic workplace injury as death has been greatly underreported (Dijkhuis et al. 1994). Yet, no set of such indicators can persuade a firm to reconfigure the shop floor or office unless there is strong management support for such a change or some form of national or local regulation or policy.

Products are a key element to be addressed in a set of sustainable production indicators, since they are linked to the environmental, social, and economic aspects of a firm. Yet, because of the diversity of products, this is often the most difficult element to address. In addition, most products become waste and thus pose significant postproduction problems for the environment. Moreover, some products, such as the automobile, impose higher stress on the environment during use rather than during production (Graedel and Allenby 1998). Integrated indicators of sustainable production can help companies better focus and utilize the various planning tools when designing and managing their products. They can be applied to a single product, a set of products, or all manufactured products at a particular facility/company.

Types of Sustainability Indicators in the Literature

The following sections explore the development and use of sustainability indicators at the three primary levels of application—national, local, and company level. The indicators correspond to these levels:

Sustainable development indicators—assessing national performance;
Sustainable community indicators—evaluating community performance; and
Company-level indicators—measuring an individual firm's performance.

A framework has been developed that may be useful at all three levels of application. The "pressure–state–response" (PSR) model was developed by the Organisation for Economic Cooperation and Development (OECD). This model "considers that human activities exert pressures on the environment and affect its quality and the quality of natural resources (state); society responds to these changes through environmental, general economic, and sector policies and through changes in awareness and behaviors ["societal response"]. The PSR model has the advantage of highlighting these links, environmental and other issues as interconnected" (OECD 1998b, 109).

Some examples of OECD indicators, developed within the PSR framework, are presented in table 1. Since this model was developed and used to evaluate environmental aspects of sustainability, it is not clear whether it is applicable for evaluation of the social and economic dimensions of sustainability. More research is needed in this area, which is beyond the scope of this study.

National Sustainable Development Indicators

Various governmental and nongovernmental organizations, such as OECD, the United Nations, and Redefining Progress[6], are developing national sustainable development indicators. These indicators attempt to cover key issues of global significance and are typically intended for use at the national level. They usually allow for comparisons across time and countries and aim at evaluating the implementation of national development plans.

In 1991, OECD began one of the earliest attempts to develop national environmental indicators. It developed a core set of eighteen environmental indicators and seven key indicators that reflect economic and population changes. It included indicators of environmental performance, some relating to environmental quality itself (e.g., river quality, protection of nature), some to national environmental goals (sustainable use of water resources, controlling generation of waste), and some to international environmental agreements and issues (e.g., SOx emissions, CO_2 emissions, and trade in forest production) (Ditz and Ranganathan 1997). More recently, OECD has developed a core set of indicators for use in its reviews of national environmental performance. The aim of the reviews was to "assess a country's efforts to reduce the overall pollution burden and manage natural resources, integrate environmental and economic policies, and strengthen cooperation within the international community" (DPIENA 1998, 9). The indicators in table 1 were developed as part of this effort.

OECD has undertaken a three-year project on indicators of sustainable development (1998–2001). Building on existing work, the project is developing a limited and manageable set of indicators of sustainable develop-

Table 1. Examples of OECD indicators within the PSR categories

	Pressure	State	Response
Issue	Indicators of environmental pressures	Indicators of environmental conditions	Indicators of societal responses
Climate change	Emissions of CO_2	Atmospheric Global mean of CO_2	Energy intensity (energy use per unit of product)
Eutrophication	Apparent consumption of N, P in fertilizers	Biological Oxygen Demand, Dissolved Oxygen, nitrogen, and phosphorous in selected rivers	% of population connected to wastewater treatment plants
Toxic contamination	Generation of hazardous waste	Concentration of lead, cadmium, chromium, and copper in selected rivers	Market share of unleaded petrol
Biological diversity	Land-use changes	Threatened or extinct species as % of known species	Protected areas as % of total area

Source: Department of Primary Industries and Energy Network Australia (DPIENA). (1998). A survey of work on sustainability indicators.

ment. OECD is promoting the idea that "rather than a single, multi-purpose list, progress can best be achieved by establishing a pyramid of indicator sets." At the top level are key indicators that are less technical, more intuitive, and provide "a practical and concrete insight into the relationships between economic, social and environmental aspects of development" (OECDa 1998, 31). Specific examples of this type of indicator are under development. At the lower levels are found indicators with more specific functions and greater technical specificity (OECDa 1998).

The United Nations is also conducting research on sustainable development indicators. At its third session in 1995, the Commission on Sustainable Development (CSD)[7] approved a work program on indicators of sustainable development that became available to national decision makers in 2000. This project developed a framework and a core set of indicators. A working list of 134 indicators of sustainable development has been prepared. These indicators include social, economic, environmental, and institutional aspects of sustainable development and are placed within a "Driving Force–State–Response" (DSR) framework (UN 1998). The DSR framework is similar to the "pressure–state–response" framework. The term "pressure" has been replaced by "driving force" in order to expand the scope of the indicators and to accommodate social, economic, and institutional indicators. In addition, the use of "driving force" suggests that the impact on sustainable development might be positive or negative, as is often the case for social, economic, and institutional indicators. There are other attempts to develop sustainable development indicators.[8]

Although many national indicators may not be directly applicable to individual firms, it is not possible to have local sustainability without global sustainability and vice versa. Thus, sustainable production indicators should aim for consistency with national sustainability indicators and goals.

Sustainable Community Indicators

Local communities are using indicators increasingly to evaluate their development. These local communities, however, tend to use a different approach from the national/international indicators approach. When applying the PSR model for environmental indicators of sustainability, communities have been more interested in the "state" of the environment, because such indicators reflect their quality of life. Examples of sustainable community indicators are "days with good air quality," "toxic chemicals released or transferred," "song bird diversity," and "number of frogs in the local ponds" (M. Hart 1998; Sustainable Seattle 1998). Typically, community sets of indicators number between twenty and two hundred fifty. Almost every set includes environmental condition indicators, which are similar to the national indicators. One classification of such indicators covers economic, environmental, social, land use, and transportation issues (Ditz and Ranganathan 1997).

As mentioned above, sustainable community indicators mainly reflect effects. Individual companies, however, are concerned with inputs, processes, resources, and outputs and would often find it difficult to correlate their activities with the environmental state of the local community. Thus, "there is a huge gap between the meaning of environmental improvement for business and communities" (Fehsenfeld 1997). It seems that corporations and communities in their attempts to cover similar issues speak two different languages. This could become a problem for companies trying to be more open about their production activities. Indicators of sustainable production should try to overcome this gap by making linkages, where possible, between environmental pressures and environmental effects as well as between economic development and social justice.

Company level indicators/benchmarks

Industry groups, investors, governments, and environmental organizations have all pushed for greater corporate accountability and public reporting of environmental practices. In response, many firms have produced corporate environmental reports (CERs). Although the quality and content of such reports varies greatly, they are an important step in opening up environmental and management practices to greater public scrutiny. John Elkington, the chairman of SustainAbility,[9] has noted that "companies are now publishing data in ways which they actively argued would be commercially suicidal as recently as the early 1990s" (Russel 1998).

In their environmental reports, companies often use environmental performance indicators (EPIs). ISO 14000 (1997) standards distinguish three types of EPIs. These correspond to the "pressure-state-response" (PSR) model, developed by the OECD to assess the environmental performance of countries:

Operational indicators measure potential stresses on the environment, such as burning fossil fuels or converting forest resources to a paper mill ("pressure").

Environmental condition indicators measure environmental quality, such as ambient air pollution concentrations or global climate change ("state").

Management indicators measure efforts to reduce or mitigate environmental effects, such as company spending on energy efficiency or environmental training ("response").

All three types of indicators can be useful to business decision makers in their pursuit of more sustainable production practices. These three, however, cover only environmental protection, one of the three dimensions of sustainability. Three examples of research related to indicators/benchmarks at the company level are presented below.

Coalition for Environmentally Responsible Economies (CERES)

A unique partnership among some of this country's largest institutional investors as well as religious and environmental organizations formed CERES in 1989. Its goal is to promote the transformation of environmental management within businesses through adherence to a core set of principles. In its dozen years of existence, it has emerged as a worldwide leader in standardized corporate environmental reporting. At the heart of CERES are its ten principles listed below. Companies that endorse these principles commit to continuous improvement, stakeholder involvement and public reporting (CERES 1998).

Protection of the biosphere	Safe products and services
Sustainable use of natural resources	Environmental restoration
Reduction and disposal of wastes	Informing the public
Energy conservation	Management commitment
Risk reduction	Audits and reports

To promote standardized corporate environmental reporting, CERES has created a list of almost a hundred qualitative and quantitative questions in ten areas (corresponding to these ten principles). All endorsers must answer these questions, and they are provided with guidance, assistance, and feedback on their draft reports (White and Zinkl 1998).

As one of the first organizations to promote the importance of standardized data and reporting, in 1997 CERES undertook the additional step of promoting the Global Reporting Initiative (GRI). The ambitious vision for this initiative is to "bring together the numerous initiatives on corporate environmental reporting that have developed independently around the world, and to help shape them into one set of coherent, consistent global standards" (CERES 1998). The GRI is intended to improve corporate accountability worldwide by ensuring that all stakeholders—communities, environmentalists, labor, religious groups, shareholders, and investment managers—have access to standardized, comparable, and consistent environmental information similar to existing corporate financial reporting. Only in this way, CERES argues, can capital markets be used to promote sustainable business practices effectively. Furthermore, the GRI will assist corporate evaluation efforts and help measure adherence to the standards set forth in the CERES principles. It is expected that the results will include the expanded credibility of corporate environmental reports, simplification of the reporting process, quick and reliable benchmarking, and risk-relevant environmental information for investment managers (CERES 1998).

The CERES environmental reporting requirements and the GRI are important efforts to advance corporate movement toward sustainability. The ten principles cover core issues of sustainable production, environment, natural resources, products, communities (but not economic viability). As an important strategy for advancing sustainability, CERES offers lessons about the applicability of its reporting criteria for creating sustainable production indicator programs.[10] A review of CERES reporting requirements reveals the following attributes and shortcomings.

CERES reporting criteria are not indicators. They are, in fact, questions covering specific areas, and usually requiring only a short answer (in many cases, just Yes or No). For example, to evaluate the use of natural resources, CERES has a whole section (section 7) in its standard reporting form, which is five pages long and includes more than thirty questions, such as "Does your company have a formal written policy regarding materials reduction, reuse and recycling (Yes/No)"; "What steps have you taken to minimize the environmental burdens associated with employee transportation for work-related or other purposes?" or, "Give some examples of techniques, practices and procurement methods employees are using to conserve materials" (CERES 1998).

Although the principles and reporting criteria cover important areas, they do not attempt to be as comprehensive as sustainability indicators. There is no attempt to evaluate the economic viability of companies in terms of R&D investment, green market development, environmental liabilities, and the like. The GRI attempts to address some of these shortcomings. It is, however, still in a process of development and pilot testing and therefore will not be discussed here.

The value of the CERES approach lies in its use of principles to guide company sustainability-performance evaluation. Development of a manageable set of sustainable production indicators consistent with their principles could enable better measurement of business success or failure in sustainability than the currently used reporting requirements.

UNEP/Sustainability Benchmarks

In collaboration with the United Nations Environmental Programme (UNEP), SustainAbility has developed an "Engaging Stakeholders" program that involves a ranking of corporate environmental reports (CERs). The project reflects what some of the leading companies are willing to report voluntarily about their practices. The evaluators used fifty ranking criteria that are grouped in six general areas:

Management policies and systems	Sustainable development
Stakeholder relations and partnership	Finance
Inputs and outputs	Report design and accessibility

The specific elements within each area are either quantitative or qualitative.[11] The survey of a hundred corporate environmental reports from sixteen industry sectors and eighteen countries found that, of the five reporting elements, the lowest scores occurred in "Sustainable development" and "Finance."

The usefulness of the UNEP/SustainAbility project lies in its demonstrating a practical way to evaluate company moves towards sustainability. Company performance can be evaluated on a yearly basis and compared to the maximum score as well as to previous scores, so that conclusions can be drawn about priorities for improvement.

Obviously, not all the fifty ranking criteria employed in the study can be used as sustainable production indicators because some are too broad. For example, an important criterion of the SustainAbility set is "health and safety"—an often-omitted component of sustainability assessment. This criterion needs additional refinement, and possibly a whole subset of indicators, to reflect the complex nature of worker health and safety. Criteria, such as material use, energy consumption, water consumption, accidents and emergency response, air emissions, and packaging, need to be further refined to reflect particular aspects of sustainable production. For instance, when evaluating material use, it is important to distinguish between toxic and nontoxic materials. Similarly, energy consumption should be divided into energy from renewable and energy from nonrenewable sources.

The SustainAbility approach is subjective, so valuable information may be lost when data are aggregated. The project researchers emphasize the need for greater benchmarking of performance measures within the same industrial sectors. Another crucial issue is the absence of information in many of these reports about the environmental aspects of core business practices, not just pollution or energy inefficiency.

The analysis also showed that third-party verification is an increasing practice in corporate reporting and can play a significant role in ensuring quality. The first survey in 1993–94 showed that just four reports had some type of verification. This grew to twenty-eight in the 1997 survey, including five of the top ten. However, the quality and scope of the verifications varies. A third-party auditor may confirm that the report is factually correct, or the verification may take the form of deeper analysis, including areas for improvement. The lessons from such uses of verification may be helpful in the development and use of sustainability indicators.[12]

ISO 14000

The ISO 14000 series emerged as a result of both the Uruguay Round of the General Agreement on Tariffs and Trade (GATT) negotiations and the 1992 Rio conference on environment and development. While GATT concentrated on the need to reduce nontariff barriers to trade, the Rio summit generated a commitment to global protection of the environment. Developed by the International Organisation for Standardization (ISO) in Geneva, the ISO 14000 series of standards addresses various aspects of environmental management. The very first two standards, ISO 14001 and ISO 14004, were published in 1996 and deal with environmental management systems (EMSs). These are management tools that enable firms to control the impacts of their activities, products, or services on the environment. According to ISO (1997, 5), "an environmental management system represents a structured approach to setting environmental objectives and targets, to achieving these and demonstrating that they have been achieved."

ISO 14001 specifies the requirements for an EMS and is the only standard in the series that requires certification. Fulfilling these requirements demands objective evidence that can be audited to demonstrate that the EMS is operating effectively in conformance with the standard. It is likely that the ISO 14000 series of standards will have a growing impact on corporate behavior. Although initially promoted primarily by large corporations, ISO 14000 may increasingly involve small and medium enterprises as a result of pressures on suppliers and distributors from ISO-certified companies. It is thus a fertile area for research on company indicators.

The draft standard ISO 14031 deals with environmental performance evaluation guidelines. It lists over a hundred environmental performance indicators (EPIs) and assists companies in evaluating their environmental performance against environmental policies, objectives, targets, and other environmental criteria. Intended mainly to support ISO 14001 and ISO 14004, this draft standard introduces environmental performance evaluation as an "ongoing process of collection and assessment of data and information to provide a current evaluation of performance as well as performance trends over times" (ISO 1998, 3).

ISO 14031 introduces the three types of EPIs mentioned earlier:

Operational indicators
Environmental condition indicators, and
Management indicators.

It provides concrete examples of indicators within each of these three categories and demonstrates how to use them in practice. While all three types of EPIs are important, operational indicators are crucial for the corporate-level assessment of sustainability (Ditz and Ranganathan 1997). The reason is that these indicators allow directly to link a facility or company's operations to the resulting environmental impacts. To assist their practical implementation, operational indicators have been divided into nine categories, as presented in table 2 below.

The strengths of ISO 14031 are that it links environmental condition indicators with operational indicators and creates a common framework for companies to use, thus promoting comparisons and integration of data. The use of such indicators will help firms introduce accountability for environmental performance. In many cases, this could lead to improvements in resource use while increasing profitability.

Since they are focused on environmental protection, the EPIs presented in ISO 14031 cover only one aspect of sustainability and thus exclude economic, social, and work environment issues. The standard notes that sustainable development indicators have been developed by governmental agencies, nongovernmental organizations, and scientific and research institutions, and that these might be used in addition to the EPIs. Yet no guidance or examples are provided, making it difficult for companies to develop a comprehensive set of sustainable production indicators using the ISO framework.

Yet ISO 14031 was not designed to encompass the larger issues of sustainable production. For example, ISO 14000 excludes occupational health and safety issues. These are interrelated with environmental issues and a key component of sustainable production. Although the original concept of ISO 14000 assumed that EMSs are intrinsically linked to occupational health and safety (OH&S), somewhere in the negotiations this linkage was compromised (Jackson 1997).[13] As a result, the draft standard ISO 14031 does not provide companies with indicators on worker health and safety.

The entire ISO 14000 series of standards was developed by business

Table 2. Some operational indicators as presented in ISO 14031 (1998)

Category	Examples of operational indicators
Materials use, including quantities and types of materials used	Quantity of materials used per unit of product Quantity of water per unit of product Quantity of hazardous materials used in the production process
Energy use, including quantities and types of energy used or generated	Quantity of energy used per year or per unit of product Quantity of each type of energy used Quantity of energy used per service or customer
Services supporting the organization's operation	Amount of hazardous materials used by contracted service providers (CSP) Amount or type of waste generated by CSP
Physical facilities and equipment	Number of pieces of equipment with parts designed for easy disassembly, recycling and reuse Average fuel consumption of vehicle fleet
Supply and delivery	Number of business trips by mode of transportation Number of business trips saved through other means of communication
Products	Number of products which can be reused or recycled Rate of defective products Duration of product use
Services provided by the organization	Amount of cleaning agent used per square meter (for a cleaning services organization) Quantity of materials used during after-sales servicing of products
Wastes	Quantity of waste per year or per unit of product Quantity of hazardous, recyclable or reusable waste produced per year
Emissions	Quantity of specific emissions per year Quantity of effluent per service or customer

representatives to serve the interests of corporations. The result is a weak start on the promotion of sustainability.[14] The application of EPIs presented in ISO 14031 is, first, insufficient to address the basic aspects of sustainable production, and, second, it will depend heavily on the willingness of firms to promote environmental improvements, without shifting the burdens of their practices either to workers, local communities, or the broader environment.

Lessons and Recommendations for Future Research

This review of issues related to sustainability indicators suggests the following conclusions and recommendations for further development, refinement, and research:

There has been marked progress in the development of sustainability indicators over only a few years, yet they are still in the early phases of development. The theoretical underpinnings continue to be developed, and indicators are still being formulated, tested, and analyzed.

Because of the increasing need for standardization in corporate reporting, it is recommended that a core set of indicators be developed that reflect significant global and local sustainability issues that are common to all companies. In addition, guidelines need to be provided for companies that are willing to develop their own production-specific subset of indicators. This approach will allow for both standardization and flexibility when evaluating company progress towards sustainability. Third-party verification is a growing trend and can help ensure quality in the use of these indicators.

No set of sustainability indicators has been identified that is applicable as a core set to evaluate sustainable production practices at the level of an individual firm. Only a few appropriate indicators have been developed, and these still need further formulation and evaluation.

Often missing in the sets of environmental and sustainability indicators are evaluations of worker health and safety, economic viability, and product sustainability. These are key elements of sustainable production and need to be adequately addressed in national, local, and company policies in order to stimulate increased indicator creation and use (see table 3 for examples of possible indicators).

Important work has been conducted regarding environmental reporting and benchmarking. These activities can be considered the first stage in corporate sustainability reporting. As corporate reports become more standardized, there will be increasing pressure to use standard indicators. At least some corporate environmental reports are broadening to include sustainability issues. More research needs to be done to define economic and social indicators of sustainability at the level of the individual firm.

Lessons from the development and use of various indicators of sustainability have led the authors to prepare the examples of sustainable production indicators in table 3. These are an early attempt—without further investigation and research, it is difficult to develop robust indicators that have many or all of the desirable qualities.

The utility of the operational–environmental condition–management indicator framework for use at the company level, corresponding to the "pressure-state-response" framework at the national level, needs further research. Lessons need to be drawn about how this framework is perceived and utilized at the level of the firm.

Beyond the theoretical discussion presented here, the strength of company-level indicator programs will depend on three factors: the strength of national and local sustainability policies, management support, and the development of specific tools and techniques for such activities as

Table 3. Examples of sustainable production indicators

Categories	Sample indicators
Community development and social justice	Community spending and charitable contributions (UNEP/Sustainability 1998)
	Community quality of life
	Number of people from the local commuinty employed at the facility (M. Hart 1998)
Economic viability	Market share
	Environmental spending (White and Zinkl 1998)
	Environmental liabilities (White and Zinkl 1998)
	R&D investment
Workers	Total lost workdays per year
	Rate of work-related injuries per year
	Average salary and benefits
	Employee training (M. Hart 1998)
Energy and material use	Energy productivity (Lowe et al. 1997)
	Materials productivity (Lowe et al. 1997)
	Ratio of virgin to recycled materials (Lowe et al. 1997)
External environment	Total water consumption
	Greenhouse gases produced
	Amount of packaging per unit of product output (in pounds)
	Vehicle miles driven for transport of raw materials, products, by-products, waste, and other services (M. Hart 1998)
Products	Percentage of products/parts reused or repaired
	Percentage of recycled products
	Percentage change in product durability

employee participation in an indicators program, creating site-specific goals and indicators, and evaluating and improving the indicators.

Environmental reporting and the development of sustainability indicators are at a crossroads. In one direction lie narrow, internally derived and focused systems such as ISO 14000. These systems provide a basic structure for compliance (to internal standards), but may do little to foster genuine progress toward sustainability. It will be difficult to show genuine progress as long as environmental or sustainability issues are seen as separate from the core business practices of the company. In the other direction lie more outward-looking, visionary approaches, including the tracking and verification of corporate policies and indicators using clear environmental, social, and economic benchmarks.

Related to the preceding point is the larger issue of how sustainability is to be interpreted and incorporated into business (or how business is to be incorporated into sustainability). A vision of sustainability as encompassing fundamental changes in economic, social, and environmental practices requires dramatic changes in the way business is conducted. Our current economic system, however, rewards only the ability to make larger profits. It is often easy to make the business case for ecoefficiency—conserving resources and energy usually saves money—but the sustainability challenge is far greater. Given current conditions, corporations that wish to redirect their practices more in line with sustainability—emphasizing equity, empowerment of workers, and the vitality of communities, for example—may feel that they are risking their economic viability. The appropriate local, national, and international policies that enable or persuade firms to make these changes is a crucial area of study.

The process of indicator creation and interpretation is subjective. Companies tend to measure and report aspects of production that demonstrate their successes rather than their failures. Academia, however, has the advantage of being a disinterested third party that may be able to introduce new approaches to old problems (Kriebel et al. 1997). Here the Lowell Center for Sustainable Production (LCSP) and the University of Massachusetts Lowell can play a key role. Although they cannot initiate changes in management behavior, they can provide valuable expertise. In 1998 the LCSP started a project, working with two local firms, to develop sustainable production indicators programs and to explore and test tools that can assist in this effort. Another related project is the development and testing of curricula and training businesses, community organizations, and governments in the use of indicators of sustainable production.

Conclusion

The use of sustainable production indicators may enable better planning and progress for those companies choosing the path toward sustainability. For other businesses, indicators can be a valuable educational tool, providing information and raising their awareness about sustainable production. For the larger society, indicators provide information about the effects of corporate practices. When this information is made public, indicators can be a valuable tool for increasing the accountability of firms. Much research, including direct experience with businesses needs to be done to increase corporate and community understanding of the creation and use of sustainable production indicators.

NOTES

1. Examples include the Responsible Care Program, the Coalition for Environmentally Responsible Economies (CERES), ISO 14000, and The Natural Step.

2. It can be argued that economic sustainability indicators have been highly

developed in the form of corporate financial reports. The relationship, however, between the short-term financial viability of a firm and more long-term sustainability is far from clear.

3. Acushnet Rubber of New Bedford, Massachusetts, manufacturer of elastomeric products, has doubled its business since 1990 and added two hundred jobs in 1996 alone (the total number of workers in 2000 was 1000).

4. In Europe the catalyst for this change has largely been the adoption of new packaging regulations and new requirements that goods such as appliances and automobiles be recycled after their use instead of being landfilled. These requirements affect not only the European companies but also many foreign companies that export to Europe, in particular the multinational corporations. As a result, the latter began to change their environmental policies and also to require their suppliers and distributors to change. BMW and General Electric are good examples here.

5. The authors recognize that work-related accidents can be an unreliable indicator owing to underreporting.

6. Redefining Progress is a San Francisco-based, nonprofit organization that has done extensive work toward developing an alternative to the GDP and is also conducting research on other indicators of sustainability.

7. The CSD was created as a response to the Rio Earth Summit in 1992.

8. The efforts of the United Nations Development Programme, U.S. Interagency Group on Sustainable Development Indicators, and NASA are among them.

9. SustainAbility is a for-profit, London-based consulting firm that promotes the business case for sustainable development. Its mission is to "help clients develop win-win-win business solutions which are socially responsible, environmentally sound, and economically viable" (White and Zinkl 1998, 13).

10. Unfortunately, CERES currently has only fifty-five endorsers. Why do many companies choose not to participate? Most of them do not understand or accept the environmental and sustainability issues, are lacking in long-term vision, or simply do not wish to make their internal operations public.

11. Points are assigned to each element with a maximum possible score of 194. As a result of this ranking system, the Body Shop's Values Report ranked highest in the last two surveys with 129 and 131 points respectively. The Body Shop International is a UK-Based cosmetics manufacturer, that has been well known for its excellent social and environmental policy.

12. An interesting observation of the SustainAbility ranking is that five of the top ten reports in the 1997 ranking were CERES endorsers. This result, however, has no straightforward explanation. It might be that CERES endorsers are much more environmentally conscious and advanced in their management practices. On the other hand, it might also mean that CERES endorsers have better prepared environmental reports but not necessarily better environmental (or even sustainability) performance.

13. In the final version, ISO 14001 states that "it is not intended to address, and does not include requirements for aspects of OH&S management, although it does not discourage an organization from developing integration of such management system elements" (ISO 1995, 6).

14. For detailed discussion of the weaknesses of ISO 14000, see Gleckman and Krut (1997).

REFERENCES

Coalition for Environmentally Responsible Economies (CERES). 1998. Boston. [http://www.ceres.org/].

Department of Primary Industries and Energy Network Australia (DPIENA). 1998. Barton. Australia. 1998. A survey of work on sustainability indicators. [http://www.dpie.gov.au/].

Dijkhuis, H., C. Zwerling, G. Parrish, T. Bennett, and H. C. Kemper. 1994. Medical examiner data in injury surveillance: A comparison with death certificates. *American Journal of Epidemiology* 139 (6).

Ditz, D. and J. Ranganathan. 1997. Measuring up: Towards a common framework for tracking corporate environmental performance. Washington, D.C.: World Resources Center.

Fehsenfeld, T. 1997. From corporate indicators to community benchmarking—How can we make the transition? Greening of Industry Network Conference. Santa Barbara, Calif. November.

Gleckman, H., and R. Krut. 1997. Neither international, nor a standard. In *ISO 14001 and beyond: environmental management systems in the real world*, ed. C. Sheldon. New York: Greenleaf Publishing.

Graedel, T. E., and B. R. Allenby. 1998. *Industrial ecology and the automobile.* Englewood Cliffs, N.J.: Prentice-Hall.

Hart, M. 1998. Sustainable community indicators. [http://www.sustainable measures.com].

Hart, S. 1997. Beyond greening: Strategies for a sustainable world. *Harvard Business Review* (January–February): 67–76.

ISO (International Organization for Standardization). 1995. Draft International Standard ISO/DIS 14001. Geneva: ISO.

———. 1997. ISO, ISO 9000, ISO 14000. 1997–09-05/4300. Geneva: ISO.

———. 1998. Draft International Standard ISO/DIS 14031. Geneva: ISO.

Jackson, S. 1997. *The ISO 14001 implementation guide: Creating an integrated management system.* New York: John Wiley & Sons.

Kanter, E. M. 1999. From spare change to real change: The social sector as beta site for business innovation. *Harvard Business Review* (May–June): 122–32.

Kriebel, D., M. Quinn, and K. Geiser. 1997. Integrating environment and health into regional economic development. Committee on Industrial Theory and Assessment (CITA) Conference. October.

Krut, R., and K. Munis. 1998. Sustainable industrial development: Benchmarking environmental policies and reports. *Greener Management International* 21 (Spring): 87–98.

Lowe, E. A., J. Warren, and S. Moran. 1997. *Discovering industrial ecology.* Columbus, Ohio: Battelle Press.

Lowell Center for Sustainable Production (LCSP). 1998. Sustainable production: A working definition. Lowell: University of Massachusetts Lowell, LCSP.

McDonough, W. 1998. The region's prosperity greater with sustainable design: A Declaration of Interdependence for the Twin Cities. *Minneapolis Star Tribune.* April 4.

Organization for Economic Cooperation and Development (OECD). 1998a. OECD work on sustainable development: A discussion paper on work to be undertaken over the period 1998–2001. Paris. [http://www.oecd.org/subject/sustdev/oecdwork.htm].

———. 1998b. *Towards sustainable development: Environmental indicators.* Paris: OECD.

Russel, T. 1998. World review—Business and the environment. *Greener Environmental Management* (Spring): 6–13.

Sustainable Seattle. 1998. Home page. [http://www.scn.org/ip/sustainable/index.htm].

United Nations. 1998. Indicators of sustainable development (ISC): Progress from theory to practice. New York. [http://www.un.org/esa/susdev/isd.htm].

United Nations Environmental Programme (UNEP)/SustainAbility. 1998. London. [http://www.sustainability.co.uk/].

Veleva, V. 1994. Development of environmental sustainability indicators for use in environmental impact assessment. M.Sc. thesis, University of Manchester, U.K.

White, A. and D. Zinkl. 1998. Green metrics: A global status report on standardized corporate environmental reporting. Tellus Institute. Boston. CERES Annual Conference, Boston. April 15–16.

World Commission on Environment and Development (WCED). 1987. *Our common future.* Oxford: Oxford University Press.

III RETHINKING SUSTAINABLE DEVELOPMENT

Technology, Business,
and the University

Introduction

Jean L. Pyle

THE chapters in this part explore the ways a university can promote sustainable regional growth by fostering relationships with firms that further the development and implementation of innovative technologies. This is a challenging endeavor that can involve fundamental rethinking of what many have assumed were the most effective ways to accomplish such goals. The authors share the results of the debates and struggles regarding effective implementation of their ideas that faculty and centers at UMass Lowell have engaged in during the past few years. These range from consideration of which forms of the innovation process and organizational collaborations are most effective in promoting sustainable growth in today's fast-changing economy, to reconceptualization of how the production process can be redesigned to be sustainable, to reflection on how we can establish viable university–industry partnerships that further mutual goals. In developing university–industry partnerships, there are often problems or tensions that have to be addressed when institutional goals differ. For example, the university typically has a longer time frame for results than firms; the university may be more focused on activities of wider social benefit than firms (which focus on products and services for which users can pay); and the university is nonprofit, while firms are market oriented and concerned with short-term profitability.

The authors do two things that others may usefully consider. First, they provide new perspectives on technological innovation and how a university can further an innovation process (and thereby promote

265

sustainable regional development) via its relationships with businesses. Second, they ground these perspectives by presenting examples of how faculty and centers at UMass Lowell have implemented them in the region. They discuss the difficult issues they have encountered, how they have worked through them, and what they have learned.

In "Innovation, the University of Massachusetts Lowell, and the Sustainable Regional Development Process," Michael Best and Robert Forrant shed light on the kinds of interactions between firms and educational institutions regarding enterprise innovation that are most likely to foster sustainable regional development in today's fast-changing technological and business environment. They present two models of innovation: the linear approach (science-led research conducted in laboratory settings with innovative applications distributed to firms later) and what they call the "dip-down" approach (innovation results from addressing the immediate needs of firms, taking place at the point of production or in interactions either among firms or between firms and the university). Although the linear approach has been dominant, Best and Forrant believe that, today, firms must innovate quickly to stay competitive in demanding markets. They need ready access to advanced knowledge and must draw on a broad network of information from other firms and institutions. They require workers and managers trained in the use of this new knowledge. The authors first discuss these models by comparing development along Route 128 in Massachusetts (during its "miracle" period from 1978 to 1986 and its "resurgence," 1992 to mid-2000) with growth in Silicon Valley. They then provide examples of structures at UMass Lowell that promote each type of innovation, discussing the advantages and limitations of the two models for the university.

Regional growth requires sustainable innovation and the continuous, rapid learning and problem solving that supports it. The authors conclude that the most useful function of the university is the promotion of a collective learning process that links firms, institutions, and communities in the region. To be effective in skill development and in generating collective learning, however, the university must consider the unique historical and social context of the region. Given the industrial structure and development in Northeastern Massachusetts, Best and Forrant argue that UMass Lowell should particularly address the needs of the hundreds of small-to-medium-sized firms that provide substantial employment in the region. Firms such as these often lack the access to resources and expertise required to innovate, remain competitive, and support sustainable growth.

In "The Lowell Center for Sustainable Production: Integrating Environment and Health into Regional Economic Development," David Kriebel, Ken Geiser, and Cathy Crumbley address the difficult issue of how firms and workplaces can meet what appear to be contradictory demands (for economic viability as well as certain degrees of environmental protection, product quality, employee diversity and training, and community good-citizenship—many of the components of "sustainability") by rethinking and redesigning the production process. Building on the growing recognition that removing costly problems, such as pollution, at their source by altering production practices is a better strategy than addressing them at "the end of the pipeline," they seek to alter the prevailing approach to production processes even more broadly. They propose the adoption of an innovative concept that is a key component of sustainable development—sustainable production. They define sustainable production as involving technological processes that are economically and environmentally sustainable and attentive to worker occupational health and safety concerns as well as economic considerations. It also involves the participation and input of workers.

Kriebel, Geiser, and Crumbley illustrate what a university center, such as their Lowell Center for Sustainable Production (LCSP), can do to foster sustainable regional development in terms of: creative rethinking of the issues and problems, designing appropriate strategies for change, mobilizing the various constituencies that must be involved, and addressing the inevitable tensions that arise. The authors provide broad-based guidelines for implementing programs in sustainable production and describe the evolving roles, perspectives, and motivations of all the major constituents who must be part of this collaborative effort, groups often at odds with one another: firms, the environmental movement, community organizations, labor unions, government, and the university. They delineate ways in which a university can promote sustainable production by building relations with these groups and provide examples of the numerous concrete ways the new approach has been implemented in projects with each group. They discuss a number of important, difficult issues they have confronted and the lessons they have learned.

The last chapter, "Effective University–Industry Partnerships in Photonics," by Michael A. Fiddy, Dik Kalluri, and J. D. Sanchez, addresses a crucial issue for universities seeking to contribute to regional development: how to engage in long-term research, support graduate students, and maintain academic programs while responding to industry needs that typically have shorter-term horizons. The authors highlight the importance of a university center's partnerships with industries for sustainable development, particularly in areas of emerging technologies where industry has almost unlimited needs for innovation (primarily via applications research but also requiring some basic research), technical expertise, and trained personnel. The authors provide an example of a center focusing on photonics, a state-of-the-art, dynamic high-tech field that combines electron physics with optical (photon) physics and involves engineers and scientists from many disciplines. To provide context, they describe the relative positions of the United States and Japan in this industry and in the key regions, centers, and consortia internationally. Developments in photonics will enable revolutionary changes in information technology, medicine, defense, energy, and manufacturing processes; the potential economic and social benefits are vast. The federal government is urging industry and government to develop the technology rapidly and encouraging universities to provide multidisciplinary education in this field.

University centers are key arenas for fostering such innovation, because they can draw people from traditional, disparate disciplinary backgrounds together to explore and develop the technology. There are problems, however, in sustaining centers that seek to nurture university–industry partnerships, given the differing, sometimes conflicting, motivations of industries and the university. University centers need stable, longer term funding to support basic research and graduate students, whereas firms often seek only to fund solutions to immediate problems. This problem is compounded by the fact that, although university centers have relied substantially upon longer term funding provided by the federal government, proportionately more recent funding comes from industry. Faced with this reality, the authors suggest a number of ways to enhance university–industry partnerships and evaluate their effectiveness. They raise issues other institutions interested in building such partnerships and balancing the diverse goals of the university and industry could consider in determining their own direction.

The approaches and reconceptualizations presented in these three chapters are thought provoking. They raise some key questions for our future intellectual and practical work:

How can a university be a force for establishing a socially beneficial balance between research that is longer term and fundamentally basic and research that is more responsive to the short-term needs of industry? Because industry is focused much more on short-term results and the profitability of its enterprises, the forces of a competitive economy may not produce the kinds of longer term research that is needed for society as a whole.

How can the university induce diverse institutions with differing short-term motivations to work together to promote sustainable development? How can a university develop relationships with businesses and also include other key groups in society—labor unions, environmental groups, state and local governments?

How can a university obtain sustainable funding to in turn conduct research, support graduate students, and engage in discussions that promote regional growth and development that is viable over the long term? How can a university alter its reward systems to encourage faculty to relate more broadly and creatively to industry and the community?

Innovation, the University of Massachusetts Lowell, and the Sustainable Regional Development Process

Michael Best and Robert Forrant

IN THIS chapter we explore the links between the University of Massachusetts Lowell and enterprises engaged in product and manufacturing process innovations and adaptations. In the early 1990s, UML began to "reinvent" itself to work systematically with enterprises and development organizations to stimulate a collective learning process, influence commercial activity, and participate in the collaborative knowledge-generation and regional-development process. Our purpose here is to offer an instructive review of UML's efforts so that other colleges and universities intent on engaging in the regional development process may benefit from what has been learned thus far through one urban university's efforts. We review this "makeover in progress" and locate it within the context of economic trends in our region to gain a better understanding of what kinds of interactions between firms and formal education institutions may foster sustainable regional economic growth. The region we refer to is Northeast Massachusetts. It is made up of forty-three cities and towns, bordered by New Hampshire to the north, the Atlantic Ocean as far south as Marblehead, and the Route 128/495 corridor from Wilmington to Chelmsford and Lowell. It includes older mill cities, such as Lowell, Lawrence, and Haverhill, as well as cities and towns that have undergone explosive growth, including Andover, Danvers, Chelmsford, and Westford.[1]

We stipulate a delineation of two innovation models, one that is science-led and another that is fashioned more out of the particular technical and organizational needs of firms. The linear, science-led model, emphasizes laboratory research. Here, a university or research laboratory pushes science and technology out to a group of receptive firms. In the second model, the "dip-down" approach, the day-to-day needs of the firm become the principal driver of innovation, not laboratory research. Production, in this approach, is the laboratory for creating, advancing, combining, and diffusing technological knowledge. The work of the University's Biodegradable Polymer Research Center and Masters in Manufacturing Science program

271

are discussed in light of these models. It is our contention that enterprises are under intense competitive pressures, and successful ones are engaged in continuous and oftentime incremental product and process developments (Lester 1998). These enterprises need help in resolving technical problems, and this help may come from anywhere in the value chain, including the firm's customers and suppliers. The dip-down model has a dynamic extra-business component, for it includes a prominent role for institutions outside the firm in the enterprise and a wider regional innovation process. We review these models and weigh their implications for UML and other colleges and universities engaged in regional development efforts.

From our perspective, innovation can occur when new manufacturing processes are utilized or new product variations are produced efficiently in a firm or a group of firms related either by product market or collaborative activities. Innovation may take place as well when a group of complementary firms, a community, or a particular geographic region fashions a successful transition from outmoded and dying industries to dynamic ones. Sustainable growth involves some combination of an expanding absolute number of firms and growing firms and emerges from and develops within an industrial infrastructure comprised of specialist and affiliated enterprises and bolstered by a set of relationships among support institutions, including knowledge-generating institutions such as universities.

Some of the factors necessary to support sustainable innovation include: an advanced pool of knowledge; recognition of technical opportunity and need; an entrepreneur who champions the innovation; financial support; managerial expertise; and continuing innovation and development over many fields. What emerges here is the perspective that a vital component of sustainable regional development is enterprise's accessibility to a pool of technical knowledge skillfully fused with an equally accessible pool of shop-floor expertise. Collaborative behavior among firms and between firms and UML is essential to deepen the technical knowledge pool continually and simultaneously to educate workers and managers in effectively utilizing new knowledge.

A Brief Note on the University and the Economy

The end of the "Massachusetts economic miracle" brought a rapid drop in state tax revenues and eventually led to the rationalization of the public higher education system as a cost-saving measure. As a result, in 1991 the University of Lowell became the University of Massachusetts at Lowell (UML), one part of the five-campus university system. Reeling from the loss of thousands of computer industry jobs in the region and realizing that there could be no easy technological fix to the problem of maintaining a healthy economy, campus leaders—faculty and administrators—set to work figuring out how UML could help the region avoid the historical pattern of economic booms and busts that had dominated its history. Today, UML has a clearly articulated development mission predicated on the notion that a sustainable economy depends upon a skilled and ever-replenished workforce, product quality, and environmental protection with worker safety. There are over thirty faculty-led interdisciplinary research centers and institutes charged with solving real-world problems by using the talents of faculty, students, and members of the surrounding business community (University of Massachusetts Lowell 1997). The work embraces the development of new products and manufacturing processes (Fiddy, Kalluri, and Sanchez in part III herein), the formulation of strategies to deal with today's manufacturing and management challenges, pollution control and environmentally sustainable production (Geiser and Greiner in part II herein), worker safety and health and job design (Punnett; Quinn et al. in part II herein), community leadership development, and community-based social research (Silka; Kleniewski both in part IV herein).

There is also a new academic department, the Department of Regional Economic and Social Development (RESD), which was established in 1996. Interdisciplinary in nature, at the graduate level it educates technologists, entrepreneurs, development professionals, and community activists to understand, analyze, and intervene in the economic and social development process. Faculty are engaged in funded research and hands-on activity on a variety of issues related to economic and social development—like many other universities (for examples, see Beck et al. 1995; Dineen, 1995; Felsenstein 1996; Harris 1996). In sum, these changes have positioned the university, in its urban and regional setting, to stimulate a collective learning process, one that has the potential to foster growth. Today, there is little quarrel over the importance of UML's engagement in and service to the region; there is, however, a vibrant discussion over what that engagement ought to consist of.

Massachusetts has a rich industrial history and Northeastern Massachusetts and Lowell have been in the midst of much of it since the early-nineteenth-century establishment of textile mills there. Two regions within Massachusetts can lay claim to being the birthplace of American industry: One surrounds the Springfield Armory, where the principle of interchangeability was first applied and which, in turn, fostered the development of the world's first machine tool industry; the other is Lowell, where the Boston Associates built a canal and water power infrastructure for America's first textile district (Best and Forrant 1996; Meyer 1998; and Gross in part I). The machine tool industry fostered the growth of a range of industrial districts in Massachusetts, such as watches in Waltham, footwear in Haverhill, furniture in Gardner, jewelry making in Attleboro, cutting tools in Greenfield, and metalworking and specialist machine making in Worcester and Springfield. The proliferation of industrial sectors, in turn, supported hundreds of small, highly specialized tool-and-die shops and foundries engaged in the production of textile machinery, fixtures, tooling, gauges, and made-to-order components in Massachusetts. Just as the consumer goods industries created a market for machine tool makers, the machine makers diffused innovations across the industrial spectrum.

In the twentieth century, Massachusetts has been on a roller coaster: caught between the decline of industries with their origins in the last century and the emergence of new industries derived from a regional competitive advantage in research and development and technological innovation, often boosted by defense spending. Between 1975 and 1980, more than one hundred thousand high-tech jobs were created in the state, and Lowell witnessed a spectacular 400 percent job growth during these years. Wang, Digital Equipment Corporation, Prime, and Data General emerged along with hundreds of small firms to establish the minicomputer industry, while thousands of metalworking, plastics, and electronics companies received lucrative subcontracts to supply computer and defense firms with components, accessories, tooling, machines, and instrumentation. Wages for the region climbed from 90 to 105 percent of the national average, and the unemployment rate fell well under the national average to only 2.8 percent. The research intensity of the region did not insulate the state from the vicissitudes of the business cycle; a sharp downturn occurred in the late 1980s as the computer industry crashed and defense spending took sharp cuts resulting in the statewide loss of 25 percent of durable goods manufacturing jobs between 1985 and 1992 (Gittell and Flynn 1995; Lampe 1988). The Digital workforce fell to 12,000 from 33,000, Raytheon's local employment dropped to 16,700 from 30,000, and Wang Laboratories skidded to 1,350 employees from 8,800. In Lowell, manufacturing employment dropped 28 percent. Symbolic of the decline was the 1992 sale of the Wang Towers in Lowell for $500,000, a building complex that had cost $23 million to construct during the Massachusetts miracle.

Nonetheless, the region has maintained a concentration of manufacturing employment above the national average. At the same time, there is a lower-than-state-average concentration of retail and service establishments, especially high-wage business, engineering, and management services. Whereas only about 15 percent of the state's workforce was employed in manufacturing in 1998, 25 percent of the northeast region's workforce (97,000 jobs) was so employed. Two industry sectors that remained strong—industrial machinery and equipment and electronic and electrical equipment—ranked first and second in the state for total dollar value of exports from 1995 to 1998. The electronics manufacturing services sector has emerged as well. Firms in this sector perform contract manufacturing of a variety of electronic assemblies for other high-tech firms. Four industries—industrial machinery and equipment (which includes computer man-ufacturing), electronic and other electrical equipment, transportation equip-ment, and instruments and related products—account for well over 65 percent of all manufacturing jobs in the region (Forrant 1999).

Hundreds of small to mid-sized metalworking, software, telecommuni-cations, and computer networking firms (under one hundred employees) have experienced impressive growth in recent years. Yet many of these firms lack information about global markets and have not invested in the internal engineering expertise needed to remain innovative in their product mar-kets. Universities have not served the needs of these firms, because the com-panies could not afford to make the large-scale contributions to faculty research that more prosperous large firms are able to make. The challenge before UML and other universities intent on stimulating job creation and economic growth is to develop ongoing working relationships with such firms.

Innovation Theory and Framework for Consideration

Two Development Models

How and why has the region been able to rebound so strongly over the past five years? Why has the region's vital manufacturing base remained so strong in the face of the more widely discussed shift to a service economy? Are there ways firms have reorganized themselves that would help to ex-plain this and point up ways for UML to engage in the continued regional development process? There are two models that can help us discern the various ways in which universities can encourage enterprise innovation: the linear model and what we term the "dip-down" model. The linear model emphasizes laboratory research work and views the university as "push-ing" science and technology out to a group of receptive firms (Hounshell 1996). Leonard Reich (1995) notes that such activities are usually per-formed in industrial laboratories set apart from production facilities and are staffed by people trained in science and advanced engineering. These individuals are somewhat insulated from the immediate demands of the en-terprise.

In the dip-down model, the day-to-day needs of the firm are the princi-pal driver of innovation, and thus firm growth. The dip-down model has a dynamic extracompany component, for it includes a role for institutions outside the firm in the enterprise innovation process. For example, in the search for solutions to particular problems, it is possible for enterprises to "dip down" into the scientific, managerial, and technical bodies of knowl-edge available in universities. Companies best at rapid new product devel-opment and/or the shop-floor integration of more effective manufacturing processes know where to access particular kinds of technical expertise and thus gain an advantage over their "less knowledgeable" counterparts. Over time, fairly stable relationships between particular university groups and particular firms are forged as these firms "dip down" into the pool of identifiable chunks of knowledge and expertise to solve their problems (Kodama 1986). In this two-way process, a continuous stream of technical and organizational knowledge is generated across firms and universities. In our estimation, it is the network character of such relations among the various parties involved in innovation that is a critical determinant of the long-term success of regional economies.

Richard Lester (1998) points out that for most of the postwar period the linear model predominated in the United States. "Research was science driven; priority was given to discovery; and the nation's achievements in basic science were outstanding." Today, enterprises are under intense competitive pressures, and successful ones are engaged in continuous and oftentime incremental product and process developments (300–301). The "one big breakthrough" strategy is less helpful to firms engaged in rapid new product development in markets where product life cycles frequently register in months not years.

The Silicon Valley Trajectory and the Innovation Process

Industry in Massachusetts has had its ups and downs over the past thirty years. Route 128, which lays claim to being the country's first high-tech region, drove the Massachusetts miracle, countering a long postwar "deindustrialization" process. But driven by the decline of the minicomputer industry and defense cutbacks, the so-called miracle was ephemeral, and it seemed as if the deindustrialization trajectory had returned with a vengeance. Although Massachusetts had extensive research capabilities and captured a disproportionate share of federal R&D funding, its business enterprises were not competitive, particularly against the surging Silicon Valley. Yet, at similar points in time in the 1980s, both regions had contained technologically dynamic firms that depended on such endogenous factors as skilled labor and renowned local research institutions to turn out goods. Why, then, did the "miracle" turn into a "mirage" in Massachusetts?

AnnaLee Saxenian (1994) argues that Silicon Valley's traditions of collective learning, the horizontal integration of complementary firms, and its sense of community provided it with a decisive edge in competition with the traditionally vertically integrated firms of Route 128. Some scholars took issue with Saxenian and argued that an industrially "fragmented" Silicon Valley would eventually lose out to vertically integrated Asian competitors. Thus far, time has been more favorable to Saxenian's explanation for Silicon Valley's success, for, based on the vision of Frederick Terman and, eventually, the spectacular growth of the personal computer industry,

Silicon Valley enjoys an unparalleled dynamism. Because the social complex and corporations within it arose from nowhere to leadership in the high-tech world, it remains one of the most admired and imitated models of regional economic development. Indeed, dozens of well-funded policy efforts have been undertaken to learn and apply its secrets, yet most of them have failed. Ironically, this is so even in the case of efforts in which Terman himself played a leading role. Saxenian's explanation for the success of Silicon Valley—the existence of a collective learning environment, a horizontal industry structure, and a sense of community—suggests criteria for evaluating these cloning efforts to fuse academia and industry.

In their review of Terman-guided efforts to replicate the Silicon Valley experience, Leslie and Kargon (1996) made good use of Saxenian's criteria and concluded that Terman's own efforts ignored horizontal integration and were driven by vertically integrated enterprises that had "never depended on collective learning" in their corporate history (471). Likewise, the derived model focused on the critical role of prestigious universities and their capacity to attract leading scholars without the development of a useful off-campus knowledge-exchange mechanism. Terman's approach to duplicating Silicon Valley was essentially static and thus failed to capture the actual dynamic processes in which Terman himself had been a key player. Silicon Valley and Stanford "had grown up together, gradually adjusting to each other and to their common competitive environment. Each helped the other discover and exploit new niches in science and technology. In the proliferation of new technical fields and new companies that characterized the early evolutionary stages of these industries, the right kind of university could make a real difference in fostering horizontal integration and collective learning throughout the region" (470).

Examples of dynamic processes include: the elaboration and re-elaboration of ideas between industry and university; the processes of discovery and exploitation of new market/product niches, which, in turn, engendered the "proliferation of new technical fields and new companies"; and the relationship between "a sense of community" and the strategy and theory of industrial development. Terman's strategy did not involve small firms and start-ups; instead he worked with large established firms such as Bell Laboratories. Similarly, he emphasized work with prestigious universities with long-established academic departments and failed to reflect the actual experience of industry and academic interaction and mutual adjustment in the early phases of Silicon Valley. Finally, he failed to capture the interplay between the organizational and production capabilities of enterprises and emerging opportunities in the market. Instead, he was attracted to the massive federal investments in R&D, especially defense-related activities and the usually nonrecurrent and noncommercial opportunities that such funding provided.

Terman's approach suggests that following the Silicon Valley model required entrepreneurial capabilities to anticipate emergent opportunities at a precise moment in time. But is such clairvoyance possible? As Leslie and Kargon note: "Then, as now, stimulating economic growth may be more a question of nurturing well-balanced, possibly unique, regional ecologies than of constructing a single model from common blueprints" (472). In other words, success is derived from an active engagement process among firms and supportive institutions and one's "luck" is gained the old-fashioned way, through hard work. Indeed, the success of Saxenian in explaining Silicon Valley's achievements in terms of "a sense of community" provoked a strong reaction around Route 128, a reaction that has engendered a policy response designed to advance a similar sense of community there (for a discussion of innovation policy, see Fountain 1998).

In fact Route 128 has enjoyed a resurgence. Can it be explained in terms of an institutional transformation making the region congruent with Saxenian's criteria? Perhaps. But it is important to ground Saxenian's analysis in an understanding of the dynamic processes identified in the experience of Silicon Valley as it was applied to the unique characteristics of the Route 128 region, of its business enterprises, and the derived dynamics of the region. Indeed, just as the relative lack of cooperation among firms along Boston's Route 128 was being blamed for the area's poor performance relative to Silicon Valley, the region's capability for systems integration was laying the foundation for revitalizing its industry around a new principle highly conducive to interfirm networking.

The two expansions (the "miracle," from 1978 to 1986, and the "resurgence," from 1992 to the present) are marked by two different models of industrial organization: vertical integration versus horizontal integration, or closed-system versus open-system. Systems integration played a role in both, but only in the "resurgence" did it lead to widespread technological diversification, a proliferation of new industries and subsectors in a process by which the establishment of new industrial niches leads to yet newer niches. CorpTech, a data-processing company, categorizes America's small and medium-sized (under a thousand employees) "technology manufacturers" (most of which are privately held) into seventeen industries as shown in table 1. The dispersion of these firms across Massachusetts and in the region is indicative of the diversity of industries associated with the Route 128/495 corridor. The mix of high-technology manufacturing in Massachusetts, with approximately 2 percent of the nation's population, is remarkable. CorpTech estimates that almost 8 percent of the nation's small and medium-sized high-tech companies are based in Massachusetts with a total of over two hundred thousand employees (CorpTech, 1999).

These data support the theme that a process of technology diversification has driven the "resurgence" of the Massachusetts economy. Technol-

Table 1. CorpTech directory of small and medium-sized high-tech manufacturers, January 1999

Companies	No. of SMEs in Mass.	% of total firms monitored in Mass.	Mass. % of total U.S. firms monitored
Factory automation	337	10.5	12.1
Biotechnology	151	4.7	3.5
Chemicals	95	3.0	4.1
Computer hardware	435	13.6	13.8
Defense	56	1.7	2.1
Energy	105	3.3	4.5
Environmental eqpt.	203	6.3	7.1
High-tech mnfg. eqpt.	421	13.1	12.6
Advanced materials	159	5.0	6.6
Medical	248	7.7	6.3
Pharmaceuticals	95	3.0	2.5
Photonics	240	7.5	5.0
Computer software	993	30.9	24.8
Subassemblies/comp	530	16.5	17.2
Test and measurement	378	11.8	11.2
Telecom and Internet	415	12.9	15.5
Transportation	92	2.9	3.6

Source: [http://www.corptech.com]

ogy management in the "Massachusetts miracle" growth industries of minicomputers and defense was locked up in vertically integrated enterprises. The downturn was critical to the upturn, as the demise of these enterprises facilitated the transition to an open-system, multienterprise model of industrial organization. It was the accompanying decentralization and diffusion of design, combined with a heritage of technological skills and capabilities to fuel the internal growth dynamic of entrepreneurial firms, that fostered regional innovation dynamics.[2]

To return to Terman, there were critical weaknesses in his "export strategy"; he failed to anchor this work in the internal dynamics of the firm and he failed to understand the history of the region he sought to emulate and the ones he wanted to work in. It is imperative that UML understand better the mutual adjustment processes in firms, on the one hand, and interfirm relations, on the other hand. The internal and external dynamics of interfirm behavior are historically contingent (historically constrained) and thus form the basis for unique organizational capabilities within regions. The mere concentration of firms in a particular industry, for example, medical instrumentation, or among complementary industries, for example, textile companies and builders of textile equipment, does not ensure an unlimited climate for enhancing firm competitiveness. After all, the existence of a vibrant textile industry in Lowell ensured only its bumpy and widespread collapse when firms began to lose market share to low-cost producers elsewhere around the world. By comparison, the development of the metalworking region in Western Massachusetts over the past fifteen years is an instructive example of how a region ravaged by job loss was revitalized through a vigorous and continuous social process, characterized by research, discussion, shared learning, reflection, and experimentation among firms, their trade association, and several public agencies (Forrant and Flynn 1998). Here, the strength of the firms acting collectively was dramatically enhanced when strict attention was paid to the realities of the region's particular history and its extant social and institutional environment. The existence of a cluster may facilitate the social construction of assets that increase learning among firms about such things as advanced manufacturing processes and technologies, but this is not an automatic process; the geographic proximity of firms only occasions the likelihood for innovative company behavior.

The growth process in knowledge-intensive industries is largely conditioned by the availability of engineering and scientific personnel required to staff rapidly growing firms. (UMass Lowell's engineering education challenge is discussed by Vedula et al. in part IV). The role that machinists and product engineers played in the diffusion of the principle of flow in nineteenth-century American industry is now taken by systems and software engineers in the diffusion of the principle of systems integration. Regional dynamics are immobile; they are comprised of intangible, complementary, and publicly induced assets, and, for good or ill, firms are embedded in this regional dynamic. Universities and colleges can play a critical knowledge-providing and -diffusion role only if they grasp this dynamic, in practice not just in theory, and become linked to the firms in a mutual learning process.

Polymer Research and the Master's in Management Science (MMS): Firm Links and Innovation

To be clear, integral to regional growth is a set of interactions between industry and formal educational institutions. The growth process in knowledge-intensive industries is limited by the available skill base. Any individual firm can attract staff members from the existing pool by offering superior pay, but the success of the region depends upon deepening the pool. Oftentime the most successful industrial policy is education policy, and this should play to the ostensible strengths of a university. Skill formation is a long-term process, including formal and informal activities within and across schools, firms, government, and the polity (for an example, see Forrant and Flynn 1998). Production capability and skill development are necessarily linked; the success of each depends on complementary developments in the other. For the process taking place at Lowell and at other colleges and universities, there is the caveat that skill-development programs designed in isolation from the region's industrial enterprises are rarely successful. The failure of most skill programs and industrial policy initiatives attests to the unfortunate fact that skill education for the sake of skills is not part of advancing production capabilities or industrial growth. We analyze two UML programs designed to enhance skill formation and firm growth to learn more about UML's efforts in the regional development process.

The Biodegradable Polymer Research Center: R&D Partnering

The UML Biodegradable Polymer Research Center (BPRC) is a representative model for conducting long-term research and development projects with firms. There, a number of America's leading materials companies have collaborated with government laboratories and researchers at UML in exploratory, fundamental, and developmental research on biodegradable polymers. The work is important, since the plastics industry in Massachusetts is large and growing with roughly six hundred sixty plastics firms, and more than two hundred additional companies, such as Gillette, which produce plastic components as part of their broader operations. The industry employs nearly fifty thousand people, and between 1988 and 1994 plastics exports from Massachusetts increased by 96 percent to over $400 million. Should research into new materials lead to significant changes in the industry, the state's firms would be well positioned to learn about such changes early on.

The university was chosen as the site for the center in part because it houses the first- or second-largest plastics engineering program in the United States and because companies were impressed with the fundamental research on polymers taking place on campus. The center is colocated with the Institute for Plastics Innovation, also a consortium structure, that offers research and new product development services to leading firms, including 3M, Ford, and GE, and many others in electronics, computers, and automation technologies. Combined with the plastics engineering department, the center and the institute offer unique advantages in R&D, education, and training for regional plastics companies. Together theirs are numerous accomplishments in terms of patent applications and patents issued, research publications, and the funding of more than thirty graduate students. Few of the participating companies are direct competitors, but all share an interest in the development of biodegradable polymers and related technologies. Taken together, the companies span much of the plastics "food chain" from grain feedstocks (Cargill), to processing agricultural feedstocks (National Starch), to petrochemical-based materials companies (Dow and Eastman Chemical), to materials developing and using companies

(Monsanto and 3M), to high-volume purchasers of plastics and synthetic rubber materials (BASF and BF Goodrich). Toward a $.5 million annual budget, the National Science Foundation contributes $50,000 annually, and each firm contributes $30,000 a year. Companies such as 3M and BASF, with R&D research budgets exceeding $1 billion can easily afford to collaborate on R&D with the center to get in on the ground floor of research on new materials.

The center's Industrial Advisory Board meets twice annually to review completed research and shape faculty proposals for future research projects. Patents are held by the university but any companies that participate in the application and support for the patent have privileged licensing rights. The companies also profit from access to dozens of students who have been involved in other research projects, while faculty and students benefit from evaluations of their research by scientists working in industrial laboratories. Furthermore, faculty, corporate researchers, and advanced students are often fellow members of professional associations and editorial boards of specialist journals as well as the international standards-setting organization dealing with biodegradable materials.

An important attraction to the sponsoring companies is the BPRC's integration of scientific research across the boundaries of three disciplines: chemistry, plastics engineering, and biology. In the evolution of biodegradable polymer research, the issues associated with plastic degradation kinetics are in the forefront. (Biodegradation involves the interaction of the physics and chemistry of polymers with the biology of natural systems.) This is so, because the development of novel materials has run well ahead of the development of processing procedures and environmental impact assessments. In other words, these exotic materials exist but without the knowledge of how to utilize the materials in the manufacturing environment and without sufficient information on how the new materials will interact in landfills and with various recycling techniques. In the first two years, the focus of the center's research was mainly on materials synthesis and the processing and blending of new materials, but with the addition of biologist Rich Farrell, the center developed a major strength in biodegradation testing.

The shift in emphasis away from material synthesis was driven largely by the interests of the companies. One industry board member indicated that his company would have left the center but for the unique biodegradation work conducted at UML. What is not yet known is the interaction of the new materials with the biology of soil systems; this will be the center's contribution. An estimated thirty companies are conducting research in biodegradable polymers worldwide, and there are research consortia in Japan and Germany looking into the links between biodegradable polymers and composting. Benefits of the research will be immeasurable and involve the repositioning of the world's plastics industry away from the proliferation of nondegradable and environmentally hazardous synthetics to materials that are environmentally friendly. For this reason, the BPRC is highly attractive to students contemplating a career in environmental sciences. However, Farrell is no longer at UML, and his departure demonstrates the difficulties that UML has in maintaining this type of scientific research without the very "deep pockets" to attract and keep researchers and provide them with the most modern equipment.

The Master's in Management Science (MMS) Program

The economic well-being of the Northeast region of Massachusetts is linked to the innovative capabilities of its goods producers. For example, the recent high growth in information-and communications-technology industries has led to strong economic growth. The resurgence has been associated with unique regional capabilities to integrate manufacturing with rapid new product development and technological innovation. At its core, the MMS program eliminates the traditionally rather separate managerial and technical views of manufacturing by offering a cross-disciplinary degree designed to give students the integrated technical and managerial knowledge required to play a leadership role in modern manufacturing. An important component of the program is a student research project conducted at a local manufacturing plant. The project provides students with an educational experience that enables them to integrate and apply their management and engineering course work and experience in a way that is also of benefit to the host enterprise. (These projects are excellent examples of the dip-down model we discussed above.) The challenge here is to focus on a particular issue that is of interest to both the student team and the company.

Modicon: An MMS Case Study

MODICON, an acronym for MOdular Digital CONtrol, was incorporated in Bedford, Massachusetts, in 1968 by Bedford Associates, a small consulting firm that saw the opportunity to marry the computer age with automatic machinery. Today it is owned by Groupe Schneider (GS), France's $12 billion global leader in electrical distribution and industrial control products with over 120 companies and 91,000 employees. At start-up, Bedford Associates had unique collective hands-on experience in designing, installing, wiring, and troubleshooting the precomputer-age automatic control systems, which consisted of complex and cumbersome series of electromechanical relays. A digital system for a new device using solid-state circuitry and ladder logic was designed there and first installed on assembly lines at General Motors in the early 1970s. Called a Programmable Logic Controller, or PLC, demand for the product led to rapid growth for MODICON. Thereafter, PLC performance advanced in lockstep with that of the microchip and translated into ever higher speeds, greater memory, expanded handling capability, and smaller size.

Typically, contract manufactures like MODICON specialized in the manufacture of a handful of products or product variations in very high volumes for consumer electronics products. Today many contract manufacturers are attempting to make a transition to the manufacture of low to medium volumes of an expanded range of products. Initially, MODICON was unable to make this switch. This was primarily because, though successful at introducing continuous flow manufacturing (CFM) to its printed circuit board (PCB) activities, the "back end," or final assembly, testing, and packing activities, were still organized for high-volume production. The production line at MODICON involves a diverse range of machines made by various companies. Flexible production in a medium-volume, high-mix environment requires that the machines interface with open-system software. Open systems are required to support, for example, various makes of pick-and-place machines and autoinsertion equipment, and various makes of machines are notoriously unresponsive to common programming. For support of both quick changeovers and rapid new product introduction, the production line requires user-friendly and reliable systems for line balancing, simulation of the assembly process, detection of error, creation of visual aids, documentation, and bills of material.

To make the shift, significant organizational and managerial changes were required, including a thorough redefinition of work assignments and a commitment to education and training for employees. Today GS's career-incentive structure is geared to lifetime learning and includes payment of tuition for advanced education including MBA and MMS degrees. From top management's viewpoint, all employees must see how important continuous learning is to the company, but it goes a step farther, for students taking courses are encouraged to apply what they learn in the company. With this support, Sean Gearin, an MMS student and Surface Mount Technology engineer at MODICON, and Joe Swadel, a fellow MMS student, did their field project at MODICON.

The goal of the project was to redesign back-end (final assembly) activities to implement a one-piece flow manufacturing pilot line within a low- to medium-volume and high-product range environment. To accomplish this, MODICON's engineers needed to link a range of machines within a standardized operating system. This challenge was addressed by working closely with two nearby companies with distinct but complementary skills, UniCam, a specialist in computer-aided manufacturing software, and Cam/ALot, a machinery builder. UniCam supplies a range of manufacturing hardware companies and has recently formed a long-term development and marketing agreement with Fuji Machine Manufacturing Company to develop FujiCam, involving computer-integrated manufacturing applications for all Fuji surface-mount technology machines (SMT). (Fuji is the world's largest supplier of SMT placement equipment). Cam/ALot, once a traditional machine shop, is expert in the integration of machine tools and electronics and builds complex manufacturing systems for many companies, including Motorola, Siemens, Cray, and Panasonic.

The application of the ideas developed in the project led to a one-third drop in manufacturing process-cycle times and improved quality, improved product traceability, a reduction in scheduling problems, and reduced new product introduction time. MODICON spent approximately $430,000 to implement the project, which had an estimated payback time of three years. These costs included the purchase of ten new pieces of equipment for $390,000 and facility improvement costs of $40,000. Total savings for the first year following implementation were estimated at $302,000, and the payback period dropped to under two years.

GS and Regional Dynamics

Groupe Schneider came to the region for its expertise in the integration of computers with automation. It has remained because it can manufacture here in a dynamic environment with partners such as UniCam, Cam/ALot, and UML. In fact, design and production have continued in Massachusetts even though the company has gone through a series of dizzying ownership changes. A new North Andover facility was built in 1995 to streamline printed circuit board manufacturing. Today the plant produces 1,000 board types, with 1,500 per month as its greatest volume and an average lot size of 34. The plant generates $200 million in revenue, with 320 employees involved in direct labor and 80 support staff. Groupe Schneider's total operation at North Andover employs more than 700 workers and generates $650 million annually in sales. Typical customers are end users in the automotive, pulp and paper, waste treatment, petrochemical, food and beverage (bottling plants), and pharmaceutical industries.

Jim Crawford, the plant manager, began his career with Honeywell when the Route 128 computer industry was emerging. After twenty years, Crawford has many friends in the industry that he shares ideas with, and his learning network has several nodes. For example, Siemens sells equipment both to Groupe Schneider and to Motorola's high-speed Internet division in Mansfield. Siemens has invited Crawford and Sean Gearin and their Motorola counterparts to visit Georgia Tech's advanced manufacturing line, and MODICON has opened its plant to Motorola, HP Colorado, Bosch, Lucent Technology, and numerous other companies as part of its "Tours R Us" program. Crawford's vision is for MODICON to become a worldwide leader in computer-integrated manufacturing, and with systems integration capabilities and a learning organization, this vision is taking shape.

The University and the Innovation Process

What are the implications of these two approaches to innovation for universities generally? Activities at UML approximate both the linear and the dip-down models, as the polymer center and MMS case studies demonstrate. One caution is necessary here: Work like this remains difficult, as is evidenced by the fact that, despite the global proliferation of modernization organizations and university–industry research centers over the past fifteen years, we still know little about whether these institutions facilitate interfirm learning and the diffusion of more efficient technologies and manufacturing processes (Shapira 1998).

To recapitulate, the linear model has dominated people's thinking for many years. Here, innovation commences with a good scientific idea in a research laboratory. This approach approximates the activities of the Biodegradable Polymer Research Center. By contrast, with the dip-down approach, product improvements and incremental manufacturing process improvements are more likely to take shape within the day-to-day activities of the enterprise itself. Firms under strong competitive pressure in demanding markets—for example, telecommunications equipment, computer networking, semiconductor automation equipment—are seeking to push ahead with product and process improvements as rapidly as possible. In these endeavors, technical problems arise that are unresolvable by drawing solely on the existing stock of knowledge in the firm. MODICON's search for a new way to speed work through its plant is an instance of this. The search for solutions leads to a sampling—a dip into the pool—of the scientific and technological bodies of knowledge that are available in universities and, perhaps, even among nearby firms. Those companies best at effective and fast new product development generally come to know where particular kinds of knowledge and expertise may be found.

What are the implications for the linear approach? It requires that universities like UML be given research funds from public sources to pursue basic scientific ideas that get published in scientific journals, and substantial efforts should be made by public bodies responsible for these funds to ensure a sufficient level of such output. The financial costs, however, of pursuing this type of work are high, as are the risks that leading researchers will leave the university. There is also the possibility that partnering firms will take their research dollars elsewhere, or a different scientific approach to a particular problem will succeed, thus derailing years of effort. The "technological fix" approach may lead to close working relationships with a handful of firms—as the polymer research demonstrates—but it hinges on the willingness of large firms to sustain the financial drain required to complete the work, when any derived and sustainable development benefits in terms of employment and new firm growth may well accrue outside the region.

For activities more in accord with the dip-down approach, there are at least three implications. First, a great deal of attention must be paid to the establishment of relationships with leading-edge companies engaged in new product development activities. The conversion of new product concepts and production capabilities into products is mediated by the creation of new firms to take advantage of the opportunities. Second, structures must exist that encourage faculty to engage in problem-solving activities with these companies, activities that will diffuse knowledge within and across firms as well as throughout the workforce. Finally, to generate greater knowledge among firms in the region, activities need to be written up, perhaps in a case study format, so that the lessons learned can be taught to others. Ideas are the raw material for the innovation process and are bound up with the skill-formation infrastructure, an infrastructure that UML must systematically learn about, embed itself in, and enhance.

Integral to regional growth is a set of interactions between industry and formal educational institutions. The growth process in knowledge-intensive industries is limited by the available skill base. One important development role of the regional college or university is responsive collaboration with industry and government in skill formation appropriate to the region. This is a time-consuming process that requires strong trust-based relationships, which cut across educational institutions and the business establishment. The activities of three universities are instructive here. The Georgia Institute of Technology (Georgia Tech), Rensselaer Polytechnic Institute (RPI), and Worcester Polytechnic Institute (WPI) have developed extensive applied research and development partnerships with industry. Each center is embedded in its regional economy and characterizes its activities as a dynamic two-way street whereby firms obtain much-needed technical and organizational knowledge while boosting the reality-based knowledge of the university. As RPI acknowledges, Technology Park companies provide faculty and students with "a living laboratory."

The Georgia Tech Research Institute (GTRI) is a nonprofit applied research and development organization that expends approximately $100 million in engineering, science, computer technology, and related research applications annually. The institute is located on the Atlanta campus and employs several hundred engineers and scientists. Its principal focus is identification of processes and procedures for clients that will improve their manufacturing procedures and reduce their costs through the application of available technologies. GTRI keeps executives aware of the latest developments in the use of information technology through its Senior Executives' Roundtable (www.gtri.gatech.edu).

RPI owns and operates the Rensselaer Technology Park in North Greenbush, New York. Park companies are able to access the extensive physical and human resources of RPI, and at the same time "faculty and students have a living laboratory in which to apply their learning." There are fifty tenant businesses that employ two thousand people. Firms in the Park can make use of the RPI library, computing center, and machine shop. RPI characterizes the Park as a dynamic resource, able to "enrich the educational environment of the university and help companies stay on the leading edge of their technologies." The park consciously links technology resources with RPI's Lally School of Management to work simultaneously on organizational change with firms. This linked approach, they contend, is fundamental to successful change in firms. (See [www.rpi.edu/dept/rtp]).

WPI's Metal Processing Institute is an industry–university partnership dedicated to the advancement of the metal processing industry. Since WPI's roots in metalworking date back more than one hundred thirty years this is, indeed, an appropriate focus. More than sixty enterprises use the institute to resolve production-related problems through plant-specific projects, attend industry-led research seminars and workshops, and host student internships. The institute operates three research consortia engaged in longer term projects to address technological barriers facing the industry. Like RPI's Technology Park, the institute is interdisciplinary in nature. The Powder Metallurgy Research Center is engaged in numerous technical research areas, but these are complemented by linked management research areas, including the development of interfirm relationships and alliances, cost modeling, and industry globalization (www.wpi.edu/Academics/Research).

Thoughts on the Future

Several questions remain that are central to the dialogue that must take place as we define UML's role in the regional development process. Are UML industry-related research activities demand-driven or are they driven more by the research proclivities of faculty members?

Can there be more frequent workshops and symposia to determine explicit ways to integrate faculty off-campus research activities with classroom settings? (This may take the form of developing an incentive structure to encourage faculty engaged in such work to prepare teaching case studies.) How much do funding considerations determine the research activities and technical assistance projects that faculty engage in? Put a bit differently, how difficult is it for small firms lacking money to dip into expertise on campus? Are incentives available to encourage faculty to pursue work with small and medium-size enterprises? Although such work may currently lack the academic prestige and financial rewards to make it attractive, it should be noted that the majority of firms in Greater Lowell are small. How are the courses of study at the undergraduate and graduate levels in the College of Management and the College of Engineering integrated?

If a key argument here—that the ability of innovators to adapt quickly to competitive pressures is paramount—is accepted, courses that prepare students to play an integrative role across technological and organizational issues are essential. Is there a way to track ongoing projects with firms through a cumulative learning process? This will help to reduce the time taken to "reinvent the wheel" with the same firm, and it will also lessen the "pestering effect" that results when too many of us call the same too few firms. Are there ways to make certain that in-plant activities are guided by an appreciation for worker health and safety concerns? Are there ways to make certain that firm-based activities do no harm to the environment? How can we learn from models already in place, such as the Toxics Use Reduction Institute, to encourage projects with positive environmental impacts? How do we evaluate our activities to determine impacts on the regional economy? To put it bluntly: How do we know if we are making a difference?

For UML to stimulate regional, sustainable growth, it must focus on knowledge creation, enterprise development, and industrial innovation. By industrial innovation we refer not simply to the number of scientists working in research laboratories, but to product development, process improvements, design and technology management capabilities, technological diversification, industrial specialization, and the nurturing of new industrial subsectors. We set out to lend our voices to the discussions taking place on the role that UML could play within our region and, by extension, we have raised issues and concerns central to a more general and extended discussion of regional development and the academy's role in this dynamic social process.

Firms under strong competitive pressure and in demanding markets are seeking to push ahead with product improvements and new products as fast as possible. In so doing they encounter technical problems that they do not know how to solve and they search for solutions, dipping down into the pool of specialist technological and scientific bodies of knowledge that are available in other firms, universities, and elsewhere. A survey of literature on innovation by Ronald Kostoff (1994) finds that the first and most important factor in regional innovation is a broad pool of advanced knowledge. According to Kostoff, "The entrepreneur can be viewed as the individual or group with the ability to assimilate this diverse information and exploit it for further development. However, once this pool of knowledge exists, there are many persons or groups with capability to exploit the information, and thus the real critical path to innovation is more likely to be the knowledge pool than any particular entrepreneur" (61).

The companies best at effective and fast new product development (NPD) have developed the capability to integrate technologies, starting with software and hardware. They know where particular kinds of knowledge and expertise can be located and how to dip down into the pool of technological and scientific knowledge and expertise to solve particular problems. It is likely that this knowledge will be in identifiable "chunks" related to the needs of the particular firms and industries and to the characteristics of the science and technology (for the importance of studying firms, see Markusen 1994).

The capacity of an industry or a regional economy to provide well-paying jobs and a broadly shared sustainable prosperity is contingent on its ability to learn new things and resolve problems as they manifest themselves. Recent studies on the textile industry (Antonelli and Marchionatti 1998; Scranton 1997), machine tools (Wieandt 1994), metalworking (Capecchi 1997; Kelley and Arora 1996), plastics (Saglio 1997), the semiconductor industry (Angel 1994), Italy (Cossentino, Pyke, and Sengenberger 1996), the Midwest (Florida 1996), and Western Massachusetts (Forrant and Flynn 1998) demonstrate that, as a systematic learning process among enterprises and institutions is nurtured, knowledge is accumulated. Kelley and Arora (1996) consider how quickly small and medium-sized metal-working enterprises adopted innovative production technologies such as computer-aided drafting equipment. Differences in "the nature of inter-firm relations" together with the "accessibility to non-market technology transfer institutions," they concluded, "hold the key to explaining differences in the rate and extent of the diffusion of new manufacturing technologies. These institutions condition the pace and effectiveness with which useful information flows to potential adopters" (268).

In "Toward the Learning Region," Richard Florida (1995) outlines the shift to knowledge-intensive capitalism and makes the case that firms need a broader knowledge infrastructure than simply that of their particular individual firm if they are to keep up with the shortened product and technology cycles that exist today. Florida explains that learning regions require an infrastructure of knowledge workers who can apply their intelligence to production. The education and training structure must facilitate lifelong learning and provide the high levels of group orientation and teaming required for knowledge-intensive economic organization (Florida 1995, 533–34). In fact, regions such as Silicon Valley and Route 128 have developed regional innovation capabilities embedded in "virtual laboratories" in the form of broad and deep networks of operational, technological, and scientific researchers that cut across companies and universities.

Although these high-tech districts are unique in terms of specific technologies and research intensity, they exhibit innovation characteristics in an exaggerated form that is common to the "virtuous circle" of regional growth we have described. The high-tech industrial district is a collective experimental laboratory. Networked groups of firms are, in effect, engaged in continuous experimentation as the networks form, disband, and reform. Both the ease of entry of new firms and the infrastructure for networking facilitate the formation of technology integration teams in real time. While a vertically integrated company may carry out several experiments at each stage in the production chain, a district can well exploit dozens of experiments simultaneously through a series of horizontal relationships with others with the requisite complementary capabilities. It is here, at the interstices of these relationships, that we envision universities insinuating themselves into the regional development process by being champions of collective knowledge creation and vigorous knowledge diffusion. As befits an educational institution, the university's most important role ought to be the persistent advancement of cross-firm, cross-community, and cross-institutional learning. In this way, the university will advance its work well beyond a random approach to development and play an informative, integrative, and innovative role in the cultivation of a sustainable regional economy.

NOTES

1. This is not an arbitrary delineation. It conforms to the regional definitions developed by the Massachusetts Benchmarks Project in its quarterly review of the state's economy. The project is directed by the Office of the President of the University of Massachusetts, and its quarterly journal *Benchmarks* is published by the university in cooperation with the Federal Reserve Bank of Boston.

2. CorpTech tracks America's 45,000-plus technology manufacturers with under a thousand employees (90 percent are "hidden" private companies and the operating units of larger corporations). Of 42,342 U.S. entities, 3,242 or 7.7 percent are located in Massachusetts. These are independent companies, subsidiaries of major U.S. corporations, and American operating units of foreign companies. Data were extracted from CorpTech/Web (www.corptech.com).

REFERENCES

Amin, A., and K. Robins. 1990. The reemergence of regional economies? The mythical geography of flexible accumulation. *Environment and Planning* D 8:7–34.

Atkinson, R. 1993. *The next wave in economic development: Industrial services in the 1990's.* Washington, D.C.: U.S. Office of Technology Assessment.

Batt, R., and P. Osterman. 1993. *A national policy for workplace training: Lessons from state and local experiments.* Washington, D.C.: Economic Policy Institute.

Beauregard, R. A. 1994. Constituting economic development. In *Theories of local economic development,* ed. R. D. Bingham and R. Mier, 191–209. Thousand Oaks, Calif.: Sage.

Beck, R., D. Elliott, J. Meisel, and M. Wagner. 1995. Economic impact studies of regional public colleges and universities. *Growth and Change* 26:245–60.

Best, M. 1990. *The new competition: Institutions of industrial restructuring.* Cambridge: Harvard University Press.

Best, M., and R. Forrant. 1996. Creating industrial capacity: Pentagon-led versus production-led industrial policies. In *Creating industrial capacity: Towards full employment,* ed. Jonathan Michie and John Grieve Smith, 225–54. New York: Oxford University Press.

Blewett, M. 1995. *To enrich and to serve: The centennial history of the University of Massachusetts Lowell.* Virginia Beach, Va.: Donning Company.

Brown, D. L., and M. E. Warner. 1991. Persistent low-income nonmetropolitan areas in the United States: Some conceptual challenges for development policy. *Policy Studies Journal* 19:22–41.

Brusco, S. 1992. Small firms and the provision of real services. In *Industrial districts and local economic regeneration,* ed. Frank Pyke and Werner Sengenberger, 177–97. Geneva: International Institute for Labour Studies.

Chancellor's Report. 1997. *University of Massachusetts Lowell Chancellor's Report.* Lowell: Office of Public Information.

Cohen, W., R. Florida, and W. R. Goe. 1994. *University–industry research centers in the United States.* Pittsburgh: Center for Economic Development, Carnegie Mellon University.

Cornish, S. 1997. Product innovation and the spatial dynamics of market intelligence: Does proximity to markets matter? *Economic Geography* 73:143–65.

CorpTech. 1999. [http://www.corptech.com].

Cossentino, F., F. Pyke, and W. Sengenberger. 1996. *Local and regional response to global pressure: The case of Italy and its industrial districts.* Geneva: International Institute for Labour Studies.

Dineen, D. 1995. The role of a university in regional economic development: A case study of the University of Limerick. *Industry & Higher Education* (June): 140–48.

Eisenger, P. 1995. State economic development in the 1990s: Politics and policy learning. *Economic Development Quarterly* 9:146–58.

Felsenstein, D. 1996. The university in the metropolitan arena: Impacts and public policy implications. *Urban Studies* 33:1565–80.

Ferleger, L., and W. Lazonick. 1993. The managerial revolution and the developmental state: The case of U.S. agriculture. *Business and Economic History* 22:67–98.

Fitzgerald, J., and A. McGregor. 1993. Labor-community initiatives in worker training in the United States and the United Kingdom. *Economic Development Quarterly* 7:172–82.

Florida, R. 1995. Toward the learning region. *Futures* 27:527–36.

Florida, R. 1996. Regional creative destruction: Production organization, globalization, and the economic transformation of the midwest. *Economic Geography* 72:315–35.

Flynn, E., and R. Forrant. 1997. The manufacturing modernization process: Mediating institutions and the facilitation of firm-level change. *Economic Development Quarterly* 11:146–65.

Forrant, R. 1995. *Metalworking in the Merrimack River Valley: Can a cutting edge be maintained?* Lowell: Center for Industrial Competitiveness, University of Massachusetts Lowell.

Forrant, R. 1997. The cutting edge dulled: The post–World War II decline of the U.S. machine tool industry. *International Contributions to Labour Studies* 7:37–58.

Forrant, R., and E. Flynn. 1998. Seizing agglomeration's potential: The greater Springfield, Massachusetts metalworking district in transition, 1986–1996. *Regional Studies* 32:209–22.

Gertler, M. 1995. "Being there": Proximity, organization, and culture in the development and adoption of advanced manufacturing technologies. *Economic Geography* 71:1–26.

Gittell, R., and P. Flynn. 1995. Lowell, the high tech success story: What went wrong? *New England Economic Review* (March–April): 57–60.

Granovetter, M. 1985. Economic action and social structure: The problem of embeddedness. *American Journal of Sociology* 91:481–510.

Harris, R. 1996. The impact of the University of Portsmouth on the local economy. *Urban Studies* 34:605–26.

Harrison, B., M. Kelley, and J. Gant. 1996. Innovative firm behavior and local milieu: Exploring the intersection of agglomeration, firm effects, and technological change. *Economic Geography* 72:233–58.

Hounshell, D. 1984. *From the American system to mass production, 1800–1932: The development of manufacturing technology in the United States.* Baltimore, Md.: Johns Hopkins University Press.

Indergaard, M. 1996. Making networks, remaking the city. *Economic Development Quarterly* 10:172–87.

Isserman, A. 1994. State economic development policy and practice in the United States: A survey article. *International Regional Science Review* 16:49–100.

Kelley, M., and A. Arora. 1996. The role of institution-building in U.S. industrial modernization programs. *Research Policy* 25:265–79.

Knudsend, D. 1997. What works best? Reflections on the role of theory in planning. *Economic Development Quarterly* 11:208–11.

Kodama, F. 1986. Japanese innovations in mechatronics technology. *Science and Public Policy* 13:44–51.

Labrianidis, L. 1995. Establishing universities as a policy for local economic development: An assessment of the direct impact of three provincial Greek universities. *Higher Education Policy* 8:55–62.

Lampe, D. 1988. *The Massachusetts miracle: High technology and economic revitalization.* Cambridge: MIT Press.

Lazonick, W., and M. O'Sullivan. 1996. *Corporate governance and corporate employment: Is prosperity sustainable in the United States?* Lowell: Center for Industrial Competitiveness, University of Massachusetts Lowell.

Leslie, S., and R. Kargon. 1996. Selling Silicon Valley: Frederick Terman's model for regional advantage. *Business History Review* 70:435–72.

Lichtenstein, G. 1992. *A catalog of U.S. manufacturing networks.* Gaithersburg, Md.: National Institute of Standards and Technology.

Markusen, A. 1996. Sticky places in slippery space: A typology of industrial districts. *Economic Geography* 72:293–313.

Moore, C. 1997. *Connection to the future: An analysis of the telecommunications industry in Massachusetts.* Boston: Massachusetts Telecommunications Council.

National Institute of Standards and Technology (NIST). 1994. *Manufacturing extension partnerships.* Washington, D.C.: U.S. Department of Commerce.

Nowak, J. 1997. Neighborhood initiative and the regional economy. *Economic Development Quarterly* 11:3–10.

Osborne, D. 1988. *Laboratories of democracy: A new breed of governor creates models for national growth.* Boston: Harvard Business School Press.

Osborne, D. 1989. *State technology programs: A preliminary analysis of lessons learned.* Washington, D.C.: Council of State Policy and Planning Agencies.

Reese, L., and D. Fasenfest. 1997. What works best?: Values and evaluation of local economic development policy. *Economic Development Quarterly* 11:195–207.

Rosenberg, Nathan. 1994. *Exploring the black box: Technology, economics, and history.* New York: Cambridge University Press.

Rosenfeld, S. 1997. Bringing business clusters into the mainstream of economic development. *European Planning Studies* 5:3–23.

Saxenian, A. 1990. Regional networks and the resurgence of Silicon Valley. *California Management Review* (Fall): 89–92.

Scott, A. 1995. The geographic foundations of industrial performance. *Competition and Change* 1:51–66.

Shapira, P. 1990. *Modernizing manufacturing: New policies to build industrial extension services.* Washington, D.C.: Economic Policy Institute.

Shapira, P. 1996. Modernizing small and mid-sized manufacturing enterprises: U.S. and Japanese approaches. In *Technological infrastructure policy: An international perspective,* ed. M. Teubal, D. Foray, M. Justman, and E. Zuscovitch. Amsterdam, Netherlands: Kluwer Academic Press.

Subramanian, S. K., and N.Y. Subramanian. 1991. Managing technology fusion through synergy circles in Japan. *Journal of Engineering and Technology Management* 8:313–37.

University of Massachusetts Lowell. 1997. *Interdisciplinary centers and institutes.* Lowell: Office of Public Information, University of Massachusetts Lowell.

The Lowell Center for Sustainable Production

Integrating Environment and Health into
Regional Economic Development

David Kriebel, Ken Geiser, and Cathy Crumbley

AS CORPORATIONS and other workplaces move into the next century, they find themselves increasingly challenged by seemingly contradictory and costly demands: in addition to remaining economically viable, they are expected to protect the environment, conserve energy, undertake quality assurance programs, provide employees with training as well as a wide range of social supports and services, be good neighbors to their communities, hire minority workers, and so on. Many of these expectations are seen as contradictory to the healthy balance sheet that owners and stockholders demand, and that is the sine qua non of any business. How can a firm sustain itself, while moving beyond narrow profitability thinking? We believe that the answer lies in broadening the fundamental design criteria for the productive activities upon which the firm is founded to include an explicit and comprehensive commitment to sustainability.

In 1995, the University of Massachusetts Lowell established a new center, the Lowell Center for Sustainable Production (LCSP), to promote systems of production that are more compatible with the ecological cycles of the earth and more protective of those who work within them. The LCSP is carrying out a series of demonstration projects whose long-term goal is to redefine environmentalism and occupational health and safety and to demonstrate how these concepts are compatible with new systems of production that are both economically and environmentally sustainable. In this essay, we present our views of the role of a university environmental and occupational health center in regional economic development and propose a strategy by which we believe the LCSP can make important contributions to environmental quality and the economic life of the region's firms.

295

Production: The Common Origin of Environmental Pollution, Workplace Hazards, and Economic Development

For nearly three decades, the United States has made an intensive and very costly effort to reduce the burdens of pollution, such as smog, acid rain, toxic emissions, contaminated effluents, and hazardous waste. Although there has been notable progress, these remedial efforts have not been as effective as originally intended. Both industry and government are increasingly recognizing that it can be far more effective and less costly to remove a chemical entirely from a production process than to try to control its release into the environment. This basic lesson, that pollution is best prevented at its source, provides the key to future improvements in environmental quality, and we believe, to economic development as well. Smog and nitrate (a major cause of acid rain), for example, originate in chemical reactions within the internal combustion engines that drive cars and trucks. Efforts to control these emissions with exhaust pipe devices have not been as successful as initially intended. Control devices are often ultimately ineffective because the technologies are costly, require continuous maintenance, and do not contribute directly to making or adding value to a product. As a result, regulators in California and the Northeastern states have been pushing the auto industry to introduce electric vehicles—thus proposing a fundamental redesign solution to the environmental degradation resulting from the transportation system, rather than attempting to retrofit a solution onto the existing technology.

Like the automobile's catalytic converter, other "end-of-the-pipe" devices have also failed to remedy industrial pollution, have proved tremendously costly, and have pitted industry against environmentalists. In a growing number of instances—the substitution of water-based cleaners for ozone-destroying and toxic solvents, for example—the high costs and ineffectiveness of traditional control strategies have pushed industries toward inherently safer ways of producing goods and services. It is this shift in strategy that the LCSP seeks to promote.

Until recently, government regulatory agencies have been guided almost exclusively by the strategy of control in their regulation of environmental health and safety. This has resulted in lengthy, acrimonious, and expensive debates over what constitutes "safe" levels of pollution or "acceptable" health risks. Slowly, government agencies are moving from command and control strategies to more preventive and facilitative approaches. The U.S. Environmental Protection Agency (U.S. EPA) and many key state environmental agencies have recognized strategies that go to the root of the problem by designing out the environmental hazard or pollutant. These strategies, called variously "pollution prevention," "cleaner production," or "toxics use reduction," are exciting new approaches for improving environmental quality while avoiding the seemingly inevitable conflict between industry and government regulators. But LCSP believes that up until now much of the work in this new direction has lacked a key component; there has been almost no consideration of the jobs that interact with the production processes that are to be changed, or the associated worker health and safety issues inherent in these jobs. This unfortunate gap between environmental and worker health and safety goals is one of the arguments for a "sustainability" perspective that overarches both of these fields.

The hazards of the workplace, including toxic chemical exposure, physical and ergonomic risks, and psychological stress, also originate from the design of production processes. Industrial hygiene, one of the core disciplines of the work environment profession, has long recognized that disease prevention is far more effective when a hazardous process or chemical is eliminated altogether than when the emissions of one technology are controlled by means of another technology. But all too often the work environment professional is not involved in the fundamental design decisions and must resort once again to "end-of-pipe" solutions. Ventilation, for example, has traditionally been one of the major tools used by the industrial hygienist to prevent worker overexposure to toxic agents. But ventilation moves the pollutants to the outside air, which must then be protected by installing pollution control devices in the vents. These control devices may reduce air pollution to some degree, but they only displace the problem. The control systems must be cleaned and the trapped toxic agents disposed of, often exposing workers—and ultimately the environment—to their hazards.

If changes in production aimed at a cleaner and healthier environment are to be effective, we must be cognizant of the social and political aspects of the environment of production. Workers, for example, are not merely residents of workplaces whose health must be protected. They can be invaluable contributors to production redesign because of their intimate knowledge of the workplace. But a worker who does not have a voice in routine work decisions, or who fears unemployment, will not be a constructive contributor to change. Sustainable forms of production must not only protect workers from occupational hazards, but must involve workers in the design of workplaces that are both productive and rewarding.

The focus on controlling pollution rather than preventing it is one of the reasons for the perceived conflict between environmental quality and economic development. The cost of a control device typically rises exponentially with its efficiency, so that a device designed to remove essentially all of a pollutant becomes prohibitively expensive. Moreover, since the controls cannot reduce emissions to zero, as economic activity expands and production increases, total emissions rise and eventually cancel the environmental value of the devices. An approach to environmental quality based on control devices can only be effective if it limits production and hence curtails economic activity. Thus, we see that control strategies inevitably perpetuate conflicts between environmental quality and economic development.

Sustainable Production: A Working Definition

The Lowell Center for Sustainable Production believes that environmental quality and safe and healthful workplaces can be achieved while at the same time enhancing the economic life of firms, through technologic and organizational changes in the ways that goods and services are produced, by incorporating ecological and human health considerations into the fundamental engineering design process at the root of all industrial processes. It is our thesis that many production processes can be safe, clean, and profitable if they are designed from the start to meet all these criteria.

Production is here defined in a broad sense to mean the creation of goods and services; whether this involves industrial processes, agricultural operations, commercial activities, transportation, or social and community services. We believe that this definition is relevant, whether production is organized as private, for-profit operations, nonprofit operations, or government functions.

By sustainability we mean to capture the sense of the term "sustainable development" promoted by the 1987 report by the World Commission on Environment and Development as development that meets the needs of the present without sacrificing the ability of the future to meet its own needs. Thus, sustainability links current organizational conditions to future consequences in a manner that assures enhancement or, at least, minimizes deterioration.

Therefore, sustainable production is defined as the creation of goods and services using processes and systems that are nonpolluting, conserving of energy and natural resources, economically efficient, safe and healthful for employees, communities, and consumers, and socially and creatively rewarding for all working people. Traditionally, these objectives are promoted independently, and, at times, even antagonistically. In contrast, sustainable production is promoted as the central organizing principle for developing new forms of enterprise that meet all of these objectives in concert. Sustainable production provides a conceptual framework for the design of production in the context of the ecological systems of the planet and the organization of work in the context of the social systems of the community, the nation, and the world. Indeed in our view, work can be organized to enhance the environment, benefit employees and communities, while at the same time leading, always in the long term and often in the short term, to more economically viable and productive enterprises.

The following nine principles serve as guidelines for implementing programs in sustainable production:

1. Products and packaging are designed to be safe and ecologically sound throughout their life cycle, and services are designed to be equally safe and sound;

2. Wastes and ecologically incompatible by-products are reduced, eliminated, or recycled.

3. Energy and materials are conserved, and the forms of energy and materials used are most appropriate for the desired ends.

4. Chemical substances, physical agents, technologies, and work practices that present significant hazards to human health or the environment are eliminated.

5. Workplaces are designed to minimize or eliminate physical, chemical, biological, and ergonomic hazards.

6. Workplace decision making is an open and participatory process of continuous evaluation and improvement, encouraging the long-term over the short-term economic viability of the enterprise.

7. Work is organized to conserve the efficiency and enhance the creativity of all employees.

8. The security and well-being of all employees is a priority, as is the continuous development of their talents and capacities.

9. The communities around workplaces are respected and enhanced economically, socially, culturally, and physically with a commitment to social equity and fairness.

University–Industry–Labor–Community Partnerships for Sustainable Production

The LCSP is organized in the Department of Work Environment and the Toxics Use Reduction Institute at the University of Massachusetts Lowell. Extensive experience working with industry, labor, and communities through these two organizations has convinced us that the university can play an important role in promoting sustainable production in this region. LCSP is building close working relationships with firms, unions, and environmental organizations with the goal of bringing technical, educational, and organizational expertise to partners interested in making fundamental changes in production that move toward a sustainable economy. LCSP requests assistance from faculty and staff from other centers and institutes and from the broader university community when specific expertise is needed on a project. In this way, LCSP can be a gateway through which the university's technical and educational capability can be brought to the region, while at the same time promoting a clear vision of environmental quality and healthful living conditions.

The Role of Firms

LCSP works directly with firms to help them make fundamental changes in production. In many industrial sectors, there are now a few forward-looking corporations that have chosen an explicit strategy of environmental quality. There are various motivations: customer demand, a "green" market niche, inspiration from European competitors, and others. These companies are often willing to consider eliminating toxic chemicals, changing technologies, or redesigning the organization of work.

We have observed that some firms have indeed made impressive strides toward reducing their environmental impacts, but these have often failed to consider occupational health and safety as a part of this effort. One of LCSP's goals is to broaden the definition of "green" to include the quality of worklife. We believe that many firms will find this in their own interest, because employees will contribute more enthusiastically to an environmental strategy that includes their own health and safety concerns.

The Role of the Environmental Movement

There are reasons to believe that a new phase is emerging in the evolution of the political movement for environmental quality. In many ways—sometimes explicitly, sometimes not—the environmental movement has traditionally focused on changing consumption and has not engaged in debates about how things are produced. But as noted above, the environmental control strategies of the 1970s and 1980s have not been particularly successful, especially considering the amount of money invested. Despite continued popular support for environmental protection, it is becoming increasingly difficult to secure the political and financial capital to continue the strategy of retrofitting high-tech environmental controls to existing technologies. This has begun to focus attention among environmentalists on pollution prevention and other strategies that change the basic technologies of production rather than attempt to clean up after them. At the same time, the environmental-justice movement has brought themes of social and economic equity to the environmental movement. This is beginning to force environmentalists to realize that a future with jobs and dignity for all requires economic activity—production—and not just reducing consumption.

The Role of Labor

Like environmentalists, labor unions too have pursued a "consumptive" rather than "productive" strategy, focusing on wages and, to a lesser extent, working conditions, while leaving entirely to corporations the decisions about what is produced and what technologies are employed in production. This coincidence of strategies with environmentalists is somewhat ironic, because it has had no discernible impact on the ability of environmentalists and labor unions to work together. Now, at least some labor unions are reconsidering their essentially defensive position of the last thirty years, and in some sectors are willing to discuss entering into more collaborative arrangements with employers in which the design and organization of work are more democratically managed.

The nature of production profoundly affects the quality of the lives of workers and their families—not only through the nature of the work and its hazards, but also by virtue of the status of workers in the workplace and in the community. As noted above, workers are often an untapped resource in developing new ideas for production and products, and truly sustainable production will need to include them as participants in the redesign process.

The Role of Government Regulatory Agencies

As noted above, government regulatory agencies have, until recently, been guided almost exclusively by the strategy of control in their regulation of environmental health and safety. This has resulted in lengthy, acrimonious, and expensive debates over what constitutes "safe" levels of pollution or "acceptable" health risks. Slowly, government agencies are moving from command and control strategies to more preventive and facilitative approaches. The U.S. Environmental Protection Agency (U.S. EPA) and many key state environmental agencies have adopted pollution prevention as the premier compliance strategy. At the same time, our experience working with many industries convinces us that, at least for the foreseeable future, compliance with essential environmental and health and safety regulations will continue to be a critical strategy by which government seeks to ensure environmental quality. Thus, we envision a dual approach; governments will seek partnerships with industry, labor, and communities promoting the transition to sustainable production, while at the same time continuing to require adherence to environmental regulations.

The Role of the University

Through the Department of Work Environment, we train masters and doctoral students as work environment professionals. The sustainable production perspective has important implications for this training. Occupational and environmental professional practice currently focuses on the development of solutions to environmental problems after they have occurred. Too often, the development of solutions begins from the identification and quantification of the harm that has been accepted as inherent in production processes, rather than from a review of the purpose and function of the production process itself. Practitioners are channeled into specializations that emulate basic sciences and thus view production processes as fixed. This approach continually reduces an end-of-pipe problem to a smaller scale in pursuit of the single, causal agent or fundamental physical principle that governs it. Such focus on first principles is useful because it can provide a sound scientific basis for controls. We are losing the ability, however, to consider simultaneously the larger-scale forces that are the social, economic, and political context of production processes—or even to see such processes as dynamic. It is as if we are looking through an ever-more-powerful microscope lens that focuses on the molecular level but obscures the whole organism.

While the fields of occupational and environmental engineering and public health are experiencing significant pressure within traditional academia to prove that they are "real science," other components of society are recognizing that the narrow solutions resulting from such an approach are not sufficient. There is growing acknowledgment, even among some sectors of private industry, that the production processes that generate the problems may need fundamental redesign. This means that the work environment professional beginning a career in the new century must be trained to be much more flexible, collaborative, and creative than in the past. The transition to sustainable production calls for work environment professionals who are not simply trained to monitor a static industrial environment, but who are agents of change.

There are three important roles for a university in the development of this new approach:

First, engaging with corporations, communities, and others on the fundamental redesign of technologies requires considerable technical expertise, which universities may provide.

Second, a university is a place where new ideas can incubate relatively insulated from economic and political realities that may prematurely frustrate their application.

Third, the university is a third party that may be able to introduce a new approach to an old problem. Traditional adversaries are understandably suspicious of each other, and may in fact have little basis for trust. The university may serve as common ground for the discussion of collaborative approaches that break down old divisions.

Sustainable Production in Action: Some Examples

One example of a firm with a demonstrated commitment to sustainable production is Malden Mills Industries, Inc., the largest textile firm in the northeastern United States, with more than thirty-three hundred employees in its four mills. In December 1995, a major fire destroyed a large portion of the Lawrence mill site. Fortunately, no one was killed in the disaster. The president of the firm, Aaron Feuerstein, immediately committed the company to rebuilding on the original site. In the two and a half years after the fire, LCSP worked intensively at the Lawrence mill site, assisting the rebuilding effort in numerous ways. This work has been in three broad areas.

1. The LCSP organized and implemented a "team building" strategy that brought labor and management together to set consensus priorities for environmental improvements in the plant. At the request of the Malden Mills Re-Design Steering Committee (in charge of building the new plant to replace the one that burned), a new team process was designed for management and union operators to identify hazards related to past production processes and to develop solutions to be considered during the plant re-design.

2. LCSP has also provided technical expertise on a variety of environmental, health, and safety issues. It provided an industrial hygienist to assist with a plantwide hazard review and to integrate health and safety into the production processes by participating in the work of multiple health, safety, and ergonomics committees as well as process engineering committees established to redesign the new plant and production processes.

3. Finally, the LCSP has helped Malden Mills to organize and integrate its numerous environmental activities.

Another LCSP project helps improve the environmental as well as the occupational health practices of hospitals. Mercury, heavy metals, and endocrine-disrupting chemicals are present in many health care products and threaten the health of patients, health care workers, and communities. Incineration of medical wastes can produce dioxin and disperse this toxin as well as heavy metals into the atmosphere. In response to these environmental and health concerns, the LCSP's Sustainable Hospitals Projects provides technical support to the health care industry for selecting products and work practices that eliminate or reduce occupational and environmental hazards, maintain quality patient care, and contain costs. The project includes: A Web-based clearinghouse (www.sustainablehospitals.org) that allows quick access to vital information about alternatives to products containing mercury, latex, polyvinyl chloride (PVC), and other potentially harmful materials. The project researches and disseminates information such as product performance specifications, infection control and other patient care issues, occupational health and safety, work practice implications, environmental quality, costs and benefits, and field reports from hospitals. Another goal of the project is to develop a comprehensive model of product and work-practice selection in health care that links pollution prevention with occupational safety and health. The researchers are working directly with several area hospitals to develop and pilot programs.

In a third example, the LCSP has developed a new program initiative to develop and test indicators of sustainable production that can be applied to individual firms. This project consists of two components: creating and testing indicators in demonstration projects with two firms—Stonyfield Farm, Inc., and Guilford of Maine. The project is developing guidelines, lessons, and case studies that can be applied in other firms. And, second, it is developing and testing curricula and training programs on indicators of sustainable production that are suitable for both industry groups and citizen organizations. The training programs will increase understanding of sustainable production indicators and use them to strengthen sustainability initiatives.

Along with the Center for Clean Products at the University of Tennessee, the LCSP has worked with national, regional, and community-based environmental, environmental-justice, consumer, and labor organizations as part of the University–Public Interest Partnership on Clean Products and Clean Production. The project recognizes that new tools and policies are needed to foster fundamental design changes that will improve occupational and environmental health. Western European governments are experimenting with entirely new policies, such as extended producer responsibility and life-cycle analysis, from which the United States can learn. Some leaders in the U.S. environmental and labor movements have begun to recognize that such product-oriented approaches are useful for changing systems of production.

National training and strategy workshops have been organized as a part of this project. These have resulted in the decision to develop a Clean Production Network. Under the leadership of a national steering committee, the university partners will provide training, research, and other technical assistance deemed strategically important for the project. The project has commissioned a "Citizen's Guide to Clean Production," which is being distributed internationally and provides technical support to two campaigns: the Clean Car Campaign and the Clean Computer Campaign. Both of these campaigns promote progressive and profound changes in their respective industries. They employ strategies of policy advocacy, direct engagement with companies, international coalition building, and increasing consumer awareness.

In another major program initiative, LCSP is partnering with the Massachusetts Breast Cancer Coalition and the Clean Water Fund to develop grassroots understanding and support for the precautionary principle. The precautionary principle says that, in the absence of scientific certainty, it is better to err on the side of caution—than to commit an act with unproven but potentially hazardous consequences. Adopting a more precautionary approach would have profound consequences for environmental and social policy. Precaution changes the question asked when decisions are made from, "What level of risk is safe" to "What are the alternatives to this activity that achieve a desired goal (service, product, etc.)?" The principle also shifts the burden of proof. Rather than presuming that a specific substance or activity is safe until proven dangerous, (a process that can take substantial time and resources), the principle sets a presumption in favor of protecting the environment and public health.

The long-term goal of the project is to change Massachusetts environmental regulatory policy to incorporate a more precautionary approach. The first two years of the project focus on such activities as building a base of support for the principle; developing model regulatory policies in which the principle is used in key aspects of environmental regulation; and training community advocates and decision makers on the limitations of science and regulations in adequately protecting public health and the environment. In addition to joint project activities, the Lowell Center role includes facilitating a series of meetings with scientists to discuss limitations on the use of science in environmental and public health decision making; how best to use science to protect public health and the environment in the face of uncertainty; and how academics can work with grassroots groups to further public health and environmental objectives.

In addition to the University–Public Interest Partnership described above, another LCSP project explicitly brings together labor and environmental representatives. The New England Pilot Project for Labor–Environment Dialogue is developing a strategy for linking labor and environmental groups there around work environment and ambient environmental problem solving. The project established a working group of labor and environmental representatives along with LCSP faculty and staff. To facilitate the process, LCSP is organizing a roundtable on labor, environment, and sustainable production and will draft a working paper on these issues that can be used for the development of further policies or strategies.

The Strategic Organizational Integration Project works with government agencies to develop innovative ways for staff to incorporate pollution prevention in the policy, permitting, compliance, and enforcement activities of environmental agencies. Each intervention process is tailored to meet the needs of each contracting agency, but involves the following four phases.

Collecting data and analyzing the extent to which pollution prevention is, should, and could be practiced in the agency at the individual, group, and organizational levels.

Developing and conducting an interactive meeting with staff to understand how they view possibilities for incorporating pollution prevention in their daily work and decision making.

Conducting an impact evaluation survey nine to twelve months later to determine the extent to which the agency has implemented recommended initiatives.

Conducting an impact evaluation meeting for participants to share successes, failures, lessons learned, and new initiatives.

The intervention process has been undertaken with environmental agency personnel in the states of Pennsylvania, Maryland, and New Mexico.

At the national level, LCSP has been working with the EPA, the National Institute for Occupational Safety and Health (NIOSH), and the Occupational Safety and Health Administration (OSHA) to develop strategies that integrate worker safety and health with environmental regulatory approaches. Agency representatives participated in a national conference to develop priorities and a research agenda in this area. Regional conferences are being planned for further elaboration of the issues.

The Community-Developed Indicators of Sustainability Project increases awareness of sustainability and assists communities in developing high-quality indicators that they can use in setting priorities and measuring progress toward a more sustainable future. The project offers community workshops on sustainable communities, trains trainers in community indicators, and provides technical assistance to groups and agencies interested in using indicators. These services are offered to all ten EPA regions. The project has produced a "Guide to Sustainable Community Indicators" and has a Web site (www.uml.edu/centers/LCSP/web_indicators.html) that contains the manual, training materials, a searchable database of over eight hundred sustainability indicators, and a reference service.

Some Issues Encountered and Lessons Learned

From our project thus far we have learned a number of lessons including the following:

1. Many of the traditional methods and tools for health, safety, or environmental management are still quite relevant for sustainability projects. What distinguishes LCSP projects from more traditional technical assistance is the explicit plan to meet the much broader sustainable production goal of integrating health, safety, and environment into production. This larger goal serves as a guide for developing the details.

2. With the important exception of Guilford of Maine, each of the firms we have worked with is not publicly traded or is a nonprofit organization, which does not have to answer to stockholders and can accomplish

larger goals than maximizing short-term profits. A crucial challenge now is to make the case for publicly traded companies' adoption of strategies more aligned with sustainability principles. The Guilford of Maine demonstration project should provide us with important insights on this issue.

3. Many of the firms we work with are on the cutting edge of sustainability practices. LCSP is working with them to develop the tools, concepts, and strategies that other companies will want to use eventually, as the strategic advantages of sustainability practices become more widely understood. Yet even in those companies that are in the forefront of sustainability, business pressures and/or cultural inertia cause many employers to continue to focus only on generating adequate financial profit. There is thus a great need for continued research and experimentation with strategies for organizational change and learning that enable a mission of sustainability to be incorporated throughout the firm.

4. In our partnerships with citizen organizations, maintaining open communication and listening to one another's needs are crucial, as is understanding the constraints under which the other operates. Because it is often difficult to understand the needs and constraints of academic research, it is important for us to clearly articulate our needs, goals, and what we can and cannot do. Citizen groups have tremendous needs for technical assistance and often assume that we have expertise or information at hand when we do not. Nor do they understand what would be involved in our obtaining it. For their needs, our research may seem to be too slow. Furthermore, they often don't understand how to articulate their needs so that we can be most helpful. Thus, an important role for us is to help them articulate their needs so that we can respond thoughtfully and to the best of our abilities.

5. The work of LCSP is inherently multidisciplinary; and multidisciplinary work is difficult and time consuming. Project personnel must be able to take the time to garner relevant information from other disciplines so that a common base of understanding and action can be developed. We are still experimenting with how to conduct this mutual learning in the most efficient way.

6. Cleaner, healthier products are a key component of sustainable production. Several of our projects have demonstrated both the importance and the complexity of influencing purchasing decisions to support the use of better products. For example, in the health care industry, a complex maze of purchasing contracts limits the entry of new products to the market on an equal footing with the established ones.

7. We are playing an important role in developing publications that provide practical applications of sustainable production principles. Handbooks, guides, evaluations, and checklists have been developed for the Malden Mills project, Sustainable Hospitals Project (SHP) clearinghouse,

the Precautionary Principle project, and the Clean Production Network. These have established a niche for us that joins science with management and policy. The fact that these materials come from a university research center lends additional credibility to their use by our project partners and others.

Future Prospects

This essay has described opportunities for new models of collaborative environmental change organized around the transition to sustainable means of production. To move in this direction, it is necessary first to demonstrate to communities, labor, corporations, and the government that collaborative approaches can be successful and that a production-oriented approach is technically and economically viable. This strategy could have a more powerful political base than the old approach to environmental problems that emphasized control of hazards rather than their elimination, because the focus on redesigning production rather than on nonproductive and only marginally effective control technologies means continued economic development—that is certainly in the interests of working people, industrialists, minority communities, labor unions, and government bureaucrats. Environmentalists often have strong knee-jerk reactions against any discussion of "growth," but if production is truly sustainable, it need not harm the environment.

For sustainable production to be a rallying cry of a broader and more effective movement, concrete examples are required: Hope depends on even small successes in the transition from our consumption-oriented economy to something more sustainable. Just as we need to build political coalitions for social change, we need to demonstrate that sustainable production can mean environmental improvement for communities, good and safe jobs for workers, and economic viability for managers. Positive examples of collaborative projects moving toward sustainable production can also serve as ammunition for those pushing recalcitrant parties to change by showing that it *is* possible. Through partnerships that yield practical successes, education and training, and dissemination of our work, we propose to promote sustainable production as the strategy for environmental quality for the new century.

REFERENCES

World Commission on Environment and Development (WCED). 1987. *Our Common Future*. New York: Oxford University Press.

Effective University–Industry Partnerships in Photonics

Michael A. Fiddy, Dikshitulu Kalluri, and Julian D. Sanchez

MANY tensions exist between providing academic programs and opportunities for students while simultaneously reacting positively to pressures for help from industry. What should a university's commitment be to research and to industry? Survival of an effective research and development effort in our institution is examined in this context, further pressured by the increasing competition for federal funding (Massachusetts Technology Collaborative 1999). The annual level of R&D funding to academia has not declined, but the annual rate of growth in federal support has been falling (NSF Science and Engineering Indicators 1998). Universities typically rely on federal support for equipment infrastructure and maintenance support in order to remain competitive. The type of organizational structure and funding necessary to ensure long-term research advances must also guarantee the resources needed to provide a timely response to the needs of industries.

CEMOS, the Center for Electromagnetic Materials and Optical Systems at UMass Lowell, specializes in modeling electromagnetic scattering and imaging and photonic materials, devices, and systems. We use CEMOS as an example to illustrate some of the issues that arise when university–industry partnerships are to be fostered. CEMOS has particular and unique expertise in modeling electromagnetic wave propagation in complex media, imaging from scattered field data (nondestructive testing, medical imaging, synthesis, etc.) and developing prototypical optical systems for automated inspection and remote sensing. Although this might seem a broad array of activities, there is a unifying conceptual and theoretical core that underpins these seemingly diverse research efforts. If it is focused on advancing our understanding of the long-term and fundamental questions at the core, the center can remain at the leading edge of its discipline. In this way, by applying developments derived from the fundamental issues, it can remain nimble in responding to requests for proposals in a wide range of applications grounded in that core. However, over the passage of time, without new equipment and support to maintain laboratories properly, the

309

R&D effort can become increasingly theoretical or simulation based. It might be argued that this is fine and that we should not try to compete with the multimillion dollar R&D efforts already in place in industry and some (Carnegie I) research institutions. Nevertheless, the long-term viability of our effectiveness requires that this point be addressed.

Industry frequently is forced to choose to settle in those geographical locations around the world where the educated workforce and other necessary technical expertise and support are found. To nurture the technological base in a region requires both a recognition of the needs of industry as well as those of the neighboring universities. An efficient balance is required between industry's need to constantly review relocation and training options and a university's ability to re-evaluate curricula and research directions. Should a university stick to "basics" and allow students to build on them once employed or continually change to try to meet the apparent needs of industries.

The anticipated impact of photonic technologies is a good illustration of some of these issues. It brings into question what is fundamental to a curriculum for the future and how intimate the relationship should be between a university and concerned industries. Understanding the propagation and modulation of light in different materials, devices, and systems defines the core for photonic technologies. This is inherently a multidisciplinary field, not widely taught at any level in physical science or engineering curricula at present. It overlaps with many traditional and established areas, such as the broad range of electromagnetic R&D, particularly involving materials, microwaves, and antenna research. These technical core competencies remain, of course, of wide interest to many companies in this region of New England. Given this setting, it would seem inevitable that the resources we have to offer at our particular center, CEMOS, would be of interest to the surrounding high-tech community, with both long- and short-term agendas.

In 1993, at the Government-University-Industry Research Roundtable and the National Science Board assembly, six specific policy issues were identified by a majority of participants (NSB 1993), and those issue areas were:

Creating and communicating priorities in research and education;
Balancing research and education activities;
Facilitating multidisciplinary research and education;
Identifying patterns of institutional support for research;
Restoring a sense of community on campus; and
Developing relationships with new partners in research.

Why Photonics Is Important

Photonics is the marriage of electron physics with optical or photon physics. The centennial of the discovery of the electron and the construction of the first cathode-ray tube was celebrated in 1997. In the year 2000 we celebrated the hundredth anniversary of the discovery of the photon by Max Planck. While the relationship between light and electricity has been exploited for years, it is only relatively recently that advances in semiconductor materials have led to an explosion of photonic applications.

As communications occur at higher and higher bandwidths, electromagnetic phenomena become increasingly significant in network and chip design. Already computers operate at greater than gigaherz (GHz) clock speeds, that is, at $>10^9$ cycles per second, which is in the microwave region of the electromagnetic spectrum. Also, optical fibers will soon bring gigabit data rates into our homes, enabling us to capture enormous amounts of information in fractions of a second. Optoelectronic or "photonic" devices will play an increasingly important role because they have the ability to handle these rates. Information technology will not reach its full potential until photonics technologies bring high-bandwidth communications into every home, business, and desktop. This is being realized in Japan, where NTT is committed to having fiber in all homes (or sites) of all their customers by 2010.

Optical fibers for communications, lasers for printing, and compact disks read by lasers for information storage are but a few examples of photonic technologies. Fueling the photonics revolution is the need (the seemingly endless need) for bandwidth for communications and information technologies. The next-generation Internet is being planned with a 100 gigabit per second (Gbps) backbone and end-user bandwidths of 1 Gbps. Switching speeds at 30 to 50 Gbps will require innovations in materials and design. Funding for this will come from both the commercial and government sectors. The Japanese government's decree that fiber be laid into every home by 2010 will stimulate the necessary innovation and generate the funding to make it happen.

The potential explosion in the communications industry can perhaps be appreciated by pondering two related facts. More than 50 percent of the world's population has never made a telephone call and, despite the fact that more than 125 million kilometers of optical fiber have been laid (equivalent to 160 round-trips to the moon), this represents only a 5 percent penetration of network installations. We are clearly still at the dawn of the information age. Photonics is already a $100–150 billion industry and is likely to grow at a rate exceeding that of the electronics industry when it was entering the transistor age (Weiss 1999).

Photonics as an Enabling Technology

A report by a committee of the National Research Council (NRC 1998) predicted that lightwave technologies will lead to a technology revolution having a "pervasive" impact on life in the new century. Optics is described as a critical enabler for technology that promises to revolutionize the fields of telecommunications, medicine, energy efficiency, defense, and manufacturing, as well as the frontiers of science. The report challenges both industry and federal agencies to ensure the rapid development of these new technologies with R&D investment in key optical areas such as photonics, sensors, and low-cost manufacturing of precision optical components. Progress in materials science and engineering is a critical part of this effort, and several agencies were identified to take responsibility for this. The report stated that universities should encourage multidisciplinary optics education that cuts across department boundaries and that they should provide research opportunities at all levels. Cutting across institutional boundaries should also be considered.

The United States has a strong base of fundamental research and basic technology in photonics, and device performance but high-volume production is low—largely as a result of the nature of past government funding. The four general areas in which the U.S. photonics industry believes opportunities exist are imaging, new information-age applications, medical technologies, and transportation (Photonics Manufacturing 1997).

Photonics is also a fast moving and an exciting field to work in. It is inherently interdisciplinary, needing the skills of materials scientists, optical engineers, physicists, chemists, and electrical engineers. The energy and size (through parallelism and low cross talk) battles between electron-based systems and photon-based systems have been fought and largely won by the photon. An industry trying to compete today in the electronics sector must think about how it should integrate optical elements or components into its products. Skills in optical engineering are not usually acquired in science and engineering curricula, making advances in electro-optics slower than they might be because of this general lack of expertise. The inevitable inertia to change is exacerbated when expertise is lacking. In a fast-moving global economy, a lack of confidence in adopting new technologies can be fatal. It results in an evolutionary rather than revolutionary growth in photonics in practice.

The biggest driver of today's photonics markets is probably components for networking, communications, and multimedia, products that are emerging as information technologies are coming together (e.g., Weiss 1999). As higher bandwidth networks are installed, applications will increase, and users will demand even higher bandwidths, more processing power, more storage, better displays, and better hard copies. If the twenty-first century becomes known as the Information Age, it will have followed the Industrial Revolution and the Electronics Age, and the dominant technology will be optics.

The present and anticipated economic impact of photonics is substantial. According to 1997 figures published by the Optoelectronics Industry Development Association (OIDA 1998), the advances in optoelectronic components (a $16-billion industry) have "enabled" a $60-billion optoelectronic equipment industry, which in turn has "enabled" a $24-trillion service industry (Photonics Manufacturing 1997). More recent data reinforce this point (*Photonics Spectra* 2000). There would be no laser printers without the semiconductor laser and similarly no CD-ROM or DVD-ROM (digital versatile disk–read-only memory). OIDA has already projected the demise of the CD-ROM, with sales declining from 1999, as DVD sales take off. Photonic disciplines include light-emitting devices, detectors, optoelectronic integrated circuits, displays, optical fibers, connectors, storage media, and several other more specialized subdomains such as optical micromechanical machines.

Photonics is already 10 percent of the electronics industry and growing faster than the electronics industry did when we entered the transistor age. Forecasts of a 20 percent to 30 percent growth in photonics are very persuasive, and we at CEMOS would like not only to be a part of this but perhaps to take a leadership role (see Jessup and Lamont 1999 for some specific examples).

Who Leads in Photonics?

According to Optoelectronics Industry Development Association data (OIDA 1998), Japan is clearly in the lead when it comes to photonic systems for which there are large markets. These include display and optical storage technologies. The United States leads in networks, sensors, and devices. Japan dominates the photonic market share, while the United States is still a technology leader (see table 1). Their roles were reversed in 1980, and the Defense Advanced Research Projects Agency (DARPA) was largely responsible for this, with photonics R&D being largely stimulated by military rather than commercial needs. The United States has to compete in the global market, however, and it has to address the lack of technicians and engineers with the skill sets needed to integrate photonic components into products. Such specialists are scarce, because there are very few programs providing an education in photonics. This subject matter is not typically found treated at any great depth in electrical engineering, computer engineering, or physics curricula. Lack of access to basic research and lack of availability of a trained workforce will certainly slow U.S. growth. Nevertheless, the United States remains very strong in communications (OIDA 1999).

Table 1. Percentage distribution of photonics patents and publications, among the U.S., Japan, and rest of the world, 1991 to 1996

	US (%)	Japan (%)	Rest of world (%)
Photonic publications	34	15	51
U.S. photonics patents approved	48	37	15
Japanese photonics patents approved	7	89	4
Total number of photonics patents	17	66	15
Photonics market share	25	60	15

Source: OIDA, *Annual report* (Washington, D.C.: OIDA, 1998).

Note: Notice the strong correlation between the number of photonics patents and market share.

There are few regional centers of photonics activity in the United States. One is Silicon Valley, another is the Research Triangle in North Carolina, and the third we have here in Massachusetts on Route 128. There is purported to be a higher concentration of optoelectronics firms within thirty miles of UMass Lowell than anywhere else in the country. Of over three thousand photonics businesses in the United States, twelve hundred are located in the Northeast and four hundred in Massachusetts. EG&G is one of the earliest and most successful high-tech photonics firms in this region. It was started in 1947 by four men associated with MIT, Harold E. Edgerton, K. J. Germeshausen, H. E. Greer (EG&G), and B. J. O'Keep. It was first housed in a garage in Boston and grew to twenty-three thousand employees. Other entities born at the beginning of this region's high-tech boom include Polaroid, founded in 1937 by Edwin Land. There is also Raytheon, once the American Appliance Corporation, started by MIT graduates Laurence K. Marshall and Vannevar Bush in the early 1940s.

The United States can still compete effectively in many areas of photonics despite the fact that Japan has the majority of patents covering enabling electro-optical technologies. The United States is still leading in innovation in many key technologies but relies almost entirely on university and government laboratory research for its ideas. The latter is in decline, leaving the universities even more central to the long-term innovation generation needed to survive in a global economy.

Photonics: Funding, Clusters, Consortia, and Partnerships

The National Science Board's *Science and Engineering Indicators 1998* comprises over eight hundred pages of facts and figures about U.S. technology (Laurin 1998). It describes the changing patterns of U.S. R&D activities in the present era of smaller military budgets. This directly impacts long-term funding for the research infrastructure in the country. However, it is apparent from the report that R&D investment hit an all-time high in the United States in 1997, with total public and private expenditures exceeding $200 billion. The present lack of growth in federal funding has been offset by industry spending. Private-sector spending now accounts for two-thirds of research spending. Not surprisingly, these funds go to applied rather than basic research. Another factor is the decline in Department of Defense spending from two-thirds of all federal R&D spending in the mid-1980s to less than half currently. Also, since 1985, U.S. industry has increased its R&D spending abroad three times faster than its spending at home. Interestingly enough, overseas companies have increased their research spending in the United States by about 12.5 percent a year, and there are now over seven hundred foreign-owned research facilities here.

In our universities, in particular, the one hundred fifty or so research universities, the traditional partnerships with federal agencies have started to decline (Lepkowski 1997). Despite their reputation for "knowledge production," these universities are beset with major problems. In many, tuitions are escalating out of control, and the rapid evolution of information technologies and advances in distance-learning make it hard to visualize the structure of the twenty-first-century university. Moreover, the traditional doctoral students in the sciences are less attractive as employees unless they have shown themselves to be interested in carrying out applied research.

Balancing the need to educate students, create new knowledge, and meet industry's demands to solve today's problems is clearly quite a challenge. As pointed out by Helms (1998), industry needs a well-trained, agile workforce, while minimizing development and manufacturing costs. Universities could help with this but lack the flexibility and resources to do so. We believe that this need can only be met through new kinds of R&D partnerships that encourage the development of broad interdisciplinary teams and the sharing of resources to make the most efficient use of them. Our center should be clearly identified as a component in a larger enterprise, with a well-defined network of connections to the other key components. An analogy might be the convenience of the shopping mall as compared to scattered individual stores. A regional "R&D mall" can, of course, be geographically scattered but still be a virtual one-stop shopping facility.

Mutual Support through Clusters

The Optoelectronics Industry Development Association (OIDA 1998) has formed a national manufacturing alliance of members to try to help sustain internationally competitive positions. Most large U.S. companies in the photonics field are members of OIDA. There is also major support at the state level for photonics clusters of industries and universities, and these include:

AOIA: Arizona Optics Industry Association
CPIA: Colorado Photonics Industry Association
CPIC: Connecticut Photonics Industry Cluster
FEOIA: Florida Electro-optics Industry Association
GRPA: Greater Rochester Photonics Alliance
MAOI: Massachusetts Association for Optical Industry

In Arizona, for example, the cluster helps to provide links to all levels of government, assists with funding by providing links to venture capitalists and "angel" investors, and provides directories for distribution. The cluster also provides a single Web site (www.futurewest.com/gtec/optics.html), resources for personnel searches, marketing via billboards and public relations, general networking for engineers and, most important, a link to education at all levels.

In Massachusetts, the Massachusetts Technology Collaborative (MTC) has worked with several other state agencies and the Massachusetts Association for the Optical Industry (MAOI) to try to characterize the photonics industry in Massachusetts (MTC 1998b). Massachusetts companies produce a wide variety of devices, from fiber-optic technologies to lasers, all with a broad range of applications, from defense to medicine to telecommunications and beyond. The goal of the MTC collaboration with MAOI is to ascertain what steps, if any, Massachusetts photonics companies and the Commonwealth should take, singly or together, to enhance the state's leadership role in photonics. A survey was distributed to some three hundred photonics-related companies; the results of the survey and subsequent follow-up interviews were published in September 1998 (Minuteman Regional Science and Technology Optoelectronics and Photonics Task Force 1998). MTC sponsored a student team from Harvard's Business School and the Kennedy School of Government, organized under the direction of Professor Michael Porter. The team conducted extensive interviews within the industry to assess the extent of competitive advantage enjoyed by Massachusetts in the photonics industry; findings were released in May 1998 (Porter 1998).

This region received a major boost to its photonics activity with the opening of the Boston University Photonics Center at 8 St. Mary's Street in Boston. The center houses thirty-five different laboratories devoted to a wide range of photonics technologies in 235,000 square feet of fully equipped state-of-the-art space. It includes clean rooms, low-vibration areas, and facilities to study femtosecond (10^{-15} seconds) switching, quantum and nonlinear optics, near-field microscopy, displays, and laser measurement. This dwarfs the other optics efforts in the region, including those at MIT, Tufts, Northeastern University, and UMass Lowell. The Boston

University center has fifty faculty members affiliated with it and thirty funded graduate students working in laboratories, which are located next to industry incubator space. The excitement generated by such a facility has already led to the formation of a New England Electro-optics Consortium, which with over forty manufacturers have joined. The consortium will provide new product ideas, market analysis, and technical expertise.

Other models for consortia include the Photonics Research Center (PRC) at the University of Connecticut with a 33,000 square-foot building and research emphasis on communication and information processing, sensors, and laser diodes. The PRC was a home for the recent DARPA-funded Technology Reinvestment Program (TRP) and for the National Alliance for Photonics Education in Manufacturing (NAPEM), an effort not unlike the Engineering Academy of Southern New England, of which the University of Connecticut and UMass Lowell were. In Huntsville, Alabama, the Alliance for Optical Technology has recently been formed, linking the University of Alabama Huntsville, with Alabama A&M, Boeing, NASA, Oak Ridge, Teledyne Brown, Nichols Research, and ten other companies. This alliance coordinates the members' technical and pre-competitive activities in optical technology. Facilities are shared and education and training are provided in applied optical technology and manufacturing. A similar consortium is based in Rochester, New York, the Center for Optics Manufacturing. This virtual center links several universities and companies and provides a valuable information hub to all members, including CEMOS. At Tufts University, the Electro-Optics Technology Center has good facilities and encourages visiting industry faculty. Their corporate-sponsors program allows affiliated industries full access to Tufts, the facilities, and faculty and students.

Looking overseas, we find a few models for successful partnering between university and industry. Riken, or Rikagaku Kenkyusho (Institute of Physical and Chemical Research) was founded in 1917 in Japan. It has several laboratories, each headed by a chief scientist who also holds a university professorship. It provides a Frontier Research Program that began in 1986, offering fixed-term contracts to assemble first-class scientists from Japan and across the world to undertake advanced research projects. An independent review body encourages continual improvement of its research programs, and to encourage fresh ideas, five of its fifteen-member advisory council are replaced regularly. The Fraunhofer Gesellschaft is the leading organization of research in Germany with over eighty-five hundred employees. Contract research with Fraunhofer provides access to experts and other linked companies that can help with product development. Small businesses that cannot afford the facilities they need can stay competitive by teaming with Fraunhofer.

What Does Industry Need?

In early 1989, an interesting paper (Boehm 1989), appeared, investigating how industry could satisfy its ever-changing requirements. There were two concerns. One was that qualified individuals could not be guaranteed to move to where the jobs were, even if they existed in sufficient numbers. A second and more relevant question was raised: whether education should continually change to meet industries' perceived needs or stick to its basics. What is a qualified individual? The idea was explored: perhaps a responsive and changing curriculum should be the responsibility of smaller, less established institutions, giving them a niche in the educational marketplace. Did industry really need specialized training for graduates or would new employees drilled in the basics be preferred? An inspiring phrase ended Boehm's remarks: "A university should provide insight and direction and then turn students loose to conquer ideas. Industry can then take these minds, set parameters for them, and let them forge ahead toward company goals" (2).

A dozen years later the economic environment has changed considerably, but many of the discussions of these issues have not. The competitive environment for industry had been transformed over the past ten years, according to Charles Duke (1997). Markets, suppliers, and partners had become globalized, and commercial competition had largely replaced military markets. From the company's perspective, Duke used the big-bang theory as a metaphor and the evolutionary universe model, when examining how research and development contributed to the success of individual firms. With the former, novel research discoveries spawn entire new industries: semiconductors, personal computers, and the Internet are examples. The latter relies on continual refinement of existing technology, based on gradual changes in the market, as with electric power, aircraft, and automobiles. A third force for change is the big-brother model, in which government investment supplies precompetitive knowledge, technology, and trained personnel; examples of this route include the university and government laboratories. Public funding promoted the development of long-term approaches to finding solutions to problems from which many enabling technologies can later be traced.

Each segment of industry has to decide for itself how best to survive, especially when companies find themselves pushed toward the interface of emerging and established technologies or emerging and established markets. Large firms, following Duke's line of reasoning, tend to adopt policies that institutionalize the pursuit of value from R&D by encouraging this process, focusing their R&D on customers' needs. The time-to-market process demands that new advances be quickly cast into the form of "technology investment options" for review of new business opportunities. These are tough decisions to make, because product design and manufacturing are much more costly exercises than R&D, and time-to-market can be critical. Only the most "basic" knowledge-oriented research is done in partnership with universities. Since commercial value is rewarded, rather than technical novelty or elegance, industry seeks partners who can help make things happen rather than those who watch things happen or follow their own purely academic agenda. In many universities, there is now a need to solve industrial research problems as well as a need to follow the more traditional pursuits of inquiry for its own sake.

The Role of a State University

As a state-supported institution, we believe that there is an obligation to provide added-value to our stakeholders in the Commonwealth through our resources, skills, and knowledge base. How best might this be done in a rapidly changing economy, one in which the advanced technological nature of many of the industries in the region demand swift action because of the highly competitive environment?

As a university unit, we have a responsibility to the tax-paying industry base surrounding us, but our primary responsibility is to our tuition- and fee-paying students and the citizens of the state. Advancing and disseminating knowledge drives our daily activities. Our students share in this both through formal classroom-based courses as well as one-on-one thesis supervision. Graduate students engaged in research have a unique opportunity to explore a subject in depth and let their creativity open up new directions of thought and understanding. Of course, this is an idealized view, but their thesis work is driven toward discovery and original work. Still, at the same time, they must receive some financial support. This is traditionally accomplished through the provision of teaching assistantships (provided by the university), followed by research assistantships (provided by externally funded research grants or contracts). A flow of graduate students through teaching assistantship positions is essential in order to provide an educated student body able to serve as research assistants.

Problems arise with this process, depending on the funding sources available for research assistants and the time frame set for the work. Usually, a compromise must be reached between the ideal environment in which to conduct research and the hard reality of having to meet a deadline to keep an industrial or government sponsor satisfied. Pressure to perform and produce in the short term and keep a center active can compromise the future of those longer term research efforts that used to be synonymous with major scientific and technical breakthroughs. The dilemma is balancing the need to support financially one's facilities and research capabilities with the need to engage in possibly long-term intellectual advances and innovation for the community. This balance has to be found while maintaining activity at the leading edge of one's research field. Only by so doing can the reputation and capability of the infrastructure of such a center be sustained.

How Can Universities Partner with Industry?

At UML, we see it as part of our mission to increase our interactions with industry and also forge partnerships with them. Partnering can be envisaged in research, development, education, and training. A university center can provide people (students, faculty, technical staff) and access to equipment, facilities for measurement, testing, prototyping, and the like as well as (customized) courses and training. In the ideal university there would be a two-way channel for the flow of innovation and knowledge. Given the objective of facilitating and reinforcing our relevance to the economic growth of the region, a first step must be to clearly define and publicize our mission and capabilities. The communication gap between academics and industrialists can be wide, and there is a healthy skepticism toward university research on the part of many in industry who can see only an ivory tower.

At UMI, there is unusual enthusiasm for and commitment to applied research. Many faculty members are actively engaged in problem solving of a very practical and relevant nature. This in no way diminishes the intellectual demands of the work, and the traditional attitude of regarding only pure research as respectable has little place here. Nevertheless, our very willingness to "muck around" and deal with practical problems can send a mixed message to the educational establishment. Are we a first-rate research establishment or are we development engineers? Is it possible to satisfy both callings? Are we dependable, that is, do we project a sustainable image? Do we understand what we are recognized externally as good at?

The tension between being responsive to industry while meeting our academic obligations to our (tuition-paying) students seems to be unavoidable. Our responsibilities to our students must take priority. Short-term pressures, however, to perform and produce in order to get support for those very graduate students can sometimes lead to difficulties. Industrial funding is almost always for a short period and for smaller amounts than the longer term, more open-ended grants once received from the federal government. A student might need a few years to become really productive in the research field, making it hard to find one with the desired skill set at short notice. Nevertheless, according to the 1997 R&D White Paper (Basic Research 1997), 73 percent of all papers cited by U.S. industry are from academia or government laboratories. Industry appears to want to see basic long-term research carried out and it wants to partner with universities, but it usually cannot find the right institution to partner with on acceptable terms. This is the situation with respect to satisfying people's needs as well as establishing appropriate intellectual property arrangements.

Education of the Workforce

In April 1998, a lead article in the magazine *Photonics Spectra* (Weiss 1998) estimated that within the next three to five years, semiconductor and electro-optical manufacturing industries would need nearly four hundred thousand trained laser and electro-optical technicians in North America alone. This represents a tremendous need that cannot be met. There are fewer than two hundred postsecondary educational institutions in the United States offering photonics courses as part of their regular curriculum (Miller 1998). It has been pointed out that, even if every college and technical school with existing programs trained full classes each year, it would take two hundred years to fill the needs that industry will have created by 2003. The United States is not alone in this problem, especially with regard to the lack of technicians. In Japan, the Labor Ministry reported that 60 percent of the twenty-two hundred businesses surveyed nationally expect to face a shortage of skilled workers in the years ahead. The ministry announced that it would be launching an initiative to preserve knowledge of skills by videotaping workers and encouraging them to lecture on their skills (O'Brien 1997). In Germany, private-sector apprenticeships have been in place for many years, but this may be a difficult approach for U.S. culture to embrace.

One reaction to long-term workforce worries in the semiconductor industry has been the initiative launched by Sematech: It actively encourages partnerships between industries and colleges through the formation of certificate programs and degree programs that provide relevant courses. This kind of partnership, even through the very publicity it generates, increases awareness among the public of the opportunities in a far wider range of science and technology careers. Funding from the companies concerned about their future hiring needs helps to fuel the program. A similar approach is required in the electromagnetics and photonics industries, with courses partly funded by private companies being offered, and many new courses offered at the high school or equivalent level to expose millions of students to the postsecondary opportunities and careers in this field.

The challenge for businesses and educational institutions is to form this sort of partnership and work toward an effective cluster. The Engineering in Massachusetts initiative of Dean K. Vedula is a step in this direction (see Vedula et al. in part IV herein). Figures released by the Massachusetts Technology Collaborative (Index of the Massachusetts Innovation Economy 1997, 1998a) indicate that the total number of engineering degrees granted in Massachusetts had declined by 35 percent between 1987 and 1997 (from 3,882 to 2,456). Nationally, in that period, undergraduate engineering degrees had only decreased by 14 percent. At the graduate level, the number of degrees in engineering awarded over those years by Massachusetts institutions rose by 14 percent (from 1,826 to 2,059). This increase was less than half the national growth rate of 32 percent. The most significant decrease was in electrical engineering, the discipline most closely allied to photonics and electromagnetics.

The shortage of qualified engineers is examined in detail by Weiss (1998). She points out that not only has higher education become increasingly expensive, but also hungry industries are hiring qualified technical employees directly out of B. S. programs or even before the degree is taken. In order to tempt the brightest students to stay on and engage in research, universities typically offer a research assistant's job with a nine-month salary of around $12,000. This does not compete with an entry-level industry job at over $40,000. Moreover, the prospects in industry, after a few years of experience are tremendous, with salaries quickly approaching six figures.

Without the education of a workforce of engineers and technicians with photonics skills, new firms and growth will be slower than need be. We can help here. The Commonwealth has many excellent schools, but it also has an image of being an expensive place to get low-skilled labor, but relatively inexpensive for high-skilled labor. Traditional handicaps, such as high energy costs, high raw material costs, and high transportation costs, are not handicaps for many high-tech firms. We can help the both the well-established firms in the Boston area, such as Polaroid Corporation, EG&G, and Raytheon, as well as the many young start-ups. Among the start-ups, two current growth areas are immediately apparent. One is volume communications services based on optical systems, and the other is the medical field, because of the wealth of high-quality medical institutions in the region.

Getting Students

Our reputation at UML is for providing the region with excellent "hands-on" engineers, but this can also give some the impression of an intellectual ceiling having been defined and perhaps reached. The upper echelons of industry, especially at larger corporations, are still not heavily staffed with graduates from UML. Perhaps we should be content with a lesser niche and a specific set of regional and appreciative stakeholders? In reality, there are many dimensions to our individual and collective aspirations, making our internal as well as external image contradictory. Our image is perhaps even unformed for many of our industry neighbors whom we believe we could assist.

Our traditional approach to planning was to try to balance long-term funding provided by the federal government, which supported more speculative and long-term research, with the need to respond quickly to small businesses needing assistance over the short term. This approach proved difficult to maintain. Sustainability of the center requires a constant flow of new graduate students employed on the active research programs, in order to maintain continuity and truly build a firm knowledge base. All too often students graduate (!) and a significant piece of our "corporate" memory leaves with them. This benefits the organizations to which the students go, but this is a contribution to the economy that hurts us, if we are not properly organized. A steady flow of graduate students is necessary to minimize these problems, but in the present economic climate that flow is being diverted elsewhere.

To attract and support students during their first year or two, while they are gaining the background knowledge needed in order to become productive, requires teaching assistantships, accompanied by other rewards, such as tuition and fee waivers. This is especially important at times when the economy is thriving and there are many jobs open. Few American students want to live on a few thousand dollars a year when they can easily earn much more in industry. The region's workforce shortage thus directly affects the university's ability to sustain its usefulness and even its core competencies. Indeed, at times like these, many students leave their academic programs before completion because of the attractive employment offers they are bombarded with. There is a crisis in the workplace, owing to the lack of skilled and talented individuals, and a crisis in the very foundations of the long-term research infrastructure of the country. What can be done?

What Resources Do We Have?

The primary mission of the center, in line with that of the university, is to be a resource for information, expertise, and education. The center has developed and offered a graduate electro-optics option and graduate certificate program for many years (see www.eng.uml.edu/~ECE). The center is also actively involved in the provision of courses for the engineering technology (evening) program as well as the courses that fulfill graduation requirements for non-electrical engineering majors, such as physics students. The center also continues to provide help in writing grant applications, especially to small businesses seeking the small business innovative research (SBIR) awards. These proposals have been submitted to various Department of Defense agencies as well as to the National Science Foundation (NSF). Grant applications range from SBIR-type proposals, designed to allow partnering between CEMOS and a company, to larger, multipartner initiatives.

Resources for industry include the expertise of faculty, post-doc, visiting research professors, and graduate students for everything from casual telephone inquiries to full-blown contracts. Equipment and facilities that are not readily available to local companies can be accessed here. This includes computing power, software, laser laboratories and optical fabrication, and testing amenities. The center also offers seminars, training courses, and courses for credit.

The center itself is well complemented by other centers in the university, especially the Center for Advanced Materials and the Submillimeter Wave Technology Lab. It also works closely with faculty from other departments. Notable among these collaborative arrangements are the ties we have with the mathematics, physics, and chemical engineering faculty and their students. These linkages reinforce the impression that we are a one-stop shop for regional industries, especially those with any kind of imaging or optics problem.

At UML we seek to increase our interactions with industry and also to forge partnerships with them. We want our particular component to be plugged in to the most effective regional cluster. The center can, in principle, provide people (students, faculty, technical staff) and access to equipment, facilities for measurement, testing and prototyping, as well as provide (customized) courses and training. We have a lot to offer.

What Has Been Done?

More quantitative measures of our output might help us assess whether we are serving our stakeholders or not. Some measurables are the many graduates who have left the center as well as its many publications. The latter are all available, of course, to interested third parties. Several patent disclosures have been filed and examples of funded research in the last five years include over a dozen industry-sponsored projects and ten government-funded (DOD or NSF) projects. The government-funded research contracts were for extended periods (of at least one year, typically for two or three). The funding from industry typically ran for six months or less, with many more progress meetings to attend and reporting deadlines to meet. The industrial funding was given for far more specific tasks, some of which would change as the work progressed. Industry emphasis was heavily on producing tangible, physical results that could be seen, rather than the more speculative endeavors, that received federal support.

If a mutually beneficial partnership cannot be accomplished through R&D ventures, then training or routine laboratory/computational work can be considered to provide the financial buffer required. The financial means to support graduate students and maintain equipment could perhaps be met by broadening the technical level of support we offer to industry. For example, we could take on fairly mundane or routine laboratory or computer work for industry based on our laboratory resources. Such work can be carried out by new students and it could provide them with a useful in-house internship during which they can earn money and take classes without too much distraction. After this work is completed, it becomes possible to move such students to longer term funded research for which they are by then prepared. Without providing funding to attract graduate students into our programs initially, there will always be a difficulty in maintaining the flow of good students through the research and doctoral programs while simultaneously maintaining our core technical capabilities. Only through an increased flow of students can we expect to build on our research advances from one generation to the next. At present, there is a serious interruption in the innovation flow as students leave, and inadequate funding persists.

An example of a service we could provide, and have been exploring with regional companies, is the provision of an electromagnetics software resource. We can help smaller industries carry out their electromagnetic or photonic device modeling by providing access to commercial or in-house developed software. Much software that was previously in the realm of defense applications has now become available. Such software allows sophisticated modeling to be done on a PC rather than a Cray supercomputer. However, the expertise in understanding the software and applying it to real world problems is a service we can provide, for it is expensive for any one company to upgrade constantly to improved platforms and the latest software releases. Some of the more sophisticated numerical techniques currently require a company to hire a consultant to model their specific problem and develop the computer simulation. The center can become the focus of a cluster of companies, each paying a reasonable amount for access to this expertise. Members could use the software and draw on our resources to the mutual benefit of all concerned.

Actual productive relationships we have been successfully involved in are varied and customized: We have been a test site and customer inspection site for a local company's pattern-recognition (joint transform correlator) system. Given our expertise and interest in correlators, we were given the responsibility of helping customers realize how an optical system with this capability could help with their machine-vision problems. Other companies do come into our labs to make use of special equipment and facilities they do not have: our beam profiler and image acquisition capabilities as well as our laser sources and power meters. Provision of optical tables space or an entire room in the center for externally funded projects is possible, if not encouraged. Recent examples of this include NSF-supported work with several small businesses. We have also built dedicated test sites and prototypes for evaluation purposes. We have provided customized courses for credit and shorter courses for immediate needs: for Raytheon, for instance, an eight-hour short course on fiber optics and courses for credit in electro-optic systems and infrared systems. We have provided both graduate and undergraduate student help for R&D work both on and off campus, and placed students at local companies as interns or, more recently, as co-op students.

With the recent changes in intellectual properties regulations for university personnel, it has become much easier for faculty and students to partner with industry and reap the rewards from commercializing technology. Institute for Massachusetts Partnering and Commercialization of Technology (IMPACT), essentially a distributed center within the UML campus, was specifically developed to nurture these relationships and facilitate technology transfer, while the more effectively supporting local industry. It is now possible for partnering companies to bring their personnel and laboratories on campus in order to be physically closer to our centers and their students, thereby increasing the flow of innovation between the two. This is clearly of great value to the students involved who can see at firsthand how industrial research and development proceeds.

CEMOS has regularly held open house for local industry. The format of these occasions has varied, but it has always allowed for a period of discussion between attendees and members of the center. Usually faculty and students associated with the center make presentations; Visitors from industry may also make presentations on their interests and needs. Some companies have exhibited their products in our labs, to be on view during our lab tours. In 1996, a much larger meeting was arranged by our center along with the Center for Advanced Materials and the Photonics and Optoelectronics Device Fabrication facility. It included invited talks by some industry partners and DOD sponsors. This was very successful, attracting over a hundred visitors from the region.

Obstacles to Successful Partnering

It is apparent that the kind of support that best suits the way in which our academic programs function is government funding, which tends to be for larger sums and spread over two or three years. It also is often based on grant proposals that are not tied to specific, stated outcomes. The funding received from industry, while very welcome, has attendant difficulties. The amounts involved are relatively small, and the short time periods over which the research is to be carried out do not easily fit in with student and faculty schedules. Indeed, it can be problematic even finding a student who can be diverted for a short time to work on some projects from industry, hence, the projects sometimes (very reluctantly) have to be declined. This is not good for our image or aspirations to help regional industries. We need to persuade companies that we are worth investing in over a longer period of time, in which expertise can be built and collectively shared by a changing but seamless body of students. This requires a larger injection of funds, however, with the need to commit to a center and develop the truly robust partnership with it that ensures that the investment pays off. If this cannot be accomplished through R&D ventures, then perhaps training or routine laboratory/computational work mentioned before can provide the financial buffer required.

Given all our activities and endeavors, it is still important to understand and measure how effective we have been as a center in serving the community. What supporting research have we carried out and what mechanisms have we exploited to diffuse this to companies? Does our education and training infrastructure really support companies in the photonics sector? What shortcomings do firms perceive in university support? Clearly the electro-optics and imaging activities and needs of the region's industry are very diverse. The center has always been receptive to getting involved in industrial problems, but our present infrastructure has made it difficult for us to respond as quickly as the industry would like. The reasons for this are simple:

Graduate students carry out a great deal of the research and development work and they need funding for at least one year and sometimes as many as four. They can not be magically conjured up out of thin air precisely when they are needed.

Graduate students have course work, and so not all of their time is available for research. In fact, they need to learn the research area in their first few months, during which time it would be improper to pay them from a grant or a contract. Institutional support is necessary for a significant number of teaching assistant positions to support incoming students during this early phase.

We need to remember that graduate students must be able to graduate, and so proprietary work needs to be handled appropriately so that their theses are publishable.

Many of these problems are diminished when government funding supports our research rather than industry. It seems clear that to sustain our effectiveness, we must constantly look for ways to increase the pool of funds that can be used to attract good graduate students (at a competitive full-time stipend). These funds must be secured on a reliable basis to support laboratory upgrades and maintenance and to maintain the flow of students and, in turn, the flow of innovation. Interruption of funding for even one year can be disastrous for the retention of students, the development of new ideas, and the center's core competency. The chances of persuading a company to invest in individual students and university laboratories for several years at a time, however, seem meager.

Conclusions

At times, depending on their priorities, a company will recognize a need for our research expertise and invest in us. For us, however, to change directions in research or teaching can take some time. Hence, given our overriding scholarly objectives, we should not be changing direction with each market-driven trend.

Some examples of university-industry partnerships we have described may help to overcome this problem. Are these good models to follow? It clearly depends on the culture of each institution, the facilities available, its pool of faculty and graduate students and their inherent capabilities. Is it better to retain an independent niche or reputation, or perhaps join forces with a regional consortium? Can we find our own unique formula for success, derived from our institutional history and mission? It would seem that having as broad as possible a perspective on one's discipline is advantageous, and awareness of the partnering models employed elsewhere becomes important information. Since no two institutions are identical, however, and, indeed, since they strive to establish their own unique personalities, it is unlikely that blind copying of successful models demonstrated elsewhere will succeed when transplanted. The awareness of other models, combined with the constant reappraisal of our own institution's role in a dynamic economy, is necessary but not sufficient. Only from developing long-term communication channels with targeted industries may some semblance of a symbiotic relationship with a university evolve. There appears to be plenty of scope for a closer convergence of goals. For universities, this must occur without compromising academic integrity and engagement in long-term research investigations, while still retaining the capacity to work closely with industry on what may sometimes be quite short-term, problem-solving ventures.

Our particular R&D expertise may best be viewed as a component in what ought to be a well-publicized and well-organized network or regional cluster of R&D providers. Under these conditions, all of the support efforts industry might need can be readily found and provided for in a more systematic fashion. Facilitating and streamlining this regional connectivity will enhance the sustainability of the region. How we accomplish this will ultimately be determined by how flexible we are prepared to be. At the moment, there is a general fragmentation and lack of information about the resources available to industry. The state is addressing this situation, but in the meantime, some state or federal support for our efforts to grow and to stabilize the pockets of photonics activity throughout the region may be essential. By building closer partnerships with companies or clusters of companies we will, in turn, be better tuned to the technology challenges facing the commercial sector—the better to provide an appropriate range of support services and educational opportunities. A broadly based relationship with a spectrum of companies may result in some fairly low-tech activities being carried out, but these, too, can help provide the funding to smooth out student support for the more speculative thesis-driven and longer term research activities. The balance between remaining a relevant center of excellence in our areas of expertise, while continuing to maintain close ties with industrial partners, will always remain the challenge, because of the fast-changing pressures and different criteria for success. For universities, however, some degree of stability, as perhaps might result from engaging with a cluster or consortium, might in turn generate a productive network of affiliated companies. Each company may need to make only a small investment to benefit from a strong partnership. The question is how to create the environment in which doing this is obviously advantageous to all concerned.

In CEMOS, the dual challenge we have is to maintain the relevance of our expertise and to ensure that it connects in a productive way with the regional agenda. Although always receptive to the challenges of industrial problems, our present infrastructure will continue to make it difficult for us to respond to industry as effectively and quickly as we and the industry would like. Some key reasons for this were discussed above: most importantly, a critical mass of graduate students is needed and support for graduate students needs to be available at the time of recruiting in order to attract the best.

REFERENCES

Bachula, G. R. 1988. Photonics: A key to the U.S. Economy. Report presented at the Second Annual Symposium of the Boston University Photonics Center, Boston.

Basic Research. 1997. Basic research white paper: Defining our path to the future. *R&D Magazine.* Des Plaines, Ill.: Cahners.

Boehm, M. A. 1989. Industry and academia: Co-operation for the future of optics. *Optical Engineering Reports* (March): 2.

Duke, C. 1997. How to get value from R&D. *Physics World* (August): 17.

Helms, C. R. 1998. Next-generation R&D partnerships. *Solid State Technology* (August): 112.

Jesup and Lamont. 1999. Compound semiconductor technologies: The optoelectronics revolution. Industry Overview. June. New York: Jesup and Lamont.

Laurin, T. 1998. Industry has taken the R&D lead. *Photonics Spectra* (August): 11.

Lepkowski, W. 1997. The troubled research university. *Optics and Photonics News* (May): 22.

Massachusetts Technology Collaborative (MTC). 1997. Index of the Massachusetts innovation economy. Boston: MTC.

———. 1998a. Index of the Massachusetts innovation economy. Boston: MTC.

———. 1998b. Photonics: The collaborative. Boston: MTC.

———. 1999. Federal R&D investment scenarios on economic growth. Boston: MTC.

Miller, E. J. 1998. Forming a partnership between public education and photonics industry. *Optical Engineering Reports* (August): 13.

Minuteman Regional Science and Technology Optoelectronics and Photonics Task Force. 1998. Optoelectronics and photonics technology in Massachusetts. Lexington: Minuteman Task Force.

National Research Council (NRC). 1998. *Harnessing light: Optical science and engineering for the 21st century.* Washington, D.C.: National Academy Press. [http://www.nap.edu/newbooks/index.html].

National Science Board (NSB). 1994. *Stresses on research and education at colleges and universities: Institutional and sponsoring agency responses.* Washington, D.C.: Government-University Industry Research Roundtable.

National Science Foundation (NSF). 1998. *Science and engineering indicators.* [http://www.nsf.gov/sbe/srs/seind98/c4/fig04–06.html].

O'Brien, T. K. 1997. U.S. programs seek to fill optics technician needs. *Photonics Spectra* (October): 82.

OIDA. 1998. Annual Report. Washington, D.C.: OIDA.

———. 1999. The optoelectronics industry: A global perspective. Presentation at Boston University's Photonics Center by Dr. F. Walsh of OIDA. September 23.

Photonics Manufacturing. 1997. Advanced Technology Program. Washington, D.C.: U.S. Department of Commerce.

Photonics Spectra. 2000. Global photonics technology forecast. *Photonics Spectra* (January).

Porter, M. E. 1998. Photonics in Massachusetts: A cluster in transition. *Report from the Competition and Competitiveness Lab.* Cambridge: Harvard Business School, 1–34.

Weiss, S. A. 1998. Help wanted. *Photonics Spectra* (April): 90.

———. 1999. Global photonics market forecasts. *Photonics Spectra* (August): 75–116.

IV UNIVERSITY–COMMUNITY COLLABORATIONS

Introduction

Robert Forrant

IN A collection of essays on the responsive university, Braskamp and Wergin state "The academy does not often believe and act as though the campus is the world and the world is the campus" (1998, 64). A goal of the University of Massachusetts Lowell is to shatter this belief system by reaching out in new ways to the community to have a positive impact on the regional economy. An iterative process has begun that involves faculty from diverse disciplines—among them criminal justice, community psychology, sociology, work environment, public health, electrical and mechanical engineering, environmental sciences, and education—and off-campus constituencies. Internal relationships among faculty and between faculty and administrators are being altered while external relationships are being reconstituted to make certain that the campus listens better and engages the community in collaborative activities. This concluding section describes the institutional challenges and difficulties encountered in the search for ways to integrate the action-oriented component of the university's mission with high academic performance.

In "Random Acts of Assistance or Purposeful Intervention," Robert Forrant, Department of Regional Economic and Social Development, contends that to have a positive impact on a region a university's activities must be guided by a reflective, continuous discourse that generates a theoretical perspective and methodological approach to direct the work. Through an examination of UML's manufacturing outreach activities and a consideration of recent literature on regional development, Forrant describes the strengths and weaknesses of the university's current development efforts. Two overarching questions are discussed: First, how can the university structure its myriad activities to

331

have a transformative and sustaining impact on the regional economy? Second, who in the external environment should participate in the conversations that guide the restructuring process? Forrant concludes that the university's most important role ought to be the persistent advancement of cross-firm, cross-community, and cross-institutional learning. The resultant confluence of approaches and perspectives can dispel the deeply held public perception that universities are averse to "getting their hands dirty" with practical activities and thus advance UML's mission-related activities beyond an indiscriminate approach to development.

Dean Krishna Vedula of the College of Engineering and other engineering faculty members describe how education is being restructured in their college to augment the university's mission. The reorganization is driven by an understanding of the importance of contextual learning; the more hands-on exposure students have to the real world of engineering practices, the better prepared they will become to assume positions in the region's high-tech economy. Vedula makes the point that one foundation stone for sustainable development is a highly educated engineering workforce capable of designing improved products that are energy efficient and environmentally friendly. And since high-tech firms dominate the regional economy, the engineering program must be agile and responsive to the needs of this industry. The college has answered one of the questions Forrant poses: Who ought to participate in the conversations that guide restructuring? Students and industry representatives are defining a shared long-term vision for the college. An advisory board consisting of leaders from industry is working with faculty and students on the redesign of courses and the development of job fairs, internships, and summer employment opportunities.

To play a catalytic economic and social development role, faculty and administrators need to behave in new ways, and various institutional structures must be reorganized to support this changed faculty behavior, Linda Silka contends in "Addressing the Challenge of Community Collaborations." Silka posits that, in the traditional academy structure, faculty very often do not see their role as having an "application component." There is an assumption that the serious academic work is completed prior to the movement beyond the classroom and research laboratory. Through a careful examination of five major interdisciplinary initiatives undertaken through the Center for Family, Work, and Community, which she directs, Silka challenges the idea that application is not the arena where real intellectual work takes place. The initiatives she describes sought to meet multiple goals of advancing research, sharing resources widely in the community, and providing rich learning opportunities for students, faculty, and community members. Active problem solving that drew on the scholarly interests of faculty was paramount, and there was a concerted effort to have community members help shape the intellectual agenda. Silka concludes with a discussion of university structures and offers concrete suggestions for changing these structures to increase activities that make a substantial difference in people's lives. Here Silka offers a cautionary perspective, for she argues that just changing the reward and accountability structures to encourage innovative faculty activity is likely to fail. Instead, the focus ought to be on the identification of the intellectual problems that bring together different campus and community constituencies; faculty responsiveness and enduring institutional change are forged through the problem-solving processes that ensue (Forrant and Silka, 1999).

Institutional change is addressed by Dean Nancy Kleniewski of the College of Arts and Sciences in "Administrative Support for Innovation." She describes the difficulties of undertaking applied community research and suggests ways that universities can support faculty so engaged. Kleniewski notes that faculty who engage in applied community research often confront debilitating structural and cultural obstacles within their institutions. Reward structures may actually provide disincentives rather than encouragement to faculty interested in behaving in "new ways." Even when research universities include community service and applied research in their missions, these goals are but one of a number of institutional priorities. Confusion remains as to exactly what applied research is, whether it should be pursued, and how it will be assessed. Community-based problem-solving research defies the silolike boundaries of academic disciplines and individualized faculty behavior, potentially exposing faculty who pursue such work to marginalization in their department and college. And when such work is undertaken, far too often a traditional expert–client relationship exists. This approach avoids the time-consuming process of forming research teams composed of community members, students, and faculty all behaving as partners, while it preserves the "faculty as expert" model. In her conclusion, Kleniewski offers suggestions on how to surmount these bureaucratic rigidities.

The establishment of interdisciplinary campus–community partnerships is central to what we discuss. Faculty responsiveness to new roles cannot be mandated, nor can community involvement be scripted for there are social processes at work (Keith 1998). Practically speaking, a healthy economy rebounds to the benefit of the university in terms of an increased budget, greater enrollments, and larger alumni donations. By extension, a thriving university can increase its activities and thus play a more dynamic role in the wider world it inhabits. The result is a virtuous circle of improvement. Herein we offer our perspective on how to mold the disparate parts of the academy into an arc on that circle's circumference.

REFERENCES

Braskamp, L., and J. Wergin. 1998. Forming new social partnerships. In *The responsive university: Restructuring for high performance,* ed. William G. Tierney. Baltimore: Johns Hopkins University Press, 62–91.

Forrant, R., and L. Silka. 1999. Thinking and doing—doing and thinking: The University of Massachusetts Lowell and the community development process, *American Behavioral Scientist* 42:814–26.

Keith, K. 1998. The responsive university in the twenty-first century. In *The responsive university: Restructuring for high performance,* ed. William G. Tierney. Baltimore: Johns Hopkins University Press, 162–72.

Random Acts of Assistance or Purposeful Intervention?

The University of Massachusetts Lowell and the Regional Development Process

Robert Forrant

THE University of Massachusetts Lowell is located in the nation's first planned industrial city. Its forerunner, the Lowell Technology Institute, was established in the mid-1890s through an act of the state legislature at the urging of textile mill owners to provide them with competent line employees, engineers, and managers. In 1991 the University of Lowell became the University of Massachusetts Lowell (UML), a part of the five-campus state university system. Thereafter, the Lowell campus began to elaborate a model for how a public university could assist in the development of a regional economy.[1] A Technology Transfer Office, over thirty research centers and institutes, and numerous certificate and degree programs are now in place to assist in the regional development process, with a particular emphasis on working with industry. This essay considers two overarching questions pertaining to the UML development mission. First, how can the university structure its myriad activities to have a transformative and sustaining impact on the regional economy? Second, who should participate in the conversations that guide the restructuring process? Whether or not the university ought to engage in this process at all has already been answered through the articulated mission.

First, several generalizable social characteristics of successful industry-modernization organizations are outlined to provide a framework for the consideration of UML's activities, followed by a brief history of the Springfield, Massachusetts–based Machine Action Project (MAP), a widely recognized successful industry-modernization effort, to derive a perspective on the role that a university can play in the development of a vibrant and sustainable manufacturing base. MAP was a network of metalworking firms and education and training institutions that emerged from a continuous process of social engagement and adjustment between public and private actors, including firms, an industry trade association, colleges, and

335

labor unions. The third section elaborates UML's development mission and reviews the work of several of its centers and institutes. In the final section, the literature on regions, place, and the importance of knowledge-creation in the economic development process is reviewed. Of particular relevance is the impact of shared learning among firms and institutions in successful regions. Here, I build on a literature review to suggest why an understanding and application of development theory is essential in defining the work of the university.

It is my argument that UML (or any other university intent on playing a strong role in economic development beyond simply the theoretical) can have a sustaining positive impact on the regional economy when its activities are guided by a reflective and ongoing, regionwide discourse. The community conversations will help to generate a theoretical perspective and methodological approach to guide development efforts. As befits an educational institution, the university's most important role ought to be the persistent advancement of cross-firm, cross-community, and cross-institutional learning. The confluence of approaches and perspectives that this process generates will act as a catalyst to the sustainable development of the region. In addition, the university will advance its work beyond a random approach to development, dispel deeply held public perceptions that universities are averse to "getting their hands dirty" with practical activities, and thus begin to play the integrative and innovative role it aspires to in the cultivation of a sustainable economy.

The Characteristics of Successful Industrial-Modernization Organizations

A Brief History of Manufacturing-Modernization Programs

In an attempt to stem industrial decline, several local and state governments in the United States, in partnership with universities, private foundations, community development corporations, nonprofit agencies, labor unions, and employer associations, established organizations in the early 1980s to improve business performance (Osborne 1988, 1989). By 1990 there existed some forty-two such programs in twenty-eight states. Typically, these programs provided direct services to individual firms in such areas as the reorganization of quality systems, market research, and new technology acquisition and utilization (Eisenger 1995; Isserman 1994). Much of this activity was inspired by European programs popularized by academics and economic-development practitioners enamored with public–private partnerships in Italy, Denmark, and Germany, and by the role of the developmental state in places such as Taiwan and Japan. The indigenous model of the U.S. Agricultural Extension Service was considered here as well (Atkinson 1993; Best 1990; Brusco 1992; Cossentino, Pyke, and

Sengenberger 1996; Ferleger and Lazonick 1993; Flynn and Forrant 1997; Lichtenstein 1992; Shapira 1990).

At the federal level, the National Institute of Standards and Technology (NIST) began to fund manufacturing extension centers in 1994, and by 1996 there were approximately forty federally funded Manufacturing Extension Programs (MEP) scattered across the country in partnership with state governments. The programs were set up to "bridge a technology gap between sources of improved manufacturing technology and the small and mid-sized companies that need it" (NIST 1994). In Massachusetts, there now exists the MEP-sponsored Massachusetts Manufacturing Partnership, with five regional offices strategically located across the Commonwealth; the office for the Northeast region is in Lowell. Several states—Massachusetts among them—also provide financial support to encourage collaborative activities among groups of firms (Shapira 1996). Such programs customarily work with firms in the same industrial sector, or among clusters of interdependent businesses linked through common or complementary inputs, innovations, processes, or products (Rosenfeld 1997).[2] Parallel to these modernization agencies and organizations there exists a vast network of university–industry research centers dating back to the late nineteenth century. Following World War II, universities, the federal government, and industry engaged in numerous research projects funded by the Department of Defense (DOD) and the National Science Foundation (NSF). And in the 1970s and 1980s, the NSF established several University–Industry Research Centers and Science and Technology Centers and funded University–Industry Cooperative Research Projects (Beck, et. al. 1995; Cohen, Florida, and Goe 1994; Dineen 1995; Felsenstein 1996).[3]

What the Record Demonstrates

In spite of the proliferation of modernization organizations and university–industry research centers, we know remarkably little about whether these institutions and collaborations facilitate interfirm learning and improve regional economies. Nor have we gained much insight into whether particular institutional arrangements are more effective than others in fostering economically vibrant regions (Harris 1996; Labrianidis 1995). With Michael Best, I have argued that this paucity of knowledge stems from the fact that policy formulators and implementers are bereft of an agreed-upon perspective on how companies change and grow and lack insight into the industries they are supposed to be assisting (Best and Forrant 1996). To discern how universities can systematically move from an individual, firm-specific approach to assistance to the promotion of cross-firm learning and the cultivation of what Florida characterizes as a "learning region," it is imperative that we advance our understanding of the organizational characteristics of successful modernization agencies (1995).

Based on detailed case studies of three organizations and a careful reading of the considerable literature on industrial modernization, there appear to be five distinguishing features that successful modernization organizations share: They must be credible, connected, catalytic, collective, and continuous (Flynn and Forrant 1997; Forrant and Flynn 1998).

For an industrial modernization organization to insinuate itself into the economic lifeblood of a region it must first off be credible with employers and with unions when they are present. One important function of modernization organizations is the provision of information on new technologies. There is simply too much material coming across a business manager's desk advising him or her on this or that new technology. It would be helpful to have a place of resort to sort through this material. But employers are reluctant to utilize the services of an organization they do not know or respect. Therefore, the establishment of credibility with managers and owners is extremely important. And since credibility is contingent on the length of relationships, organizations that develop from existing relationships with firms are the most likely to succeed most expeditiously. Rather than creating new institutions, existing institutions with successful records of working with firms, such as trade associations or technical schools, should be used whenever possible.

Second, the organization must be connected to the training institutions, business and trade associations, financial institutions, and technical colleges and universities that exist around it in order to network services and benefit from the institutional learning about industry that takes place. Regular meetings to discuss industry issues, a newsletter to communicate with firms, and the hosting of occasional conferences and workshops on issues of concern to firms are ways to stay connected. This minimizes the "reinventing the wheel" syndrome and maximizes the use of scarce resources. It also makes it easier for firm representatives to access services without fighting through a frustrating bureaucratic maze. This is particularly important when working with small enterprises that may lack a management group. The owner of the firm may indeed be the human resources director, the chief production engineer, computer programmer, and the liaison to outside agencies and organizations such as the university. Efficiency in the delivery of services is thus essential.

Third, the organization has to be catalytic: It has to make things happen. It is essential that the energy built up through meetings and one-on-one firm visits progresses beyond the simple telling of "war stories" to problem solving and knowledge diffusion. The learning exchanges that work best are those that take place among industry peers. Such encounters, however, almost never occur in the normal discourse among firms or in the course of a modernization organization's typical daily routine.

Fourth, it is incumbent on modernization organizations to pursue a variety of collective strategies along with their one-on-one efforts. One-on-

one consulting and technical assistance services are important in order to establish credibility and facilitate specific firm-level changes. But to reach a scale of activity commensurate with the number of firms in any given region, group work is imperative. Therefore, organizations need to articulate and then consistently implement strategies that promote interfirm learning and create opportunities for firms to act on issues of common concern.

Fifth, organizations need to be continuous, that is, to have staying power. For owners to have confidence that timely assistance will be available to them, stable leadership, long-term funding, and a strategic vision are indispensable.

A Vibrant Approach: The Western Massachusetts Metalworking Model

Background

In Massachusetts throughout the 1980s, a dramatic wave of lay-offs and plant closings led to rapid industrial decline. As a result, a consensus formed among state and local policymakers that the manufacturing sector was devastated. Economic revitalization would require a transition from manufacturing to service-based employment. Today, however, Western Massachusetts is home to a thriving and competitive agglomeration of small metalworking firms, many of them at the forefront of technological and organizational innovation. In an industrial sector known for cutthroat competition, cost-driven strategies, and fierce privacy, Western Massachusetts stands apart, as firms throughout the region routinely open their doors to provide training to employees from other companies. A four-year apprenticeship program is in place that rotates apprentices among the firms to cultivate the next generation of skilled workers. There is a product-development group engaged in rapid prototyping projects with companies, such as Pratt & Whitney, General Electric, and Ford Motor Company. Hundreds of workers and managers have participated in group training sessions and seminars on topics including blueprint reading, computer numerical control machine tool programming and repair, work-flow management, and ISO 9000 certification. What brought this about, and what lessons can be gleaned from the development strategies utilized by the Machine Action Project (MAP)?

The Machine Action Project as Social Mobilizer

MAP was an experimental program established in 1986 in Greater Springfield, Massachusetts, in the aftermath of a series of devastating plant closings. For over a hundred years, Springfield was the hub of a prosperous metalworking manufacturing region stretching up and down the Connecticut River Valley between Hartford, Connecticut, and White River Junction,

Vermont. However, this rich legacy failed to stem the tide of more than a dozen plant closings in less than ten years in the late 1970s and early 1980s. In response, the Commonwealth's Industrial Services Program (ISP) allocated $100,000 a year for three years to MAP.[4] MAP's original intent was to transition the regional economy from metalworking toward the service sector and tourism; however, business-based research revealed that there were close to three hundred small metalworking firms present in Greater Springfield that supplied tooling and parts to the machine tool, aerospace, defense, and electronics industries. Rather than retrain displaced workers for something other than metalworking, it was determined that the program ought to nurture this small-firm economy vigorously.

To embed itself in the region, a project board of directors was assembled, comprised of five managers, five union officials, three local economic development officials, two banking representatives, the president of the local two-year technical college, instructors from technical high schools in the area, members of the business press, and an economics professor from the nearby University of Massachusetts Amherst who was familiar with successful international examples of regional economic development. Thus, MAP was engaged in collective behavior by being connected to the relevant institutions, trade associations, and labor unions keenly interested in the revitalization of the metalworking industry. The board met monthly to discuss what it believed were the industry's problems and to devise a research agenda and work plan for the MAP staff. In this way, MAP acted as a catalyst to get things done. Eventually, members of the local chapter of the National Tooling and Machining Association (NTMA) assumed a lead role on the board, giving it more credibility among firms.[5]

A paucity of information on the metalworking sector as a whole led to several research initiatives. Close to one hundred firms were visited to determine their strengths and weaknesses and to assess whether owners and/or managers and community actors were predisposed to work together to revitalize the industry. On the basis of the knowledge accumulated, a strategy slowly took shape, predicated on the belief that, to have a significant impact on the region, it was important to work with groups of firms rather than attempt to provide one-on-one assistance to hundreds of small companies. Finally, of paramount significance for the project's longevity, the visits identified a nucleus of industry representatives who became an active core that would eventually provide leadership continuity through the transition from a public sector–led to an industry-led program.

Analysis of MAP and Implications for UML

MAP received widespread attention in policy circles for its activities as a social mobilizer. It was awarded a Rockefeller Foundation prize in 1988 for Innovation in State and Local Government and an Arthur D. Little award for development innovation in 1990. And several influential academic researchers noted that MAP was a "best-practice" model of regional industrial and training policy (Batt and Osterman 1993; Fitzgerald and McGregor 1993). These analysts observed two particularly innovative features of MAP, both related to the embedded and social nature of its strategy and activities.

First, its strategy evolved from a detailed analysis of the local industrial base and through local actors' discussions of their problems and needs. The conventional wisdom was that the machining industry in the region was devastated. The research dispelled this perspective, however, and a course of action emerged that central state planners could never have designed on their own. As Fitzgerald and McGregor point out, firm-based research was notable throughout MAP's existence. The second feature was the broker role that MAP played among firms, educational institutions, and state agencies. It acted as a catalyst to translate the demand for higher-skilled workers into specific training courses and programs and to improve the local education and training infrastructure to make it more responsive to the real needs of industry. For Batt and Osterman this approach went well beyond "the simple provision of training funds or technical assistance to that of creating new forms of organization and cooperation within the private sector and between public and private sector organizations" (1993, 57).

What can be learned from the MAP case study that is relevant to universities considering their particular role in regional development? MAP's success was predicated on three things: (1) knowledge creation, the application of industry-based research to identify the issues and problems that confronted firms; (2) the establishment of industry leadership to guide the entire effort; and (3) the decision to promote interfirm and cross-institutional and institutional learning. In an article on knowledge creation and economic development, Richard Florida (1995) discusses the importance of what he terms "learning regions." Just as there is a burgeoning theory of learning organizations and knowledge-creating firms, he contends, there needs to be a comparable theory of regional-knowledge creation. "Regions must adopt the principles of knowledge creation and continuous learning; they must in effect become learning regions. Learning regions provide a series of related infrastructures which can facilitate the flow of knowledge, ideas and learning" (532).

The development of the Western Massachusetts metalworking sector is reflective of this. MAP grew out of a social process characterized by research, discussion, shared learning, reflection, and experimentation among firms, their trade association, and numerous public agencies. Through continuous interaction with industry and economic development leaders, MAP revealed sectoral dynamics that public policymakers were unaware of. MAP provided industry leaders with an opportunity to participate in the process of problem definition, and as a result they assumed ownership of the entire program as it developed. Shared learning reinforced a strong sense of partnership and provided an incentive for private and public actors to accommodate one another to meet agreed-upon common goals. And the commitment to continuous improvement—at both the firm level and program administration level—enabled participants to learn from their experiences and keep moving forward. It is my contention that it is the regionwide learning dynamics established by the intersection of firm-specific activity (important for the development of precise knowledge of a firm's problems) with cross-firm activity (important to promote regional and sectoral development) that provide the justification for public expenditures on manufacturing modernization and economic development programs.

The University of Massachusetts Lowell and Regional Development

The Economy as Context

From the end of World War II through the 1960s, the Greater Lowell economy was influenced by the decline of the textile industry. Then, there was a spectacular economic rebirth in the 1970s and 1980s brought about by the emergence of nearby Route 128 as a high-technology center and the Reagan buildup of the defense industry; between 1975 and 1980, 100,000 new high-tech jobs were created in the state. Wang, Digital Equipment Corporation, Prime, and Data General emerged along with hundreds of small firms to create the minicomputer industry, while thousands of metal-working, plastics, and electronics companies received lucrative subcontracts to supply these computer and defense firms with components, accessories, tooling, machines, and instrumentation (Gittell and Flynn 1995). A lack of industry diversification, however, resulted in a sharp downturn in the late 1980s when the computer industry crashed and there were sharp cuts in defense spending (Forrant 1995; Lampe 1988).

Remarkably, even with this decline, the economic base of the Northeast region still mostly rests on its manufacturers. Whereas only 15.3 percent of the state's workforce was employed in manufacturing in 1995, 21.7 percent of the Northeast region's workforce was so employed. Electronic components and accessories, communications equipment, industrial machinery and equipment, transportation equipment, and measuring and testing instruments are the largest employment sectors. There is a lower-than-state average concentration of retail and service establishments and especially high-wage business, engineering, and management services in the region. Two industry sectors that are strong in the region—industrial machinery and equipment and electronic and electrical equipment—ranked one and two respectively in the state for total dollar value of exports in 1995. And there are a number of small to mid-sized metalworking, software, telecommunications, and network firms in the region that are currently experiencing impressive growth. A challenge before the university is to shape a role for itself in the nurturing of these firms.

The University Restructures

The end of the "Massachusetts economic miracle" brought a rapid drop in state tax revenues, and led to the rationalization of the public higher education system. In 1991, the University of Lowell became the University of Massachusetts at Lowell (UML), one part of the five-campus university system. Almost immediately, UML began to construct a model for how a public university could marshal its research, educational, and outreach capacities to help a region develop. The mission is predicated on the notion that "a sustainable economy includes the development of a skilled workforce, product quality, and environmental protection and worker safety" (Blewett 1995, 31–32, 196).[6] To implement the mission, faculty and administrators established over thirty interdisciplinary research centers and institutes between 1992 and 1998, charged with solving real-world problems by utilizing the talents of faculty, students, and the surrounding business community. The work of the centers and institutes includes the development of new products and manufacturing processes, the formulation of strategies to deal with today's manufacturing and management challenges, pollution control and environmentally sustainable production, worker safety and health and job design, and community leadership development and community-based social research (Chancellor's Report 1997; UML 1997).

Examples of the richness of the work now taking place are the Department of Plastics Engineering and its related centers and research institutes, the Institute for Massachusetts Partnering and Commercialization of Technology (IMPACT) and the Department of Regional Economic and Social Development (RESD). In Plastics Engineering, fully equipped laboratories are available for area firms to use in designing new products and testing the properties of new materials. IMPACT was set up in 1995 with two purposes: first, to develop partnerships with industry to ease the transfer of technology from university labs to enterprises, and, second, to develop partnerships to perform generic research with groups of firms. RESD was established in 1996. An interdisciplinary academic department, at the graduate level it educates technologists, entrepreneurs, development professionals, and community activists to understand, analyze, and intervene in the economic and social development process. Faculty are engaged in funded research and hands-on activity on a variety of issues related to economic and social development.

Social Cohesion, Social Structures, and Theories of Development

In the late 1980s and early 1990s, UML leaders recognized there was no easy technological fix for the problem of maintaining long-term, healthy growth in the Merrimack River Valley. To understand the region's history of boom-and-bust cycles required a deeper analysis than had heretofore existed. The university needed to construct a variety of bridges across the river that traditionally had divided social scientists and historians from engineers and scientists. The Committee on Industrial Theory and Assessment (CITA) was established in 1993 in part to explore these issues and develop new theoretical and pragmatic ways of thinking about regional development and UML's role in it.

While the university continues to make strides in its restructuring endeavors, the long-term success of its centers and institutes remains inhibited by two deficiencies. First, there has not been a consistent campuswide discussion on what a sustainable economy is, and whether and how the various academic and technical programs on campus can support it. As Kleniewski and Silka point out (in the next chapters herein), there is no unanimity when it comes to understanding the importance of community collaborations and applied community research. Nor have the structural barriers that make it difficult for this type of work to be evaluated— emblematic are tenure and promotion guidelines—been altered. As Kleniewski notes: "Although all complex organizations are resistant to change, a few institutional barriers within universities deserve special mention as hindering the growth of applied community research. They are the following: the narrow focus of academic disciplines; a lack of fit between institutional missions and faculty reward structures; difficulty accessing resources for applied community research; and strong boundaries between the university and the community" (528).

Second, there has been no vigorous effort to involve firm owners, unions, community groups, and others interested in the success of the regional economy in discussions about the forms a meaningful university role might take. As Silka queries, "Consider, for example, the issue of who originates the intellectual questions to be made the focus of endeavors. Traditionally the direction of influence has been from the researcher to the community" (in part IV herein). But what would be truly innovative would be a shift so that regional actors increasingly help to shape the intellectual agenda. Without these discussions, there will be no inclusive set of strategies to guide the economic development mission. Thus, there is no widely shared agreement on what the concept of sustainable development evokes (Knudsend 1997; Reese and Fasenfest 1997).

Does it imply an expanded economic pie—for instance, a larger tax base, more jobs, new firms—with little regard for who gets the additional slices? Does it foster an increase in the stability of the economy, an end to the boom-and-bust cycles that Lowell has witnessed for much of the twentieth century, a general increase in the levels of income of the population, an increase in the economic empowerment of citizens, a nurturing of start-up enterprises that may require very little advanced technologies or materials, more jobs for Lowell residents, greater energy efficiency, and less toxics usage among producers? A more encompassing definition of development, one that moves beyond a firm-specific focus, will require programs and policies that simultaneously promote a more equitable distribution of new jobs and income (Brown and Warner 1991) while boosting the region's capacity to innovate (Beauregard 1994). To be clear, a too-narrow focus on firms will fail to link Lowell residents to employment, while a too-narrow focus on neighborhood development, as the thirty-year history of community-based development demonstrates, will not foster sustainable economic growth (Nowak 1997). The linkages between these development paths need to be explored, but absent the construction of a commodious room for discussion—one with many seats at the table—this will not take place.

Social Cohesion and Social Structures

Who ought to be at the big discussion table? Why does it make sense to think about universities within their regional space? Numerous articles have advanced the perspective that the clustering or geographic concentration of similar types of firms—e.g., metalworking, computer circuit boards, plastic molded parts—portends a variety of scale advantages in the delivery of technical assistance and education and training services (Amin and Robins 1990; Cornish 1997; Florida 1996; Gertler 1995; Harrison, Kelley, and Gant 1996; Indergaard 1996; Markusen 1996; Saxenian 1990; Scott 1995). One additional benefit is the ability of firms to share in the costs and risks associated with new product or process research and development. As we have seen, the geographic proximity of some three hundred metalworking firms in Western Massachusetts engendered the opportunity for innovative firm and institutional behavior to occur. The early history of the Lowell textile industry also demonstrates this point. When the mills boomed, there were machine shops, dye houses, specialized design firms, and other firms that supported the industry. A similar argument can be made on the community development side. Today, in Lowell, there are numerous agencies and organizations working to provide job training to residents. Yet without a shared perspective on the economy and a collective sense of the jobs and their requisite skills, each organization continues to roll its own rock up the hill, and the advantages that collaboration might bring are not actualized.

In a review of regional development case studies, Cornish (1997) determined that the fast pace of innovation and the agglomeration of certain types of firms is important, "because closeness encourages the informal exchange of information and tacit knowledge" (144). This perspective recognizes that, for the economy to grow, there must be continuous investments in the preparation and preservation of a broad and deep skill base. Granovetter makes the point that the behavior of economic actors is "embedded in concrete, ongoing systems of social relations" and that it is these social relations that are "mainly responsible for the production of trust in economic life" (1985, 482, 491). And Indergaard (1996) notes that a successful flexible manufacturing network among Toledo, Ohio, metalworking firms depended on local institutions to act as "resource and social mobilizers" in order for the competitive advantages that might be gained from their proximity to be realized. Without a similar collaborative approach, UML's centers and institutes will continue to do good work, but their activities will lack cohesiveness, and thus their impact on the regional economy will be marginal.

Learning from One Another

There are manifold reasons for firms and the university to work together for the good of the economy. Universities desire healthy cities to make themselves presentable to perspective students, and city leaders and employers recognize that universities can provide jobs in their own right and serve as a source of educated employees. UML is in the region to stay, as are most of the hundreds of small and medium-sized firms that make up the regional manufacturing base. Both draw upon the same population base: the firms for their employees, and the university for their students. Approximately 85 percent of UML students come from Massachusetts, and 89 percent stay and enter the workforce (Chancellor's Report 1997). A healthy economy rebounds to the university in terms of an increased budget, more enrollments, and larger donations. And a thriving university can expand its technology holdings and thus offer more assistance to firms engaged in such endeavors as new product development.

It must be noted that there are distinct differences in the ways that businesses and the university operate; these need to be acknowledged and reviewed to determine how the university can adapt to the rhythm of business and the community. Two examples may be helpful here. First, if the university is to engage in the economic development process in a consistent manner, the semester culture must be scrapped. Firms are open for business year-round and urban problems do not abate for the summer. There must be a degree of confidence that the university will be a credible and timely source for assistance year-round. Therefore, structures must be put in place that do allow for continuous assistance. Second, the university culture too often revolves around interminable meetings and discussions. A firm with a problem, or a group of firms seeking some understanding of new market opportunities, has a limited amount of time to spend in discussions and wants meetings to be action oriented (see Kleniewski and Silka for a discussion of these issues in the next chapters).

Random Activities or Purposeful Intervention?

Since the early nineteenth century, no other state has been the seedbed of so many vibrant industrial sectors; unfortunately, the Commonwealth has a poor record in sustaining them. The textile, apparel, metalworking, machine tool, papermaking, boot and shoe, furniture, and jewelry industries all have illustrious pasts in the state, and the decline of each has left its mark in the form of rundown areas. The old competition, based on relatively secure and lengthy production runs of standardized goods, succeeded because it standardized manufacturing processes and thus drove down the costs of production (Hounshell 1984; Rosenberg 1994). This model is inappropriate in the fast-paced, high-tech world of Massachusetts manufacturing where competitive advantage is achieved through the fusion of productivity and innovation. Firms are engaged in the rapid development of new products, the continuous refinement of existing ones, and the regular improvement of their production processes. The high-quality and low-cost mantra is not a formula for sustained success. Firms must be able to produce faster and develop new products more quickly.

In his study of the Massachusetts telecommunications industry, Craig Moore (1997) concluded that the state has the nation's highest concentration of people employed in telecommunications-equipment manufacturing and telecommunications software/network integration. The highest concentrations of firms in these two sectors are within a fifty mile radius of UML. Indeed, across a wide spectrum of manufacturing sectors, there has been a rapid diffusion of technological knowledge. Under these conditions of intensified competition, continuous innovation is a necessity. Why does any of this matter to the university? Intrinsic to this concept of innovation is the accumulation of tiny, but unflagging production and organizational improvements at the shop-floor level, which, when summed, lead to superior performance. For a high-wage region to sustain itself, its firms must stay ahead of their rivals from other regions on these learning and innovation curves. This can be done only if firms gain increasingly higher degrees of what Lazonick and O'Sullivan call "organizational integration." The term describes the social relations that provide participants in a complex division of labor with the abilities and incentives to apply their skills and efforts to the innovation process (Lazonick and O'Sullivan 1996).

All employees must be able to contribute to the cumulative learning process within the firm. The firm must invest in workforce skills in order to create an atmosphere where workers are able to provide their input on the production process, and managers must be committed to a management style that continually solicits and utilizes this input. This is a social process, whereby, within an industry sector, or among firms in different but complementary sectors, high levels of cooperation are achieved in order to share the costs associated with such things as technology-related research and development and managerial- and worker-skills enhancement. The university should do everything possible to create numerous venues where this cooperation and systematic learning can flourish.

Centers, Institutes, and Interdisciplinary Activity

The phenomenal post-1960s success of the Japanese machine tool industry amplifies the importance of utilizing a theoretical and strategic approach to regional development. In 1971, a law was passed in Japan that called upon machinery and electronics firms to work together on product development. Federally sponsored business consortia made numerous advances in machine tool design and operation, which boosted the Japanese machine tool industry to a position of world-export leadership and helped domestic manufacturers reap significant productivity advantages (Forrant 1997; Kodama 1986). Ten years after the passage of the law promoting firm collaborations, the Ministry of International Trade and Industry's (MITI) Small and Medium Enterprise Agency established the Technology Exchange Plaza Project (1981) to encourage firms to increase their exchange of information and technology. And in 1988, the Technology Fusion Law[7] was passed to promote different industry sectors working together in much the same way that electronics and machine tool firms did (Subramanian and Subramanian 1991).[8]

The successes generated by cross-firm and cross-sector approaches to technology development demonstrate that the establishment of explicit linkages across various research centers and institutes and the promotion of linkages among small and medium-size firms is essential. How can the university advance collaboration as a consistent development strategy? In *The New Competition* (1990), Best notes that Japan's manufacturing success "derives from the effective joint public-private articulation and coordination of long-term strategic analysis at the sector level. Sector strategic analysis is a public good in that it costs no more to provide it freely to all than it costs to provide it to one consumer" (183–84). It is incumbent on the university to become more strategic in its development focus and more expansive in the discussions it engages in to shape that strategy if it is to play a leadership role in the regional development process.

A nascent strategic focus is evident in the work of centers and institutes engaged in activities related to energy conservation, biodegradable materials, cleaner and safer production, and the environment. Geiser and Greiner provide an analysis herein (in part II) of the work of the Toxics Use Reduction Institute (TURI). TURI is a multidisciplinary research, education, and policy center, working with companies to introduce the use of nontoxic materials in their manufacturing processes. It sponsors and conducts research in the development of cleaner manufacturing technologies and safer materials. Their work relies on the integration of the life sciences with engineering and economics. The case studies they report on demonstrate the potential of and the obstacles to the diffusion of industrial process and products that prevent pollution and minimize long-term risks to the human population and the environment. The Center for Environmental Engineering and Science Technologies has a research agenda that includes wastesite characterization and containment, soil remediation, and air, water, and soil sampling. It provides direct technical assistance to firms working to develop innovative technologies for solving environmental problems. The Lowell Center for Sustainable Production—discussed in Quinn et al. (in part II) and Kriebel, Geiser, and Crumbley (in part III) in this volume—focuses on the development of explicit linkages between the promotion of economic activity that is conserving, nonpolluting, and efficient with the promotion of worker and community well-being. The challenge it faces is to establish an innovative agenda for helping industry make the shift to production processes that simultaneously prevent environmental pollution, enhance the health and safety of workers, and augment productivity. The center fosters collaboration across several academic disciplines, including Work Environment, Engineering, Regional Economic and Social Development, and Public Health, as well as among several constituencies including the environmental movement, trade unions, the state and federal governments, and academic departments and groups that often are at odds with one another.

Why do these interdisciplinary activities matter? They are significant because they more aptly mirror what is taking place in the regional economy as firms collaborate across several "manufacturing disciplines" to produce innovative products. For example, in the region there are several manufacturers engaged in such things as the design and production of environmental measuring and testing devices (Forrant 1995). And there are high concentrations of firms engaged in the manufacture of computer and communications equipment, medical equipment, electronic and electrical components, and measurement and testing devices for the environment. To support these companies, there are hundreds of small fabricated metal and industrial machinery firms concentrated in a ring around Greater Boston, along Routes 128 and 495. They provide design, engineering, and prototype production services to the state's high-tech firms. If we consider that five Massachusetts high-tech manufacturing sectors consisting of approximately one thousand firms—computers and office equipment, telecommunications equipment, electrical components, measuring and control devices, and medical equipment—depend on precision metalworking, we can see the importance of learning about interfirm linkages and how the university might enhance such relationships.

Even this cursory analysis reveals numerous synergies between metal-working firms, design and engineering companies, manufacturers of environmental and pollution-control devices, and the university centers that are engaged in the search for environmentally appropriate materials, cleaner manufacturing processes, and the tools to clean up past environmental degradation. The region's manufacturers may well be in the forefront of the global environmental effort to cut toxic wastes, reduce energy consumption, and increase recyclable products. But to remain global leaders, the diffusion of new technical knowledge and the rapid development of products and manufacturing processes is essential. With its present expertise and the potential it has to investigate and learn even more about these issues, the university should insinuate itself into this knowledge-creation and diffusion nexus. In this way it will contribute to the development of a sustainable manufacturing economy.

Conclusion: A Regional Development Framework

Based on several years of economic development work in various settings Best and Forrant have developed a ten-point methodology for conducting regional, production-led, industrial policy initiatives (1996, 248–52). The points are:

1. Develop a sector-strategy agency capability guided by the world "best-practice" methods. Industrial policymaking agencies must be guided by an awareness of the competitive dynamics that both drive and confront enterprises.

2. Conduct a strategic-sector analysis. The starting point here is an outline of a strategic competitor analysis that accounts for world-class best practices at the enterprise level. The ultimate task here is to track the performance of the best firms in the world that compete with key enterprises and sectors in the region.

3. Conduct an audit of industrial capabilities by sector. This audit starts with a compilation of manufacturing censuses and other available data to learn as much as possible about the firms in a particular region. Directories of firms, geographical-density maps, and business patterns among firms are constructed to learn as much as possible about interfirm and cross-sector relationships.

4. Visit businesses to generate data and develop case studies. Firm visits were used to great advantage by MAP. The visits have a way of bringing survey and other data to life.

5. Promote technology research and development and its rapid diffusion. For Massachusetts firms to remain successful, they must gain access to new materials, new manufacturing processes, and new technologies in order to achieve world-class performance standards.

6. Develop a strategic orientation to technical and educational services. The technical and educational capabilities at the university need to be linked to and informed by a continuously upgraded base of knowledge about the firms being served.

7. Promote discussion among industry insiders on the results of the sector analysis. If a sector analysis is prepared but is not discussed, firms cannot benefit from it, and the study's authors receive no reality check on their work. Hence, no plan of action to improve the sector will be forthcoming.

8. Develop performance and transition indicators. The performance of firms can be measured against similar indicators for world-class companies. This will help to identify specific things to work on with firms, for examples, set-up time reduction, scrap and rework rates, energy usage, and toxics-use reduction. The March 1997 Massachusetts Toxics Use Reduction Program evaluation is an excellent example to consider as a model.

9. Establish group implementation plans. Workshops can be set up to attack common problems in firms within a particular sector. For example, several precision machine shops may need to learn how to improve their use of computer-aided design software. Rather than work on a case-by-case basis, the firms can be brought together to do this. Costs to the firms are shared and thus minimized, and a faculty member's time is spent working with several companies. This establishes a forum for peer learning, as participants discuss how to introduce what they have learned on the shop floor.

10. Diffuse the restructuring ideas widely in the sector. Case studies of successful firms, newsletters, workshops, and forums are needed to facilitate the diffusion of the new production practices that are being introduced.

Presently the university is engaged in the first five activities on our agenda; however, absent an articulated and overarching theoretical and strategic approach to its work, and a thoughtful, systematic, and ongoing evaluation process to inform industry and university participants about what transpires (points 6 through 10), cumulative universitywide and regionwide learning will not be manifested. Only when the activities of the centers and institutes are informed and influenced by rigorous discussion and evaluation among faculty members, community leaders, students, and industry representatives will the university be able to upgrade its regional development role continuously; thus the university must direct its attention to points 6 through 10 in the methodology discussed above. This is necessary for three reasons.

First, simply providing one-on-one help to firms is an inefficient way to carry out an agenda to improve the capabilities of significant numbers of firms across the region. To reach a critical mass of firms and to make the most efficient use of laboratories and staff time, group activities are essential. Second, one-off activities will not enhance a strategic approach to development. For example, a strategic focus on making a range of training and educational programs available to groups of firms helps to resolve the problem of scale, while it simultaneously invigorates peer learning. This approach also makes assistance affordable to small firms, because the costs are shared. Thus, a publicly supported university should place its emphasis on building these kinds of cross-firm learning partnerships. Finally, one-off activities will not generate cross-collaboration among firms with different technical capabilities; yet, as was demonstrated by the purposeful fusion example of Japanese machine tool builders and electronics firms, it is this kind of collaboration that fosters economic innovation and growth.

The greatest "values added" that modernization organizations provide firms occur through the opportunities that are established for firms to discuss and assess their problems and learn from one another. Left to their own devices, small and medium-size firms typically do not discuss common and important market failures, such as the lack of skilled employees, the demise of key customers, product and market shifts that eat away at their customer base, and the demands of their customers to step up their quality assurance programs. From the perspective of broad-based regional development, too little good can be derived from a one-at-a-time approach to company assistance. In addition, unless the centers and institutes regularly evaluate their work and widely disseminate its lessons, little on-campus learning will take place. There is a solid technical infrastructure in place. Now the university must turn to nurturing the social relationships required to derive widespread and long-term benefits for the region and the campus from this technical system. Simultaneous learning needs to occur across three broad fronts: among firms, among the various centers, institutes, and academic departments on campus, and between firms and the university community. The university's continuous promotion of this collective learning dynamic will allow it to play a credible and catalytic role in the elaboration and advancement of a sustainable regional economy.

NOTES

1. The region that the university is most concerned with is bounded by New Hampshire to the north, Route 495 to the west, and the Massachusetts Turnpike to the south. It stretches to the Atlantic Ocean on the east, and includes communi-

ties located in a ring along Routes 128 and 495 to the north of Boston. These boundaries conform to what the Massachusetts Office of Business Development calls its Northeast Service Delivery Area. Unless otherwise indicated, when I refer to "the region," or "the Northeast," this is the area I am discussion.

2. In the literature, sectors, clusters, and networks are variously defined. Here I use the terms in the following way: A sector is a group of firms that perform a similar manufacturing function. For example, firms that machine, bend, and shape metal are referred to collectively as the metalworking sector. Clusters expand the universe. A cluster is made up of firms in a variety of industry sectors that perform a complementary set of activities in order to bring a product to the market that no single firm in a sector is capable of accomplishing. A specific example here is the Hickory–High Point region in North Carolina, where woodworking, design, upholstery, and machine-building firms congregate to build furniture (Rosenfeld 1997). When the textile industry dominated the Lowell economy there were a textile sector of many mills and also a cluster made up of the various machine shops that built textile equipment, dye houses, specialty design firms, marketing agents, and the Lowell Textile Institute. Other examples of clusters include: ceramic tiles in Sassoulo, Italy; upholstered furniture in Tupelo, Mississippi; electronics and computer graphics in Silicon Valley, California; and precision metalworking and plastics injection molding in western Massachusetts. A network is a formal or informal arrangement of firms that represent a sector or a cluster that interacts with a variety of public and private institutions—like training providers—to resolve common problems that inhibit success.

3. It is not my intention to discuss these university–industry research institutions in great detail, but to acknowledge their existence and to make the point that UML's industry-funded research should be evaluated against these national activities. Cohen, Florida, and Goe (1994) consider the short-run implications of university–industry research relationships for dissemination of research findings and conclude that the free flow of information is inhibited. This appears to be offset, however, to a degree, by the "more effective mechanisms for advancing commercial technology" that result (30). Nonetheless, the question remains as to whether there is a broader economic development value to such technology-based research arrangements.

4. MAP was one of five Industry Action Projects (IAPs) established across the state through the Regional Labs program. Each project was designed to involve displaced workers in industry-specific economic development projects to revive their communities. Industry Action Projects (IAPs) were set up in Greenfield, North Adams, Fall River–New Bedford, Springfield, and Worcester. On each site, IAPs were independent, nonprofit public–private partnerships that empowered local boards to design and implement strategies to strengthen local industry. A defining characteristic of the IAPs was their recognition that labor had a large role to play in this process. The IAPs that achieved the greatest success were MAP and the Needle Trades Action Project in the Fall River–New Bedford area. The entire program awaits a systematic historical evaluation.

5. The National Tooling and Machining Association (NTMA) in Washington, D.C., is the national representative of approximately three thousand custom precision manufacturing companies throughout the United States. There are fifty-five

local chapters of the NTMA located throughout the country. Member firms pay fees directly to the national chapter, and local chapters receive a 20 percent return on these dues. Local chapters operate thirteen regional training centers for toolmakers and machinists. Since 1970, the Western Massachusetts chapter of the NTMA has operated the Western Massachusetts Precision Institute.

6. UML's primogenitor, the Lowell Textile School, had a similar mission. The Massachusetts Legislature established Lowell Textile in 1895 to train workers and improve the technological sophistication of the region's textile mills. The legislature was hopeful that Lowell Textile could simultaneously enhance the overall competitiveness of the United States in the global textile market and maintain Greater Lowell as a high-wage, fine-textile goods producer.

7. The term "fusion" is borrowed from physics and signifies the confluence of several inputs. For example, in music it is used to describe a sound derived from a blend of two or more distinct styles, such as jazz and rock. In politics, it is a term used to describe a coalition of various parties or factions, as in a "fusion ticket."

8. Gertler (1992) examines the technology development process in the machine tool industry and puts forward the working hypothesis that, under specific conditions, "closeness between user and producer appears to be crucial if the machinery/system developed is to be truly leading-edge. Machinery system producers seem to gain important advantages by being able to participate in regular and deep interaction with their primary customer base" (271–72). The ways in which the fusion of the interests between users and builders contributes to the development of a regional economy is very important to understand and adds further credence to the argument I am making that theory and research must be an essential part of the work of the centers.

REFERENCES

Amin, A., and K. Robins. 1990. The reemergence of regional economies? The mythical geography of flexible accumulation. *Environment and Planning* D 8:7–34.

Atkinson, R. 1993. *The next wave in economic development: Industrial services in the 1990's.* Washington, D.C.: U.S. Office of Technology Assessment.

Batt, R., and P. Osterman. 1993. *A national policy for workplace training: Lessons from state and local experiments.* Washington, D.C.: Economic Policy Institute.

Beauregard, R. A. 1994. Constituting economic development. In *Theories of local economic development,* ed. R. D. Bingham and R. Mier, 191–209. Thousand Oaks, Calif.: Sage.

Beck, R., D. Elliott, J. Meisel, and M. Wagner. 1995. Economic impact studies of regional public colleges and universities. *Growth and Change* 26: 245–60.

Best, M. 1990. *The new competition: Institutions of industrial restructuring.* Cambridge: Harvard University Press.

Best, M., and R. Forrant. 1996. Creating industrial capacity: Pentagon-led versus production-led industrial policies. In *Creating industrial capacity: Towards full employment,* ed. Jonathan Michie and John Grieve Smith, 225–54. New York: Oxford University Press.

Blewett, M. 1995. *To enrich and to serve: The centennial history of the University of Massachusetts Lowell.* Virginia Beach, Va.: Donning Company.

Brown, D. L., and M. E. Warner. 1991. Persistent low-income nonmetropolitan areas in the United States: Some conceptual challenges for development policy. *Policy Studies Journal* 19:22–41.

Brusco, S. 1992. Small firms and the provision of real services. In *Industrial districts and local economic regeneration,* ed. Frank Pyke and Werner Sengenberger, 177–97. Geneva: International Institute for Labour Studies.

Chancellor's Report. 1997. *University of Massachusetts Lowell Chancellor's Report.* Lowell: Office of Public Information.

Cohen, W., R. Florida, and W. R. Goe. 1994. *University-industry research centers in the United States.* Pittsburgh: Center for Economic Development, Carnegie Mellon University.

Cornish, S. 1997. Product innovation and the spatial dynamics of market intelligence: Does proximity to markets matter? *Economic Geography* 73:143–65.

Cossentino, F., F. Pyke, and W. Sengenberger. 1996. *Local and regional response to global pressure: The case of Italy and its industrial districts.* Geneva: International Institute for Labour Studies.

Dineen, D. 1995. The role of a university in regional economic development: A case study of the University of Limerick. *Industry & Higher Education* (June): 140–48.

Eisenger, P. 1995. State economic development in the 1990s: Politics and policy learning. *Economic Development Quarterly* 9:146–58.

Felsenstein, D. 1996. The university in the metropolitan arena: Impacts and public policy implications. *Urban Studies* 33:1565–80.

Ferleger, L., and W. Lazonick. 1993. The managerial revolution and the developmental state: The case of U.S. agriculture. *Business and Economic History* 22:67–98.

Fitzgerald, J., and A. McGregor. 1993. Labor-community initiatives in worker training in the United States and the United Kingdom. *Economic Development Quarterly* 7:172–82.

Florida, R. 1995. Toward the learning region. *Futures* 27:527–36.

———. 1996. Regional creative destruction: Production organization, globalization, and the economic transformation of the midwest. *Economic Geography* 72:315–35.

Flynn, E., and R. Forrant. 1997. The manufacturing modernization process: Mediating institutions and the facilitation of firm-level change. *Economic Development Quarterly* 11:146–65.

Forrant, R. 1995. *Metalworking in the Merrimack River Valley: Can a cutting edge be maintained?* Lowell: Center for Industrial Competitiveness, University of Massachusetts Lowell.

———. 1997. The cutting edge dulled: The post–World War II decline of the U.S. machine tool industry. *International Contributions to Labour Studies* 7:37–58.

Forrant, R., and E. Flynn. 1998. Seizing agglomeration's potential: The greater Springfield, Massachusetts metalworking district in transition, 1986–1996. *Regional Studies* 32:209–22.

Gertler, M. 1992. Flexibility revisited: Districts, nation-states, and the forces of production. *Transactions in Institutional British Geography* 17:259–78.

Gertler, M. 1995. "Being there": Proximity, organization, and culture in the development and adoption of advanced manufacturing technologies. *Economic Geography* 71:1–26.

Gittell, R., and P. Flynn. 1995. Lowell, the high tech success story: What went wrong? *New England Economic Review* (March–April): 57–60.

Granovetter, M. 1985. Economic action and social structure: The problem of embeddedness. *American Journal of Sociology* 91:481–510.

Harris, R. 1996. The impact of the University of Portsmouth on the local economy. *Urban Studies* 34:605–26.

Harrison, B., M. Kelley, and J. Gant. 1996. Innovative firm behavior and local milieu: Exploring the intersection of agglomeration, firm effects, and technological change. *Economic Geography* 72:233–58.

Hounshell, D. 1984. *From the American system to mass production, 1800–1932: The development of manufacturing technology in the United States.* Baltimore, Md.: Johns Hopkins University Press.

Indergaard, M. 1996. Making networks, remaking the city. Economic Development Quarterly 10:172–87.

Isserman, A. 1994. State economic development policy and practice in the United States: A survey article. *International Regional Science Review* 16:49–100.

Knudsend, D. 1997. What works best? Reflections on the role of theory in planning. *Economic Development Quarterly* 11:208–11.

Kodama, F. 1986. Japanese innovations in mechatronics technology. *Science and Public Policy* 13:44–51.

Labrianidis, L. 1995. Establishing universities as a policy for local economic development: An assessment of the direct impact of three provincial Greek universities. *Higher Education Policy* 8:55–62.

Lampe, D. 1988. *The Massachusetts miracle: High technology and economic revitalization.* Cambridge: MIT Press.

Lazonick, W., and M. O'Sullivan. 1996. *Corporate governance and corporate employment: Is prosperity sustainable in the United States?* Lowell: Center for Industrial Competitiveness, University of Massachusetts Lowell.

Lichtenstein, G. 1992. *A catalogue of U.S. manufacturing networks.* Gaithersburg, Md.: National Institute of Standards and Technology.

Markusen, A. 1996. Sticky places in slippery space: A typology of industrial districts. *Economic Geography* 72:293–313.

Moore, C. 1997. *Connection to the future: An analysis of the telecommunications industry in Massachusetts.* Boston: Massachusetts Telecommunications Council.

National Institute of Standards and Technology (NIST). 1994. *Manufacturing extension partnerships.* Washington, D.C.: U.S. Department of Commerce.

Nowak, J. 1997. Neighborhood initiative and the regional economy. *Economic Development Quarterly* 11:3–10.

Osborne, D. 1988. *Laboratories of democracy: A new breed of governor creates models for national growth.* Boston: Harvard Business School Press.

———. 1989. *State technology programs: A preliminary analysis of lessons learned*. Washington, D.C.: Council of State Policy and Planning Agencies.

Reese, L., and D. Fasenfest. 1997. What works best? Values and evaluation of local economic development policy. *Economic Development Quarterly* 11:195–207.

Rosenberg, Nathan. 1994. *Exploring the black box: Technology, economics, and history*. New York: Cambridge University Press.

Rosenfeld, S. 1997. Bringing business clusters into the mainstream of economic development. *European Planning Studies* 5:3–23.

Saxenian, A. 1990. Regional networks and the resurgence of Silicon Valley. *California Management Review* (Fall): 89–92.

Scott, A. 1995. The geographic foundations of industrial performance. *Competition and Change* 1:51–66.

Shapira, P. 1990. *Modernizing manufacturing: new policies to build industrial extension services*. Washington, D.C.: Economic Policy Institute.

———. 1996. Modernizing small and mid-sized manufacturing enterprises: U.S. and Japanese approaches. In *Technological infrastructure policy: An international perspective,* ed. M. Teubal, D. Foray, M. Justman, and E. Zuscovitch. Amsterdam, Netherlands: Kluwer Academic Press.

Subramanian, S. K., and N.Y. Subramanian. 1991. Managing technology fusion through synergy circles in Japan. *Journal of Engineering and Technology Management* 8:313–37.

University of Massachusetts Lowell. 1997. *Interdisciplinary centers and institutes*. Lowell: Office of Public Information, University of Massachusetts Lowell.

Addressing the Challenge of Community Collaborations

Centers as Opportunities for Interdisciplinary Innovation

Linda Silka

> The constraint of tradition on innovation explains much of the history of the university.
>
> —J. T. Bonnen

THIS volume concerns approaches to sustainable development, with specific emphasis on the question of how innovations can contribute to sustainability in regional approaches. As noted in the introductory chapter, this volume considers what members of various academic disciplines mean when they talk about innovation in social, institutional, and organizational arrangements, how innovation affects the world around the university, and how innovation in a region surrounding a university can in turn lead to change on campus. Three important questions about innovation were raised in the call for chapters:

How do university centers introduce their research to organizations off campus in order to stimulate new ways of thinking and acting?

How do faculty members who are engaged in research draw the lessons of that research into their classes in order to encourage innovative ways of resolving a variety of economic and social development problems?

What are the barriers to the flow of social, institutional, and organizational innovation, both on campus among faculty and departments and from the campus out into the business community?

This chapter considers these questions through an examination of the activities of one center, the University of Massachusetts Lowell's Center for Family, Work, and Community. The center's initiatives are instructive because its goal is to promote community–university collaborations that enhance regional economic development, broadly construed. Our interest here lies in the degree to which a "mechanism," such as an interdisciplinary university center, can be an engine for innovations through its relationships with the community, other centers, and the university's various departments.

We examine three major undertakings at the center in detail. These

358

three multiyear initiatives link different departments and faculty, approach research in varied ways, and seek multiple outcomes. Discussion of these approaches sets the stage for a consideration of the university's ongoing analysis of the concepts of innovation, regions, and sustainability. The study concludes with a critique of recent calls in higher education for changes in reward structures so that urban universities can sustain the kinds of outreach initiatives described here. I maintain that if community–university initiatives are to become an integral part of the intellectual agenda pursued by universities, the focus should be placed on how these innovations tie into existing reward structures rather than on a need to create new rewards. Woven throughout this discussion is a concern for how we can overcome the commonly held view that application is not an arena for innovation. Too often the actions of faculty members suggest that they view the real intellectual work as having been completed by the time attention turns to application and community collaborations. Throughout I argue that this standard view is problematic. It keeps university faculty from recognizing the intriguing problems of application that are in need of solution and from seeing the intellectual challenges that lie at the heart of outreach scholarship.

The changes taking place at University of Massachusetts Lowell are not isolated ones. To provide some perspective on the analysis offered here, I review recent national trends among urban universities as they have begun to struggle with the need for developing different kinds of relationships with the urban areas in which they are located.

Emerging Trends around the Country: Urban Universities Face Their Communities

Over the last decade, universities are once again coming under attack (Bowen and Shapiro 1998; Kennedy 1997; Lerner and Simon 1998b). They are seen as increasingly elitist and contributing little to the public good. The forms of these attacks have been varied, but most call universities to task for returning too little to the larger society. While these external criticisms have been mounting, urban universities have also begun to recognize the extent to which their fates are, in fact, linked to those of their surrounding areas. The decay of urban areas is making it more difficult for urban universities to attract qualified students and faculty; unlike other enterprises, universities cannot move elsewhere. As a result, they have begun to look more closely at ways to work directly with their local communities to improve these urban areas.

The nature of the response has varied in different locales, but one important approach has been for universities to take stock of their resources that might be used in partnership with communities to bring about change.

Some universities have set up centers and outreach arms or have developed other forms of collaboration that transcend disciplinary boundaries. Although these efforts draw on the land-grant approach (Bonnen 1998), they are decidedly urban: They focus on the constellation of problems common to urban areas. Various universities are attempting to devise new strategies capable of being sustained in their institution, particularly approaches that emerge from the universities' goals as scholarly institutions, which also can be shown to have a measurable impact on local communities. Urban institutions, such as Boston College (Brabeck et al. 1998), have begun modeling more effective approaches by which interdisciplinary centers and departments can work together. The University of Pennsylvania has modeled ways to create a centralized office (the Center of Community Partnerships) that spearheads new methods of applied research throughout the university (Harkavy 1998). These universities are seeking out niches that reflect the particular combinations of disciplinary strengths that exist in their institutions and that blend with local need and context.

The University of Massachusetts Lowell joins this effort in its own unique way. As a part of the university's mission, faculty members have been called upon to focus on regional economic development, to make a measurable difference in the local region, and to do so by using the intellectual resources of the university in innovative ways. Just as other universities are testing out various strategies and then attempt to learn from their hands-on experiences, so, too, is the University of Massachusetts Lowell. Faculty have begun a close examination of what it means to make regional economic development a focus of our efforts. Leaders in the academic community are increasingly asking pointed questions about the roles different centers and various departments might play in advancing this mission. Within the university, people have begun to examine locally directed university efforts closely and determine whether they are optimally organized to contribute to the university's goal. An examination of the work of one center, that of the Center for Family, Work, and Community, provides an opportunity to look at these issues in closer detail.

Outreach through the Center for Family, Work, and Community

The Center for Family, Work, and Community was established a decade ago by the chancellor of the University of Massachusetts Lowell with the intent of broadening the university's outreach mission. This center does not limit its focus to a specific area of research, but, rather, attempts to break down disciplinary barriers in encouraging collaborative research, creating learning opportunities for students, and acting as an access point for groups and organizations outside the university. The center brings together groups of faculty, graduate students, and undergraduates to work in partnership with the community in solving problems. Our focus is on application and outreach scholarship, and this work is carried out through grant-funded technical assistance, applied research, program evaluation, and capacity-building initiatives.

Our work includes national and state initiatives, but we place a special emphasis on work with the local community. This diverse community of over one hundred thousand people is home to many different ethnic groups, and there are more than four hundred minority-owned businesses in the city. Lowell's Cambodian community is the second largest in the country. Poverty is prevalent. In the central part of the city, the median household income is under $10,000, many households are female-headed, and the educational-attainment level is low. According to the HUD report, "Now Is the Time: Places Left Behind in the New Economy," Lowell's estimated 1995 poverty rate was 23.4 percent, giving it a rank of 71 out of 170 cities on a list of what HUD had determined to be high-poverty central cities (U.S. HUD 1999). And of the 180 mid-sized cities (population 50,000 to 100,000) the HUD report analyzed, only 49 had poverty rates in excess of 20 percent; Lowell was one of those communities. According to federal and state data, approximately 30 percent of Lowell's children live below the poverty line. The city is one of only 64 municipalities in the country with a HUD Enterprise Zone, which has brought several million dollars of federal funding into the city directed at improving the social infrastructure. Lowell also has a history of environmental contamination and has been designated a "Brownfields Showcase Community," that is, a location where new practices for bringing economic renewal to abandoned contaminated industrial sites are being tested. Much of the recent economic effort (e.g., construction of a new hockey arena, baseball stadium, and river walk) spearheaded by city government has been directed at altering Lowell's economic fortunes by turning the city into a destination site for recreational activities.

Three of the center's major initiatives will be described here to indicate the range of the approaches that have been undertaken by center faculty, staff, and students in partnership with this highly diverse and environmentally contaminated community. These efforts are being expended as innovative, experimental attempts at bringing the university and the community together in productive, sustainable relationships. These three initiatives focus on environmental justice, leadership-capacity building with newcomer groups, and community economic development. In each direction, the center attempts to meet multiple goals of advancing research, sharing resources, and providing opportunities for students, faculty, and community to address jointly pressing urban problems.

Environmental Justice

The center's Southeast Asian Environmental Justice Partnership was begun with four years of funding from the National Institute of Environmental Health Sciences (Alston et al. 1996; Silka, Benfey, and Khoeun 1996), with the goal of bringing together researchers, primary health care providers, and an underserved community to identify ways in which research can be rigorous and still meet the needs of communities. This branch of the National Institutes of Health (NIH), over the last few years, has begun to raise concerns that the research they fund is often carried out in ways that fail to include the community's perspective, concerns, or knowledge. The work is too often done "to" rather than "with" the community (Quigley 1996). NIEHS has awarded several grants to groups around the country who have devised innovative approaches for bridging these gaps, and the Southeast Asian Environmental Justice Partnership is one of these.

Lowell's partnership brings together the Cambodian and Laotian communities, health care providers, and environmental and health researchers in the university. This collaboration acknowledges Lowell's unenviably long history of environmental contamination (perhaps the longest in the United States) and juxtaposes that history with newcomer families' lack of experience in confronting urban environmental problems (e.g., lead in homes and garden soil, mercury in fish).

A number of features of this collaboration have been unusual (Silka 2000). The institutional arrangements, for example, involve situating partnership staff not just on campus, but also in community agencies and primary health care centers. This arrangement is intended to reduce barriers to communication and the sharing of knowledge and resources. The partnership has also aggressively sought out groups in the university that are generating new knowledge and looked for ways to build a structural linkage with those groups. Thus, for example, a Southeast Asian staff person has become a toxics-use reduction "translator" who is working to find bridges between the University's Toxics Use Reduction Institute and the community's concerns with urban contamination. We have also looked for fresh ways to use technical tools to heighten communication across literacy levels. For example, we use geographic information systems (computerized mapping) as a way to combine information about environmental problems with community strategic planning. Even something as small—but still symbolic—as the nature of our time line has been redesigned, made circular rather than linear to combine seasons, culture, and environmental issues and convey the understanding that tasks need to be completed, but that time does not end at the close of each calendar year.

Early in the environmental justice partnership, it became clear that information about community knowledge of and exposure to environmental hazards was insufficient. The partnership aimed to find ways to collect rigorous information, train graduate students, and at the same time build linkages to the community. Over several semesters, we involved teams of graduate students who, as a part of their capstone experiences in the Department of Nursing, developed the protocol for a community health survey carried out in three languages, Khmer, Lao, and English. These graduate students not only collaborated with the community in designing the survey using participatory research methods, but also, with the three dozen community leaders and students, went into homes to collect the interview data. All the activities of the community survey had multiple goals: to collect data, to impact student training, to provide information to the community, and to have the community and university work together in closing an important information gap in the community. The data are now being used in the community for policy purposes, and by university faculty for planning additional collaborations and research.

The Southeast Asian Water Festival represents another striking method for advancing multiple environmental-justice aims (Silka 1997). As the partnership hosted community conversations to learn of community concerns about the environment, we quickly discovered that water was the salient environmental concern. People were fishing and swimming in local ponds and rivers, such as the Merrimack River, but they were uncertain of the environmental hazards associated with these activities. According to pollution advisories, there was reason to be concerned, and industrial pollution continues to be a problem. At the same time, Southeast Asian newcomers helped the partnership understand the importance of water in Southeast Asian cultures. For hundreds of years along the Mekong River, water festivals had been important traditional celebrations of the purity and importance of water. This history provided an opportunity to build a natural link to the Merrimack River and life along this very urbanized river that has been the center of traditions in this region of Massachusetts. The partnership decided to recreate a Southeast Asian Water Festival that would bring together environmental concerns and cultures. Youths served as River Ambassadors, carrying out water-quality testing and engaging others in looking at factors that impact environmental quality. Nearly ten thousand people (mostly of Southeast Asian background) participated in the first festival in 1997, and twenty thousand people from around the country attended in 1998 and in 1999. Such a festival, by shifting the focus to newcomers as experts on traditions, also shifted the balance between researchers and community and encouraged university participants to look to the community as a resource.

Offshoots of the festival have been varied: the U.S. Department of Agriculture has become involved with the community and the center in supporting urban farming initiatives in which clean soils are a central focus.

A partnership, composed of university people (the CFWC, the Center for Sustainable Production, and the Center for Chronic Diseases), community leaders, and nonprofit organizations, has emerged as a consequence of the water festival to undertake an urban aquaculture project that combines developing new economic opportunities, carrying out research on nutritionally enhancing fish, and designing aquaculture ventures that can be sustained environmentally. Each of these aspects serves as a national model for addressing long-standing research and application questions in the field of aquaculture.

The environmental-justice program attempts to maintain its overall goals, while actively seeking out new opportunities by which these goals can be met (Silka 2000). We looked for opportunity points to involve university specialists in environmental-risk assessment, for example, and to have this experience contribute to their research and impact on the community. One of the partnership's Southeast Asian organizational partners had been offered a massive old mill building with an unknown history of use, but one that was likely to have resulted in contamination. University faculty provided hands-on training in risk assessment to key leaders in the community, including members of an executive board (all of whom had expressed little interest in this topic until it directly affected the most significant fiscal decision the organization had yet faced). This training flowed back into research, classroom teaching, and into the writing of additional collaborative grants to create more opportunities for research and environmental education.

Refugee and Immigrant Leadership and Empowerment (CIRCLE)

A second example of the initiatives undertaken by the Center for Family, Work, and Community was its empowerment training of immigrant and refugee leaders throughout the Merrimack Valley (Archer et al. 1997). This initiative, entitled CIRCLE and begun under the auspices of the Massachusetts Office for Refugees and Immigrants (MORI), was an outgrowth of MORI's discovery that their policy-based programs were insufficient to meet the needs of the newcomers who were emerging as leaders in their communities. MORI staff viewed universities as having the resources to design innovative strategies that would incorporate recent research on community empowerment and would bring newcomer groups into the development and evaluation of nontraditional programs. The challenges to effective program design were many. Often refugees and immigrants arrive in the United States with many skills that lack transferability (particularly if they move from an agrarian society to an industrialized region such as the Merrimack Valley). Refugees and immigrants typically have little in common with one another, because their home countries differ greatly from one another, and men and women frequently respond in sharply different ways to the immigration experience. Hence, any attempt at generalized training is likely to be unsuccessful.

Working in collaboration with the University of Massachusetts campuses at Boston and Amherst, the CFWC designed innovative strategies in their program for community empowerment. Community leaders and faculty from various departments (e.g., education, psychology, economics, and political science) came together to design multiweek training sessions on topics of strong interest to the community and for which expertise could be found within the university. These topics included political leadership, economic development, and educational partnerships. Faculty members initially led the empowerment training classes, but soon CIRCLE graduates and community experts representing the different refugee and immigrant communities were able to assume these roles.

A central CIRCLE focus has been and continues to be community economic development. CIRCLE leaders have struggled with what it means to try to increase the economic prospects for their community and how to place these struggles within the context of global economic trends as well as the history of immigrant contributions to the Greater Lowell economy. CIRCLE continues to develop its programs and practices to respond to changes in the university (e.g., how to tap into the newly available intellectual resources of the Department of Regional Economic and Social Development) and changes in the community (e.g., changes in the welfare laws that are particularly punitive toward the immigrant community and changes in the local infrastructure—such as the creation of a small business assistance center—that present new opportunities to immigrant entrepreneurs).

CIRCLE's educational leadership program also shows how institutional arrangements can be organized. The core of this training is based on the problem of how to draw on current research (such as analyses of parent–school partnerships) and translate these principles into a set of empowerment experiences that consistently meet the needs of the immigrant and refugee community in the Lowell area. Leaders from different refugee and immigrant communities come to the training sessions with strongly held and highly distinct views of the proper relationship between schools and parents. Those views can include the belief that parents show their respect for teachers by making few demands on the system. The CIRCLE training is built on the premise that CIRCLE's role is to strengthen parent–teacher relationships that are now quite fragile. Refugee and immigrant leaders look to CIRCLE to learn more about how schools work in the United States and about the rights parents have in a democratic society to insist on a quality education for their children; teachers and school administrators look to CIRCLE to learn how to negotiate relationships with newcomer leaders that will create partnerships for improving educational opportunities. Through CIRCLE, Dr. Joyce Gibson of the Graduate School of Education has integrated CIRCLE training into her courses, with the result that the teachers in these graduate classes experience at firsthand the kinds of successful collaborations that are possible. Dr. Gibson has gone on to build connections between CIRCLE training and the school system's struggling parent-liaison initiative (a primary means by which the schools reach out to parents), again as a way to increase links between parents and the school system.

Community Outreach Partnership Center (COPC)

A final CFWC initiative that indicates the range of outreach activities it carries out is the Community Outreach Partnership. Under the auspices of a three-year U.S. Housing and Urban Development grant, the University of Massachusetts Lowell has developed this collection of outreach innovations (Forrant and Silka 1999). Through its University Partnerships Office, HUD allocated funding to bring about changes in how urban universities work with their surrounding communities. This HUD program is intended to make universities more effective partners in community building. From the very outset, the proposal development process undertaken by the University of Massachusetts Lowell was an interdisciplinary effort carried out by a newly formed department of Regional Economic and Social Development. The proposal was structured to assist faculty in the department in identifying common interests and transcending disciplinary boundaries in outreach scholarship. The CFWC was one of two centers (along with Center for Industrial Competitiveness) affiliated with the department that led this universitywide initiative, which has as its ultimate goal the institutionalizing of a set of outreach activities central to the university's mission.

A key feature of this project concerns the methods for drawing faculty into effective outreach scholarship. What we now call the "University in the City Scholars" program provides course-release time for faculty from across the university to design programs collaboratively with the community (Forrant and Silka 1999; Silka 2000). The goal is to have these programs become a regular part of university functions (i.e., incorporated into classroom activities and research initiatives). Early on we faced the problem of how to encourage faculty to see these efforts as structurally different from simply extending their individual research programs into the community; that is, as Lerner and Simon (1998a) note, to see the community as research partners rather than research pools. Even the introduction of the title "University in the City Scholars" turned out to be an important step in signaling that the expectations for involvement were different here. The planned initiatives needed to emerge from a shared focus with the community, they had to be grounded in scholarship, and these collaborations could not simply occur at university locations.

We developed a variety of infrastructure elements through the Community Outreach Partnership Center (COPC): a Community Breakfast series that brought faculty and community leaders together so that faculty could identify the policy concerns at the forefront in the community; the Community Archives in which information about interventions, data gathering, and the like could be housed at one location, thus making the information accessible to the community and to scholars; and a Community Request for Partnership process to enhance community–university collaborations. Under the latter initiative, community groups could make a "request for partnership" (RFP) to work with teams in graduate courses such as Community Mapping. A major goal here was not only to change the way students are taught, but also to make certain that useful new technologies were readily available to the community so that leaders could incorporate these resources into their activities and programs. And it was not merely the method of setting up the technical assistance relationship that was important. Having seen firsthand that community groups often face the unfamiliar and daunting task of writing grant proposals, we wanted to design a RFP process that would assist people in learning the skills necessary to seek external resources.

Another emphasis within the COPC is the collection and dissemination of data that will serve the broader community in its strategic planning. Under the Access to Jobs program (i.e., a local welfare-to-work jobs initiative), we gathered data from local employers about the availability of public transportation. We also undertook a survey of minority businesses in Lowell. In this instance, it became apparent over our first year of COPC interaction with the community that many structures were in place to work with the minority business community, but no one was certain who these business leaders were, what their strengths were, and what kinds of support they might need. Some four hundred businesses were identified, and a survey of a sample of these businesses was used to gather information about the contributions these companies make to the local economy (Sastry, Bader, and Garcia-Barragan 1999). The vertical integration of training that took place as a part of carrying out the survey was as important as the actual data themselves. A diverse team of graduate students and high schoolers worked with faculty to design, collect, and analyze the data.

The University in the Neighborhoods initiative is one final example of how COPC efforts have been organized to advance multiple goals (Nordstrom 1999). Here, we worked with the city and neighborhoods to design a set of trainings that would benefit community residents and involve faculty, students, and others in offering the training. From the university's perspective, we hoped not only to create a university presence in the neighborhoods but to increase the community's understanding of the city's own economic-development history and opportunities by offering trainings that drew on recent research. We quickly discovered that our intended offerings were at odds with what city gatekeepers had envisioned. The challenge was to find a compromise that still left an opportunity to focus on research in a community-responsive fashion.

Entering into the Dialogue about How to Change the Traditional Practices in Universities

The summaries of these three initiatives should illustrate the ways in which this hands-on work opens up opportunities for combining different disciplines, for involving students in different roles and combinations of roles, and for creating new configurations of relationships between the community and the university. Multiple links between teaching, research, and service have been tested out. Many different groupings of faculty and students have been encouraged. New opportunities for collaboration have been aggressively sought out. Active problem solving that draws on the scholarly interests of the faculty has been paramount.

The hands-on work that I have just described is, in significant respects, at odds with traditional practices in universities. As Lerner and Simon (1998a) point out, "Too many of our faculty, in all of our disciplines, are far too insulated, too isolated, and in fact and perception are seen as indifferent to worlds other than their own" (3). We are reminded by much recent commentary of how common it has been for universities to absent themselves from the daily life around them: "In defense of its interior intellectual life, the university tends to produce a culture that, in an extreme form, rejects as inappropriate all direct involvement with the affairs of the world, which leads some individuals even to deny that the university as an institution has any social role" (Bonnen 1998, 31).

Do centers represent a way to break with these traditions of disengagement, yet still hold to the scholarly standards by which universities measure themselves? Many universities are now exploring this possibility (Brabeck et al. 1998; Harkavy 1998; Silka 1999). Lerner and Simon (1998a) argue that centers are ideal vehicles for these purposes. Because they transcend disciplinary boundaries and ways of framing problems, centers can be effective mechanisms for advancing outreach scholarship, particularly as issues and concerns undergo shifts and change. And centers have a structural advantage over departments in that centers offer flexibility in terms of community connections, in terms of who works with whom, and with regard to how students become involved. Centers allow for more than the "mere assembling of researchers from different disciplines or the simple layering" of approaches (Lerner and Simon 1998a). They encourage the integration of investigations. And centers can provide a structure for and continuity to pursuing these collaborative, integrative activities.

At the Center for Family, Work, and Community, we are frequently asked what we are: Are we an outreach center? Is this a center for students who want to become involved in the community? Are we a location where faculty do their research? Such questions are telling, because they point to an expectation for the separation of activities rather than their integration. The work of the Center for Family, Work, and Community as exemplified in the three initiatives (environmental justice, CIRCLE, and COPC) poses a model for outreach scholarship that attempts to bring these different roles together (Silka 1999; Silka 2000). At a metalevel, the goal of the initiatives I have described is to keep combining and recombining teaching, research, and outreach functions in order to test out new structural arrangements and see if they succeed or fail. These initiatives provide opportunities to experiment with combining different disciplines. We can test out different strategies for obtaining a "fit" with the university's mission. These initiatives allow us to try different combinations of relationships with the community. They provide rich opportunities to try out various strategies for maximizing the involvement of junior and senior faculty in interdisciplinary endeavors. Various approaches to involving students can be investigated. Eventually, university involvement and responsiveness to community opportunities can be optimized.

Much of what has been attempted at the Center for Family, Work, and Community speaks to the analyses of innovation, regions, and sustainability, as these terms have been used in ongoing debates over the university's mission of regional economic analysis. At the same time, the university–community collaborations promoted by the center also raise questions about gaps and limits in the extant analyses of these concepts. In the sections that follow, we consider how the experiences of the center might contribute to the university's ongoing analysis of the concepts of innovation, regions, and sustainability.

Reflections on the Concept of Innovation

With the intent of promoting innovation, the university's Massachusetts Council for Federated Centers and Institutes (CFCI) has offered a set of defining criteria for center initiatives; these six criteria include promoting collaboration, involving multiple disciplines, addressing significant questions, pushing new frontiers, encouraging intellectual risk taking, and presenting sustainable solutions. Each center is charged with developing practices that show how these criteria can be operationalized within their area of scholarship.

The work of the Center for Family, Work, and Community, as illustrated through these three initiatives, pursues all six criteria in an application-driven context. Consider the urban aquaculture initiative as one outgrowth of the environmental-justice initiative. This initiative promotes collaboration among multiple disciplines by bringing together a surprising but effective mix of nutrition researchers, specialists in economic development, community researchers, and faculty who study the sustainability of new economic practices. On the national scene, urban aquaculture is a growing business, but one for which many technical, economic, and environmental problems remain to be solved. The questions that are driving CFWC's work reflect a focus on significant issues and sustainable solutions: For example, how can this work be done in a sustainable fashion that integrates this initiative into the regional economy? Can such an initiative demonstrate how to move away from harmful environmental practices, such as reliance on the commonly used polyvinylchloride piping that produces toxic waste through its manufacture? Can the fish that will be produced in the aquaculture venture be nutritionally enhanced so they retain nutrients that are often lost? And what kinds of research and technical-assistance collaborations will optimize the involvement of community farmers who learned their fish-farming practices in the very different context of Southeast Asia? Thus, the urban aquaculture work attempts in an integrated fashion to address significant questions and push new frontiers.

Other aspects of the environmental-justice work are also driven by a focus on innovation that combines research and application. The faculty involved in the project, for example, are in the process of completing an analysis of environmental justice as a movement and intellectual frontier and are asking questions such as: What has been gained by reframing urban environmental issues as ones of justice? How do such reframings alter the terms of the dialogue, change the participants in that dialogue, and affect how new policies emerge from those discussions? The CFCI criteria for innovation have served as a guide and a goad. That said, any set of criteria inevitably leads us to focus on some characteristics while ignoring others. It should be recognized that current criteria may well overlook the most important criteria in a mission-driven, discipline-diverse university. We have found that perhaps the most important missing criterion is that of fit (for example, the degree of "fit" with the university mission). Research in the psychology of creativity examines the tendency of groups or individuals to generate original ideas or practices (Sternberg 1999). These findings indicate that the mere number of new ideas is not an adequate criterion by which work can be judged as truly creative or innovative. Innovation is reflected in the degree to which the work represents a creative and unexpected but apt solution to a seemingly intractable problem. The solution must represent a *good fit*. As center directors, we are constantly struggling in one way or another to develop creative solutions to problems of fit. For example, to what degree do our practices represent a good fit to the need to draw in the particular combinations of disciplines that are strong at the university? To what degree do our practices create a good fit for combining competing university and community interests? How well have we optimized our strategies for using scarce university resources to enhance regional economic development? Appropriateness of fit may well be what ties the different elements of innovation together, providing direction to otherwise disparate initiatives.

Evaluation of the criteria that mobilize our work can also result in an expanded analysis of existing criteria for innovation (Silka 1999). When we attempt to translate the innovation criteria into terms that are meaningful for the kinds of community–university collaborations being attempted by the Center for Family, Work, and Community, we begin to see the emergence of new issues. Consider, for example, the issue of who originates the intellectual questions to be made the focus of endeavors. Traditionally, the direction of influence has been from the researcher to the community. Researchers have tried to organize activities so that they generate hypotheses, gather data to answer the questions, and so they are the ones to shape how the findings are shared with the community. Innovations in collaboration—such as we have attempted with the COPC—shift the balance so that the community increasingly shapes the intellectual agenda for which answers will be sought. That agenda may take on a different form from one originated by the university. Instead of the next step following from the literature, it may come from community practice (e.g., the community may insist that economic revitalization efforts take a particular form that speaks to the concerns of a city that has twenty-three countries represented in four hundred minority-owned small businesses, with one-quarter of these businesses having been in operation less than one year). One consequence of such changes is that the commerce with the community over ideas and their application is increasingly an area for innovations.

Discussions of innovation also call attention to what many people have regarded as the oxymoronic concept of "innovative application." In many discussions, application is contrasted with innovation, with innovation and application implicitly being placed at opposite ends of a continuum of inventiveness. Innovation is seen as occurring earlier, as taking place in the breakthroughs in the laboratory or through a line of research designed to test a series of increasingly refined hypotheses. Then, at the end of the process an intervention—fully formed on the basis of research—is brought to the community. The serious scholarly work is complete; what remains is mere tinkering to get the details of the application correct. The examples given of the work of the center point out the possibility that innovation (with its demanding intellectual puzzles) is at the heart of effective applications. So much application work fails so miserably, not because it lacks theoretical interest, but because so little attention has been given to what that theoretical base for application might be and how it might integrate a variety of disciplinary perspectives. If applications are to be maintained, it is important to be cognizant of the fact that innovation is often the engine that drives faculty interest. When faculty become involved in outreach scholarship, they often are driven by the need to engineer innovations that can be reported to the scholarly community through publications. In the case of community work, faculty hope to identify a problem that they can then analyze and perhaps ameliorate. But the problem has to be one that has not been previously "fixed," and it cannot be entirely local in its significance. The more the problem is a concern in other communities or has been identified in the research literature as posing questions of theoretical significance, the more advantageous involvement will be from the perspective of faculty members, because there are opportunities for innovative solutions. The goal of the faculty member is often to use the local experience as an illustrative case study, or to "mine" the experience for raw material that can be brought back to the laboratory for study (Mawby 1998). In short, a focus on innovation is often central to maintaining faculty interest and commitment in the kinds of projects we carry out at the center.

Reflections on the Concept of Regions

The value of a regional analysis has been eloquently argued in political and economic development circles. Such an analysis is as useful for the questions it raises as for the solutions it provides regarding how best to define regions. No ready answers are provided to the question of what constitutes "the region" in the university's mission of regional economic development, but such an analysis draws our attention to these important issues. This attention to regions and how they are defined has yet to become the basis for thoughtful analysis in many of the areas in which the center works. A focus on regions has not surfaced, for example, as an explicit theme in community and social psychology. Indeed, paying attention to regions has been anathema in many social science disciplines, because such a focus has been associated with a narrowing of perspectives. As Lerner and Simon note: "historically the more decontextualized the knowledge the higher its value" (1998a, 5). It is better that work not appear to be contingent on extant sociocultural context (Lerner and Simon 1998a). To begin to focus on regions—especially in an application context—raises the specter of parochialism. Centers that are working with their local areas are at risk of being seen as engaging in activities that lack national significance. This dearth of attention to regions raises the question of how those who are engaged in outreach scholarship through a center such as the Center for Family, Work, and Community can draw from the economic and political analysis so as to have this work inform and shape their activities.

Much of the work of the Center for Family, Work, and Community and other centers like it has its region defined for it by default. Nearly all of the external funding that the center receives carries with it at least some implicit definition of a region. The requirements of the COPC grant stipulate that we are to work within the confines of the city, but primarily with groups and organizations in those zip codes that fall within the city's Enterprise Zone. The environmental-justice funding targets the city as a whole, but also includes any sources of pollution that are upstream or upwind from the neighborhoods in which Southeast Asian residents live and work. And the CIRCLE work encompasses the region of the Merrimack Valley for some of our training and the entire state for others. At the center, we have not reflected on these differences, nor have we examined closely the decisions that we have made to pursue various funding initiatives that constrain our regional focus.

In a sense, what the center has done is to acquiesce passively in the definitions of regions that have been imposed upon us. We have allowed the funding streams to determine what those regions will be without giving much thought to a regional analysis, or shifting our efforts so that they are explicit in terms of their regional implications. This arbitrariness in our approach may limit how well we interface with others who are working to further the university's mission of economic development.

As we begin to think more systematically about the question of regions, an important issue is what might colloquially be called the "drop in the bucket" question: How do centers organize their interactions with a "region" so that these collaborations have more than a minimal impact? Is there a way to build "regions" into the work so that the impact of the interventions will become more predictable? Underlying these questions is the issue of what might constitute a "natural kind" of regions in the various outreach areas in which the centers works. Questions about regions also raise issues about interlinked systems. Many of the systems in which the center works are interlinked: Although in some respects they end at the edge of the city of Lowell, these groups of organizations and individuals are also parts of elaborate statewide networks that can constitute a kind of learning region. Our collaborations are often intended to increase learning, so how do we foster such learning efforts on a regional basis? In his work with regional machine shops, Forrant (2001) has shown the benefits of organizing interventions by working with groups of organizations rather than one-on-one. The effectiveness of this approach is not simply a consequence of saving time. The success results from the shared learning and development that occurs among regional competitors (Granovetter 1985). In Forrant's work we have an implicit view of regions as associated networks of businesses that share distribution networks, suppliers, and the like. Incorporating an explicitly regional focus into our interventions with groups of nonprofits and community organizations is likely to bring useful new ideas that enhance collaborations and the use of existing techniques.

Although the center staff members have not been explicit in their attention to questions of regions, we have relied heavily in all of our efforts on a specific conceptual and application tool—geographic information systems (GIS)—that fosters a focus on regions (Silka 1996). By its very nature, the GIS provokes and structures a focus on regions. This computerized mapping technique enables the researcher to compare and contrast multiple types of data for which geographical locations are available. To begin with, GIS makes us confront differences in how different groups "parse" geographic regions. A "region" is no longer an abstract topic when we consider how different groups geographically organize the information they collect, compare, and contrast. Consider some of these differences and their implications for the various kinds of work that the CFWC does with our community partners. At the local level, information about health, for example, is available in terms of what are called CHNAs, or state-defined community health network areas. The regional unit of analysis for crime information, on the other hand, is the local precinct. City schools are divided by zones of school choice based on past desegregation rulings. The Regional Employment Board, in contrast, encompasses the entire city in its catchment area. Yet a different region is indicated by the important Access to Jobs initiative that analyzed job growth along eight suburban corridors reaching out of the city. Some regions are even larger, such as the networks of small businesses displayed on city maps that include import and export markets encompassing the countries from which newcomers emigrated.

Across these different topic-areas, then, the regions are varied and overlapping. These definitional differences in regions become apparent when seen firsthand on maps that are used for display, analysis, and strategic planning. The different contours of regions become readily apparent once the varied regional analyses are mapped. Using this GIS tool has the added benefit of providing a window into disciplinary differences in the analysis of regions. The tool is used in different ways by community psychologists, environmental scientists, sociologists, and economists. The result for an analysis of regions is that (1) framing differences become much more apparent, and (2) discrepancies become more explicit and can serve as a possible topic for investigation.

As we cast our net broadly to locate social science research that includes a regional analysis and combines that analysis with an explicit focus on application, we often find useful perspectives in perhaps unexpected places. Within the social sciences, for example, program evaluation has been an interdisciplinary arena where a surprisingly large degree of thinking about regions has occurred. In this assessment of approaches to program evaluation, Scriven (1991) offered a colorful analysis of what he termed the "big footprint" metaphor, as a way to encourage systematic thinking about regional impact:

> The Big Footprint approach unpacks into a homely measure of merit by looking at six dimensions of program impact in terms of footprint parameters: depth, length, breadth, number, direction, and location. . . . The depth of the project's footprint is the importance of the effect on the average individual involved. . . . The breadth is the number of individual affected. . . . The length is the duration of the effects. . . . Number of footprints reflects the number of discrete efforts within the project. . . . Direction—forward or backward—gives us the difference between progress and regress. . . . Location involves two considerations. First—and of particular importance to the planner—there is the question of whether the program is being located in territory where, with the resources available, it is feasible to leave a big footprint, perhaps because the ground is relatively untrodden—or, on the other hand, whether the ground is heavily compressed by the footprints of others or is on rocky ground, on which footprints are hard to make. (70–71)

As the above example suggests, Scriven's work on program evaluation is guiding us toward a regional analysis that links impact to intervention.

What should be apparent in the points made here about regions is that various confusions remain with regard to how the university will approach this area (see also Forrant 2001). Are we studying the region? Are we working with the region? Is our intent to introduce new forms of regional comparisons? Are we attempting to have an impact or are we trying to study regions and use the fine nets of our conceptual analyses to predict outcomes and trends? These differing goals—particularly impacting the regional economy versus standing at a remove in order to study that same economy—will often collide. The links to collaboration in working with local groups to change the dynamics of regional economic development remain obscure.

Reflections on the Concept of Sustainability

Again, as with regions, much has been written about the concept of sustainability and its multiple meanings. At UMass Lowell, we see these uses intersecting in discussions of sustainable regional economic development. In their outreach, different centers focus on different and multiple meanings. There is sustainability as understood in terms of the concept of sustainable prosperity (see, for example, Lazonick and O'Sullivan 1997), represented in the work of the Center for Industrial Competitiveness. Here, the focus is on businesses being able to maintain themselves (and therefore their regions), because they invest in skill development throughout the corporation including the shop floor. In this analysis, regions where costs are high (such as Massachusetts) will never be able to sustain long-term prosperity by resorting to the familiar low-skill, low-wage strategy. The maintenance of a robust industrial base will depend on skill investment.

A second approach to sustainability being explored at the university is represented by the analysis being carried forward by the Center for Sustainable Production. Here researchers have focused on the environmental impacts of production processes on a region. This analysis is intended to remind us of the limited usefulness of environmental remediation as a strategy for sustainability. Long-term sustainability, instead, calls for the redesign of industrial processes so that fewer pollutants are generated by manufacturers, hospitals, and other large polluters. The Center for Sustainable Production has been leading the way in changing the terms of the discussion about sustainability at the university.

A third, very different, use of the term "sustainability" is intended to focus attention on what the centers, their associated departments, and the university in general must do to sustain their efforts in regional economic development. How can new initiatives become regularized through incorporating them into our research programs, our classrooms, and our interdisciplinary efforts? How does work of this sort become a core of university efforts rather than remain on the periphery? How can such initiatives be incorporated into many different university functions rather than merely a few? How do we develop institutional champions for this work that will enable organizations like the Center for Family, Work, and Community to sustain themselves?

Varied as the uses of the concept of sustainability are, they share in common a focus on the future, on avoiding the depletion of resources, on overturning practices that create rather than solve problems (e.g., aggressively pursuing new businesses that will come into this region but pay low wages; thinking only after the fact about ridding systems of pollutants; depending on isolated, individual faculty to sustain and carry out what should be a systemic approach to the university's economic development mission). All of the aforementioned uses of sustainability concern themselves with changing the terms of the discussion about which characteristics or practices will create sustainability. All three discussions take place both at the level of theory (e.g., what research questions need to be addressed) and practice (what we should do in policy terms).

For integrative centers such as the Center for Family, Work, and Community, concerns with sustainability can operate at all three levels (Silka 2000). In terms of the concept of sustainable prosperity, how do we draw this idea into discussions as we work with local people on economic development issues through the COPC, environmental justice, and CIRCLE initiatives? With regard to sustainable production, how do we create conditions whereby people working in environmental justice and urban aquaculture, for example, can consider these themes? How do we do this integrative work in ways that build sustainable relationships and practices within and among centers, departments, and the community?

Consider questions of the sustainability of centers, which to some degree rest on questions of who should be involved and who will drive these efforts. Contradictory views are often espoused. New ideas, energy, and flexibility are frequently said to be key to effective collaborations, and thus junior faculty are sought out to play central roles in these activities. These faculty members are more likely to have had relevant interdisciplinary training (Higgins-D'Alessandro, Fisher, and Hamilton 1998) and are not steeped in outmoded, individualistic perspectives. Yet, once junior faculty are involved, they are also pointed to as exactly the individuals who should avoid these activities. Seasoned scholars are needed, they say, who understand that collaborations involve multidisciplinary research. Junior faculty lack tenure and hence they are too vulnerable to engage in the highly political process that partnerships represent (Beaulieu, Mullis, and Mullis 1998) and they lack the clout within the university to involve others in the interdisciplinary initiatives that underlie center collaborations.

Complicated views also exist about whether a direct or indirect approach will result in more impact on sustainable regional economic development. For some projects, the links to economic development are clear. These ventures teach about economic development, involve the collection of data on access to jobs, provide information to the community on minority businesses, and host summits on economic development, but many center activities have little direct focus on economic development—their goal is to bolster the social infrastructure. Yet, it may well be that the center activities having the most impact on economic development are those which do not have economic development as their stated goal. The environmental-justice partnership, for example, makes no reference to economic development. Yet, as a part of this initiative, we highlight environmental risks in this region and continue to raise questions about how a strong manufacturing economy can be maintained that will not harm those living near industrial sites. Clearly these initiatives contribute to the local community, but including them under an economic development umbrella also raises questions about where we place the boundaries around what will count as an economic development initiative.

How Universities Can Change to Increase Innovative Applications

At the outset, I noted that pressures are increasingly being felt by urban universities in terms of how they pursue their mission-related research goals. These pressures, far from abating, are increasing (Kennedy 1997).

The model of scholars working independently to pursue "ethereal knowledge" (Lerner and Simon 1998b) has come under renewed scrutiny, with this criticism becoming especially pronounced toward publicly funded universities. As Beaulieu, Mullis, and Mullis (1998) point out: "The reality is that stakeholders are no longer content to know that cutting-edge research is being conducted at their institutions of higher learning. They want to know how these investments are making a real difference in the lives of people in the state" (155).

Universities have an explicit role to play that draws on their research function, but they will need to learn about their community if such efforts are to have any hope of success. Lerner and Simon (1998a) have been leaders in efforts to point out the contributions that are possible as universities begin to change: "Universities have a critical role to play in such collaborations. They can act as agents of technical assistance, knowledge development, demonstration, training, and dissemination. However, to make such contributions, universities must change from their currently perceived status as enclaves for ethereal elitism and become agents in community engagement and empowerment" (4). Others remind us that we need to "seek to collaborate with community members in understanding the forces that shape their development and to work with policymakers to ensure that community goals can be achieved" (Higgins-D'Alessandro, Fisher, and Hamilton 1998, 158). As I have suggested here, universities such as the University of Massachusetts must give considered attention to how they will organize such efforts to ensure their sustainability and fit. Land-grant universities have been attempting such efforts for decades, but many past efforts remain piecemeal. They have not impacted the community nor have they greatly changed the culture of the universities. How will the movement aimed at encouraging sustainable community–university collaborations be different this time? Several strategies have been emphasized in current discussions in higher education, but are unlikely to have the intended effect because their focus is misplaced.

Much discussion in higher education circles focuses on the need to change the reward system fundamentally (Lerner and Simon 1998b). It is said that we must find new ways to reward those individuals who engage in these labor-intensive outreach efforts. Only with an overhaul of the reward system, it is argued, will the prodigious amount of work required for applied research be adequately recognized. By changing the reward system, it is assumed community collaborations will be increased and widespread skepticism in higher education circles about the value of these efforts will be reduced (Lerner and Simon 1998b). Yet, this focus on rewards is misplaced (Silka 1999). What is missing from such an analysis is an understanding that application provides opportunities—perhaps unparalleled opportunities—for innovation, and innovation is already rewarded when it serves as the basis for journal articles submitted for peer review. The reward system should not be changed simply because applied research activities are labor intensive. The rewards should follow because of the ways in which this work is innovative in resolving long-standing problems of interest to scholars. Just as with other types of scholarship, sometimes incentives will be needed in outreach scholarship to assist in getting the work underway (e.g., teaching release time, coverage of start-up costs), but the focus on rewarding the identification of and communication about cutting-edge approaches should be no different from the norm and does not call for a separate set of rewards that are inimical to the knowledge-function of a university.

Erickson and Weinberg (1998, 199) use the phrase "value added uniqueness" to highlight this idea that innovation is central to cutting-edge applied research. They note that when applied work is innovative, it already is rewarded. New systems of rewards are not needed because the application has already increased opportunities for innovation. Applied research brings together many forms of innovation. Grants for applied research are formed around a core innovative idea. And not only are proposals for collaborative research innovative in concept, but the principal investigator often must also be effective at devising innovations if the implementation is to be effective. Thus, the rewards will come from highlighting how this work advances the scholarly aims of the university (Silka 1999).

A second common but unhelpful approach to encouraging collaborations comes out of a barrier analysis that focuses on identifying and eliminating impediments to collaboration. Thus, commentators pinpoint turf battles (such as disagreements between departments and centers) as factors that stand in the way of interdisciplinary work. If these internecine wars can be eliminated, then (it is assumed) interdisciplinary research will proliferate. Again, this emphasis is misplaced. Turf battles cannot be eliminated by making their eradication the focus of attention. Exhorting people to stop interdepartmental warfare rarely leads to a cessation of hostilities. Turf battles are a by-product; indeed, they are best seen as epiphenomenal. They coexist with a problem, but are not its cause. The focus should instead be placed on innovation, on identifying those intellectual problems that bring together different disciplines, departments, and centers.

Erickson and Weinberg (1998, 200) note that the "the time is right to identify the common knowledge and competencies that cut across disciplines, to tear down (or at least cut windows in) the wall that separate disciplines." I have argued that centers can be one means for accomplishing this interdisciplinary integration and outreach. If centers are to be successful mechanisms, the initiatives that occur within them must be tied to the intellectual agenda of the university. A center cannot be held captive by a single discipline, nor can centers operate merely as university outreach arms. They need to integrate research, teaching, and service, and they must serve as a place where the puzzles of application can be analyzed and addressed.

NOTE

My thanks to Laurence Smith and to the volume editors for insightful suggestions regarding the material in this chapter. Portions of the work described here were funded by National Institute of Environmental Health Sciences Environmental Justice Partnership Grant ES07718–04 and U.S. HUD Community Outreach Partnership Grant COPC-MA-96-073, and by the Massachusetts Office for Refugees and Immigrants and the Massachusetts Jobs Council. More information about University of Massachusetts Lowell's Center for Family, Work, and Community can be found at their Web site [http://www/uml.edu/centers/CFWC/].

REFERENCES

Alston, S., J. Benfey, R. Khakeo, S. Khoeun, C. pierSath, L. Silka, and N. Som. 1996. The Southeast Asian Environmental Justice Partnership. Paper presented at Meeting Community Needs: Improving Health Research and Risk Assessment Methodologies National Conference, Worcester, Mass.

Archer, J., M. Darlington-Hope, J. Gerson, J. Gibson, S. Habana-Hafner, and P. Kiang. 1997. New voices in university–community transformation. *Change* 29:36–41.

Beaulieu, L. J., A. K. Mullis, and R. L. Mullis. 1998. Building university system collaboration. In *University–community collaborations for the twenty-first century: Outreach scholarship for youth and families,* ed. R. M. Lerner and L. A. K. Simon, 139–56. New York: Garland.

Bonnen, J. T. 1998. The land-grant idea and the evolving outreach university. In *University-community collaborations for the twenty-first century: Outreach scholarship for youth and families,* ed. R. M. Lerner and L. A. K. Simon, 25–70. New York: Garland.

Bowen, W. G., and H. T. Shapiro, eds. 1998. *Universities and their leadership.* Princeton: Princeton University Press.

Brabeck, M., J. Cawthorne, M. Cochran-Smith, N. Gaspard, C. H. Green, M. Kenny, R. Krawczyk, C. Lowery, M. B. Lykes, A. D. Minuskin, J. Mooney, C. J. Ross, J. Savage, M. Smyer, A. Soifer, E. Sparks, R. Tourse, R. M. Turillo, S. Waddock, M. Walsh, and N. Zollers. 1998. Changing the culture of the university engaged in outreach scholarship. In *University–community collaborations for the twenty-first century: Outreach scholarship for youth and families,* ed. R. M. Lerner and L. A. K. Simon, 335–64. New York: Garland.

Erickson, M. F., and R. A. Weinberg. 1998. The Children, youth, and family consortium. In *University-community collaborations for the twenty-first century: Outreach scholarship for youth and families,* ed. R. M. Lerner and L. A. K. Simon, 185–201. New York: Garland.

Forrant, R. 2001. Random acts of assistance or purposeful intervention? The University of Massachusetts Lowell and the regional development process. In this volume.

Forrant, R., and L. Silka. 1999. "Thinking and doing—doing and thinking": The University of Massachusetts and the community development process. *American Behavioral Scientist* 42:814–26.

Granovetter, M. 1985. Economic action and social structure: The problem of embeddedness. *American Journal of Sociology* 91:481–510.

Harkavy, I. 1998. Organizational innovation and the creation of the new American university: The University of Pennsylvania's Center for Community Partnerships as a case study in progress. In *University–community collaborations for the twenty-first century: Outreach scholarship for youth and families*, ed. R. M. Lerner and L. A. K. Simon, 335–64. New York: Garland.

Higgins–D'Alessandro, A., C. B. Fisher, and M. G. Hamilton. 1998. Educating the applied development psychologist for university–community partnerships. In *University-community collaborations for the twenty-first century: Outreach scholarship for youth and families*, ed. R. M. Lerner and L. A. K. Simon, 157–83. New York: Garland.

Kennedy, D. 1997. *Academic duty.* Cambridge: Harvard University Press.

Lazonick, W., and M. O'Sullivan. 1997. Investment in innovation: Corporate governance and employment: Is prosperity sustainable in the United States? The Jerome Levy Economics Institute of Bard College Public Policy Brief No. 37.

Lerner, R. M., and L. A. K. Simon. 1998a. The new American outreach university. In *University-community collaborations for the twenty-first century: Outreach scholarship for youth and families*, ed. R. M. Lerner and L. A. K. Simon, 3–23. New York: Garland.

———, eds. 1998b. *University-community collaborations for the twenty-first century: Outreach scholarship for youth and families.* New York: Garland.

Mawby, R. G. 1998. Mobilizing university expertise to meet youth needs. In *University-community collaborations for the twenty-first century: Outreach scholarship for youth and families*, ed. R. M. Lerner and L. A. K. Simon, 367–88. New York: Garland.

Nordstrom, J. 1999. University in the neighborhoods. Paper presented at the Eastern Psychological Association Annual Convention, April, Providence, R.I.

Quigley, D. 1996. Meeting community needs: Improving health research and risk assessment methodologies. Conference report, September 20–22. Available from the Center for Technology, Environment, and Development (CENTED), Clark University, Worcester, Mass.

Sastry, M., T. Bader, and G. Garcia-Barragan. 1999. Profiles of immigrant-owned businesses. Paper presented at the Eastern Psychological Association Annual Convention, April, Providence, R.I.

Scriven, M. 1991. *Evaluation thesaurus.* Thousand Oaks, Calif.: Sage.

Silka, L. 1996. Transforming the psychology of risk: From social perception to the geography of communities. In *The psychology of adversity*, ed. R. Feldman, 219–39. Amherst: University of Massachusetts Press.

———. 1997. The Southeast Asian water festival: A civics lesson for all. *Community Catalyst* 6:1–2.

————. 1999. Paradoxes of partnership: Reflections on university-community collaborations. In *Community politics and policies,* ed. G. Rabrenovic and N. Kleniewski, 335–59. Greenwich, Conn.: JAI Press.

Silka, L. 2000. Strangers in strange lands. *Journal of Public Service and Outreach* 5 (2): 24–30.

Silka, L., J. Benfey, and S. Khoeun. 1996. Southeast Asian environmental partnership for communication. Paper presented at the American Public Health Association Annual Convention, November 20, New York.

Sternberg, R. J. 1999. *Handbook of creativity.* New York: Cambridge University Press.

U.S. Department of Housing and Urban Development (HUD). 1999. Now is the time: Places left behind in the new economy. Washington, D.C.: HUD.

Administrative Support for Innovation

The Case of Applied Community Research

Nancy Kleniewski

THE central question of this volume is: How can universities contribute to innovation in their region? One important path by which university leaders can support innovation is by encouraging their faculty and students to engage in applied community research. This essay identifies several barriers to the institutional acceptance of applied community research and offers a number of suggestions for overcoming such barriers. These suggestions are based on the experience of a growing number of institutions that are attempting to support enhanced community-outreach elements within their missions.

What Is Applied Community Research?

Applied community research is research that focuses on the needs or issues arising from the multiple communities in which the university is situated: residential communities, business communities, political communities, religious communities, and so on. Often interdisciplinary in nature, applied community research strives to answer questions about "the real world" and propose and test solutions to problems. It is one element of university–community involvement that has come to characterize "the engaged campus" (Edgerton 1994).

Applied community research often receives its impetus from outside the university or college. It may be initiated by local groups who seek the expertise of academics to help them solve immediate problems. Alternatively, university faculty may initiate projects by reaching out to incorporate community groups in their research. When applied community research is initiated by faculty, it sometimes addresses those local issues as defined by the faculty, without the participation of community members as full partners—a traditional expert–client relationship. Ideally, however, it takes the form of research teams that include faculty, students, and community members, with the community members acting as full research partners rather than as clients or subjects.

There are at least two variations on applied community research. One

383

is collaborative research, as described by Philip Nyden and his colleagues (1997): "In the collaborative research model, academics and nonacademics work together in identifying the research issue, developing the research design, collecting the data, analyzing the data, writing up the results, and even working with policy makers and practitioners in designing programs and policies. . . . [It is] research done with the community, not to it" (Nyden et al. 1997, 5–7). Nyden and his colleagues describe many different forms of collaborative research (some of which are considered below), but several characteristics unify the category. Whether the specific research project is evaluation research, policy research, or an intervention to meet an immediate problem, collaborative research must involve the community in setting the research agenda and must tap knowledge in both academia and community. When used instead of traditional research, collaborative research can demystify the research process, spread expertise, and make research tools accessible to the community (Nyden et al. 1997).

The other type of applied community research is community-based research, as described by Richard Sclove and Madeleine Scammell. In their view, community-based research is "research that is initiated by communities and that is conducted for—and often directly with or by—communities (e.g., with civic, grassroots, or worker groups throughout civil society)." Community-based research differs from "the bulk of the R&D conducted in the United States . . . which . . . is performed in response to business, military, or government needs or in pursuit of academic interests" (Sclove and Scammell 1998, 2).

I have chosen to use the broader term "applied community research" to encompass both collaborative and community-based research. That is, it includes projects initiated by a local need, whether from the business, government, or nonprofit sector; it is carried out with the cooperation and/ or direct participation of community members; and it results in an intervention for a positive change that is implemented at the grassroots level and is consistent with community needs and wishes. In my view, it also, likely but not necessarily, implies research done not by individual scholars but by teams composed of academics drawn from multiple disciplines, students, possibly experts from outside academia, and community members.

Applied community research in the university helps us answer the question, "Knowledge for what?" posed by many analysts of higher education. The values that support applied community research provide principles stemming from the best impulses of the university tradition. One value is based on a commitment to serve others: The ends of research and learning can be altruistic rather than self-centered. Another is the desire to use the institution's expertise to improve the conditions of life: Research and learning can be applied to making the community and the world a better place. A third value is the benefit to students: Research and learning applied to real problems have a powerful and lasting intellectual and developmental impact on young people. These notions are both idealistic and pragmatic, in the sense in which John Dewey (1916) used the term. They help to provide a beacon by which metropolitan universities can navigate the myriad demands and opportunities that surround them. In evaluating the multiple and conflicting interests, activities, and needs that our faculty and students propose, the principle that knowledge should be produced and shared in service to social needs is a very powerful criterion that administrators can use in allocating resources and rewards.

Why Should Universities Promote Applied Community Research?

There are several compelling reasons why university staff should foster applied community research in their institutions. Among them are furthering the institutional mission, creating mutual linkages between the university and the community, providing regional leadership, and supporting student learning.

Furthering the Mission

Applied community research is probably beneficial in every type of institution and every type of community setting, but it is particularly useful in metropolitan universities. Metropolitan universities are not only located in metropolitan areas, but they also have symbiotic relationships with those areas. They directly address the instructional needs of the population of the region, they link theory and practice in professional education, and their research addresses regional issues (Lynton 1995). The Declaration of Metropolitan Universities states, "Our research must seek and exploit opportunities for linking basic investigation with practical application and for creating synergistic interdisciplinary and multidisciplinary scholarly partnerships for attacking complex metropolitan problems, while meeting the highest scholarly standards of the academic community" (Johnson and Bell 1995, viii).

Metropolitan universities often invoke the model of the land-grant university in fashioning their mission and programs (Cordes 1998; Johnson and Bell 1995). They have increasingly addressed community issues, particularly as it has become obvious that cooperation and attention to social and economic issues benefit both communities and universities. In recent years, these universities have fielded many discussions and publications about community capacity, or "the interaction of human capital, organizational resources, and social capital existing within a given community that can be leveraged to solve collective problems and improve or maintain the well-being of that community" (Chaskin et al. 2001, 7). Metropolitan universities explicitly or implicitly include in their mission the goal of increasing such capacity in the various communities with which they interact.

Creating University–Community Linkages

Since the publication in 1995 of Robert Putnam's article "Bowling Alone," academics have been discussing the unraveling of civil society. But like the weather, we are more prone to discuss it than do anything about it. Building social capital and social relationships is a complement to the discovery and dissemination of knowledge. This is not the role simply of the social sciences, although this essay concentrates on examples from those disciplines. Any interactions we academics have with firms, government agencies, nonprofit organizations, or other community groups have the potential to increase what Georg Simmel (1955) called "the web of group affiliations" in building a more cohesive community.

In addition to providing service, the community partnerships created by applied community research are beneficial both to the universities and to the communities of which they are a part. Most obviously, a weak community base hampers a university's attractiveness to students, employees, patrons (e.g., of sports and the arts), and funders. The partnerships created by applied community research, although requiring a substantial investment of time on both sides, can result in long-lasting relationships between the university and community groups that can be transferred to other specific issues (Nyden et al. 1997, 25). More significant, the links that universities build are not simply two-way ties between the university and a single external constituency. Rather, universities can have even more impact by helping to create webs of relationships in which companies, nonprofit organizations, and government agencies collaborate to produce changes and to learn from one another (Kanniyappan et al. 1998).

In addition, creating partnerships through research opens the door to new information from the outside. Many researchers acknowledge that "the flow of information from the outside in is as important as the transfer of knowledge from the inside out" (Lynton 1995, xii), and that applied community research affords opportunities for the researchers to learn as well as to serve. The relationship between the university and the community is one of reciprocal learning. Planning professor Pierre Clavel argues that real innovations most often come from outside the university, not from within. He continues, "Disciplines, academics, and many professionals lag behind the real innovators. But the university is better suited to observe, codify, compare, and feed back what the innovators do. This can speed up diffusion, bringing regional innovation to a new level, perhaps" (Clavel 1998).

Collaborations with community groups are also a good way to respond to a changing and uncertain economic and social structure. Elsewhere in this volume, William Lazonick points out the benefits to both academic institutions and the community in establishing collaborative efforts to promote innovation. Because the future is always uncertain, he argues, it is difficult to determine the precise types of research that will be useful to a community at a future time. Thus, he urges academic institutions to foster a cumulative and collective learning process in which participants can identify and act on critical issues as they arise. The important innovation, according to Lazonick, is the creation of an organizational foundation and learning dynamic that facilitates identifying and responding to change.

Providing Regional Leadership

Universities often seek to provide leadership in research. They may seek to become nationally ranked research institutions, although only a tiny faction of schools can hope to gain national prominence in more than one or two fields. But most institutions have the potential to become leaders in their own regions. If institutions choose to focus on this role (and many metropolitan universities now see it as a priority), they can gain recognition as prominent and positive forces within their regions.

Mary Ellen Mazey argues that universities have an advantage over other types of institutions in performing this leadership role. She holds that existing entities, such as local governments, businesses, or community organizations, may be less able to institute regional initiatives (such as grassroots planning), because any actions of theirs might be perceived as "a means to enhance [their] own well-being" (Mazey 1995, 196). Universities, however, often have the prestige to act as brokers, bringing together different interests around a broader agenda than any one of them might envision.

Enhancing Student Learning

In recent years, academics have rediscovered the principle of active learning, expecting students to take a role in their own learning process. Such active learning is useful, not only for building the intellectual skill base of the students but in preparing them to become active participants in society. Marlynn May and Robert Koulish (1998) argue that applied community research, and problem-based learning in general, helps students learn to work in teams to analyze and solve problems. They find that this pedagogical tool enhances students' agency in their own education and helps them learn social responsibility. In discussing the impact of community research on undergraduate learning, Barbara Lieb, of the U.S. Department of Education, says, "What we're finding is that for many college students, this kind of learning gives meaning to their educations" (Cordes 1998, A38).

Applied community research supports the teaching/learning mission of universities and particularly supports the increased pedagogical emphasis on experiential education. By creating teams of faculty, students, outside experts, and community members engaged in real–world issues, students are forced to become active, engaged learners and responsible members of problem-solving teams. These are exactly the skills and characteristics that accreditation bodies and educational experts (as well as legislators) have recently enunciated as the desired outcomes of undergraduate education.

Applied community research also affords faculty the opportunity to create coherence in their own and their students' activities by integrating teaching/learning with research and community service. The creation of such research teams helps create coherence in education and congruity between the real world and the academy. This type of research, which is widely practiced in professional schools, is also applicable to the sciences, humanities, and social sciences. Charles Hathaway and his colleagues suggest that the university can "transform itself by empowering the entire campus to utilize the metropolitan area as a living laboratory" (Hathaway et al. 1995, 11). Such experiences may well pay off in student recruitment and retention. Faculty and students engaging in collaborative community research claim that it is more interesting and enjoyable than other learning and research environments (Nyden et al. 1997).

Another advantage of applied community research is that it lends itself to approaching problems from interdisciplinary or multidisciplinary perspectives. Many observers (e.g., Haaland et al. 1995) have argued that, as the problems of science and society become more complex, we need to broaden our perspectives beyond the borders of the disciplines, thereby creating new knowledge. The research-team approach inherent in applied community research is interdisciplinary by its very nature, thus exposing students to ideas from a broad array of perspectives.

As we have seen, the reasons for promoting applied community research are many. Despite these advantages, however, we still find many barriers to instituting aggressive programs of applied community research. The traditional barriers to applied community research in universities and how we can avoid or surmount these barriers concern us next.

What Are the Barriers to Applied Community Research?

Universities are complex institutions. They contain a huge number of smaller entities, each of them oriented toward its own specialized role, with sometimes only the faintest recognition of the links between its work and the mission or goals of the whole. Thus, barriers arise to changing the focus of work that the smaller units (such as academic and administrative departments) carry out. The barriers may be structural, that is, created by overt policies such as organizational charts, flows of resources, and job descriptions, or they may be cultural, that is, related to such intangibles as tradition, prestige, and networks of friendship.

Although all complex organizations are resistant to change, a few institutional barriers within universities deserve special mention as hindering the growth of applied community research. They are the narrow focus of academic disciplines; a lack of fit between institutional missions and faculty reward structures; the difficulty of accessing resources for applied community research; and the strong boundaries between the university and the community.

The Narrow Focus of the Disciplines

The academic life of most universities and colleges is centered on the disciplines. Faculty are hired to teach and conduct research within a discipline, students major in a discipline, and academic departments are, for the most part, congruent with disciplines. Disciplines are the "homes" of faculty and students, the intellectual communities within which they gain their expertise, know the language and key players, and aspire to make their mark.

Tensions arise, however, between the disciplines in the arts and sciences and the nature of applied community research. The disciplines define the nature and goals of research that the faculty and students must do to be recognized and published. Whereas the faculty in professional schools are expected to engage with "real world" problems, the disciplines in arts and sciences can and often do experience major gaps between theory and practice. The foremost goal of research in a scholarly discipline is to move the discipline forward, not to solve problems. "Successful" scholars are most often those who narrow their focus and address esoteric theoretical questions. Applied research is often considered inferior to theoretical or basic research. In some institutions, applied research is seen as "merely service." In short, the disciplines are a construct of the academy, and disciplinary values can deter faculty from engaging with the needs of outside groups (Ziegler 1995).

In addition, the disciplines tend to be relatively segregated. Scholars read and publish in the journals of their own discipline, attend conferences of their peers, and have their own established standards for quality work. A disciplinary focus can create a kind of tunnel vision in which the interests and norms of one's disciplinary colleagues are the paramount influence on choosing scholarly and teaching projects. Such an orientation (to which the vast majority of faculty are socialized as graduate students) leads faculty in most departments to aspire to be part of a national conversation of their peers, not to make a priority of research that meets the needs of the local community.

Problem-solving research, such as applied community research, is usually not the focus of single disciplines but of the integration of applied knowledge derived from different fields. This integration, while more reflective of the real world than of the university, can deter faculty and students from engaging with the community. An applied research focus, particularly an interdisciplinary one, can limit the graduate programs to which students might apply, the faculty positions for which new Ph.Ds might be hired, and the funding sources to which researchers might apply for support. Individual faculty can be marginalized within their institutions if they do not publish in the proper journals, collaborate with recognized partners, or contribute to the departmental (read, disciplinary) workload.

The Faculty Reward Structure

Different institutions of higher education have different missions. The largest categories are the undergraduate liberal arts colleges, the comprehensive institutions that encompasses both liberal arts and professional programs up to the master's level, and the Ph.D.-granting universities. Universities can be further divided into those that have basic research as their primary mission and those that primarily serve the needs of their region, the latter group including land–grant institutions.

For some time, researchers have pointed to a mismatch between the missions of many institutions and their faculty reward structures. Faculty rewards, chiefly tenure and promotion, are governed by faculty and administrative groups on each campus and theoretically should reward faculty for contributing to the specific mission of each institution. Yet in recent decades, institutions have increasingly adopted a series of national norms for promotion and tenure based on nationally recognized scholarship. In the process, both teaching and service have been de-emphasized in favor of published research. As a corollary, applied research has often been de-emphasized in favor of basic research that is published in a small number of flagship journals (Diamond and Adam 1993).

The former president of Stanford University, Donald Kennedy, explains why this convergence has occurred. He points out that the vast majority of Ph.D.s are trained in the largest research universities, absorbing the norms and values of those institutions. When new Ph.D.s leave their graduate institutions, however, only a small fraction take jobs in research universities; the vast majority find positions in other types of institutions. Kennedy argues that faculty learn the norms of the professoriate in research universities, and they carry them to the institutions in which they subsequently work, whether or not the norms are appropriate for their new settings. Thus, research institutions have an inordinate influence on setting the norms for the professoriate (Kennedy 1995).

Rice (1995) argues that national norms for scholarship are often inappropriate and counterproductive for the majority of faculty, yet because they are the national norms, they are difficult to challenge. He argues that the expectations for faculty have been ratcheted up, and the range of approved activities narrowed by several recent trends. One is the trend toward accountability. As institutions have tried to raise their standards, they have adopted more stringent measures of quality, and because publications in refereed journals can be easily quantified, they tend to become the primary criterion of success. Another trend that has contributed to changing faculty norms is downsizing. As institutions have become smaller (or stopped growing), they have had to differentiate more stringently among faculty members being considered for tenure. Ratcheting up standards of scholarship (as opposed to teaching or service) can make it easier to deny tenure to junior faculty whose credentials would have been sufficient for tenure in the past (Rice 1995).

A certain amount of "mission creep" has occurred among institutions in recent years. Rather than adhering to a focused mission, institutions have tended to broaden their scope to encompass the missions of disparate institutions. The American Association for Higher Education (AAHE) has addressed this phenomenon by calling on institutions to realign their faculty reward structures with their institutional missions. Metropolitan universities have missions that include scholarship as service to the community. Yet a faculty reward structure that stresses pure research is often driven by institutional attempts to gain the prestige and visibility of research universities. "By believing themselves to be what they are not," say critics, "these institutions fall short of being what they could be" (Lynton and Elman 1987, 13).

A final problem of the faculty reward structure is that the individualistic nature of faculty rewards can be a barrier to engaging in applied community research. Although such research need not be collaborative, the best examples normally involve multifaceted teams of collaborators. Nonfaculty team members provide not just footsoldiers or subjects but input into the conceptualization, study design, data collection and analysis, and even the authorship and dissemination of the results. A tenure and promotion process that devalues multiple authorship, particularly coauthorship with students and nonacademics, can discourage faculty from collaborating with students or community groups.

Research Funding Access

Another significant barrier to faculty's engagement in applied community research is the nature of funding available to support research. A great deal of the funding available is earmarked for support of basic research. This is the result of the research buildup in the post-Sputnik era, which led to an international race to gain basic knowledge in the sciences. Although professional schools can easily receive funding for applied projects, faculty in colleges of arts and sciences find it more difficult to fund applied research. Support for applied community research may come from the same sources that fund basic research: private firms looking for research and development projects, government agencies, and foundations. The problem with these funding sources is that their goals may not mesh with the goals of the community partners that need assistance; conversely, community groups rarely have funding sources available to finance their necessary investigations. In addition, government agencies that have money available for funding applied research often do not pay overhead, thus making the projects they fund less attractive for the university. The third source of funding, foundations, often have narrowly defined and highly targeted requests for proposals that allow little breadth for many of the specific investigations that community groups request. In short, identifying funding sources for applied community research projects can be difficult.

Institutional Boundaries

The university as an institution has not always been friendly or accessible to outside groups. Faculty and administrators speak their own language, address themselves to fellow academics, and seem inscrutable to those not initiated into the mysteries of academe. Outside groups that might benefit from collaboration with the university can find the institution unfriendly and inaccessible. Thus, access into the university from the outside can seem difficult.

Even when universities initiate contact, their overtures can be met with skepticism because of past experiences community groups have had with academic institutions. Kornreich (n.d.) describes the hostility among groups in Phoenix in response to Arizona State University's announcement that it had formed a Community Outreach Partnership Center, funded by the Department of Housing and Urban Development: "ASU had been here before; many times before, in fact. And each time was the same. Professors and students would ask a lot of questions, write a report, get published in a journal, and disappear from the community, leaving little behind in the way of change or progress. This time, when news of the grant for over half a million dollars hit the streets, community activists were united in their response: 'Give the money to us. We know what to do with it. ASU doesn't have to tell us what our problems are or how we should solve them'" (Kornreich n.d., 1). Thus, the current motives of the university are scrutinized because of previous experiences. Researchers following the collaborative or community-based models may have to invest many months in the process of establishing trust with people and differentiating their particular project from "The University."

How Can the Barriers Be Overcome?

Despite the difficulties and barriers cited above, applied community research is well-grounded in a number of universities and colleges, including the University of Massachusetts Lowell. Administrators and faculty who want to support such efforts can take lessons from the institutions that have successfully integrated applied community research into their programs and can be creative in reevaluating the established procedures on their own campuses. Five areas for change that can support a broad range of research in service to the community include:

Clearly articulating the mission and supporting it with rewards;
Assisting faculty to participate in applied community research;
Creating interdisciplinary structures as an alternative to disciplinary departments;
Actively supporting faculty seeking funding for innovative projects; and
Creating a gateway into the institution for community groups.

Aligning the Faculty Reward Structure with the Mission

Many institutions have missions that are compatible with applied community research. Land-grant institutions were formed specifically to aid agriculture and industry in their areas. Religious institutions frequently hold service as their highest value and integrate it with teaching and research. Metropolitan universities have a mission to assist their local communities and regions. Within such institutions, however, top administrators must clearly articulate that priority because of the host of constituencies competing for institutional recognition.

In recent decades, the missions of metropolitan universities have been contested by various groups of administrators, faculty, students, trustees, government officials, and community members. Different constituencies argue for the desirability of cutting-edge research, prominent athletic programs, open admissions, a full array of doctoral programs, continuing education for adult learners, an impressive physical plant, and other goals. In the mid-1980s, presidents and senior administrators, often through their national organizations, initiated discussions on many campuses to define coherent, focused missions to which university resources would be directed.

A consistent theme in the missions of metropolitan universities is service to the regions of which they are a part (Johnson and Bell 1995). To make that priority a reality, campus administrators must frequently, clearly, and concisely articulate what the university is about. At the University of Massachusetts Lowell, for example, the central mission is service to the region, particularly framed as "university-community interaction to promote sustainable development." This dedication is repeated in the campus newspaper, alumni magazine, annual report, and other publications. It is the subject of an outreach video that the advancement office is using to recruit corporate partners and donors. Both internally and externally, but primarily internally, audiences must hear these priorities repeatedly before they can be expected to embrace them.

In addition to articulating the mission clearly and repeatedly, the next most important step administrators can take is to align the reward structure with that mission. If it is service to the region, then the metropolitan university should clearly value and reward teaching and research that provide such service. Kornreich (n.d., 4) says bluntly, "If outreach and applied scholarship are to become integrally woven into the fabric of university practice, institutional endorsement must be unambiguous and support explicit."

To align scholarship with the institutional mission of outreach requires explicit attention from administration and faculty. The problem with traditional scholarship is that it separates theory and practice (or application). Even when some research findings might be useful for a community, they are too often buried in scholarly journals where they are presented in incomprehensible jargon and with insiders' descriptions of the methodology. University administrators need to support faculty and students who could engage in research that results in both good analysis and practical solutions to problems.

In 1993, the AAHE inaugurated its Forum on Faculty Roles and Rewards to broaden the view of scholarly work in higher education. As a key participant in the AAHE forum, Professor Robert Diamond of Syracuse University emphasized the need for institutions to align their reward structures with their missions. To assist in this project, he developed a set of criteria for broadening the definition of scholarship that institutions use to evaluate faculty work. Diamond (1995; Diamond and Adam 1993) began with the notion of "the scholarship of application," as advanced by Ernest Boyer (1990), and worked with groups of faculty and disciplinary associations to operationalize the concept. According to Diamond's studies, applied community research can be evaluated under criteria that academics routinely use to judge basic research. He demonstrates that it is possible to document and evaluate the scholarship of application (and the scholarship of teaching), but that institutions must take conscious steps to do so.

In another study, Barbara Holland (1997) analyzed practices at a number of campuses and created a matrix of institutional commitment to community service. She found that those campuses that made service a "central and defining characteristic" of their mission emphasized applied community research as a key element in faculty hiring and in promotion and tenure decisions.

The institutional reward structure is not the only incentive that affects how faculty use their time and expertise. The intrinsic rewards of doing interesting work, of collaborating with others, and of making a difference are powerful motivational factors. Elsewhere in this volume (in part IV), Linda Silka argues that institutional leaders should focus on "innovations rather than rewards if community–university initiatives are to become an integral part of the intellectual agenda pursued by universities." Intrinsic rewards are always powerful forces motivating faculty. But if the university reward structure does not clearly support scholarship in service to community groups, the faculty receive a clear signal that it is not important.

Assisting Faculty Participating in Applied Community Research

In addition to a clearly articulated mission statement and a supportive reward structure, administrators can provide tangible assistance to faculty who are engaged in applied community research. Administrators can give mission-based projects priority for receiving such resources as release time, project space, start-up or seed money, and research assistance. The University of Massachusetts Lowell channels such resources to applied community research through several competitive programs that provide release time for various mission-related projects and make small grants available for community research and other public service projects.

To effect changes in academic culture, it is necessary, but not sufficient, to provide a reward structure and even tangible supports as described above. It is helpful to provide many ways of educating the faculty, both new and continuing, in the importance of applied community research. One way of educating faculty is through peer mentoring. Seasoned faculty with records of successful applied community work can mentor junior faculty just establishing their research programs. They can help them avoid the pitfall of doing research that may not be transportable because it is perceived as parochial. Brown (1997) suggests that junior faculty can be advised to ground unorthodox research in recognized approaches, to use good science, and to write for publication in a straightforward manner, as a way of integrating a community study with more general academic questions.

Tenured faculty are often drawn to applied community research by the opportunities it presents for their professional growth and development. They may discover that, as Kornreich (n.d.) argues, applied research and outreach can help expand the knowledge base of their disciplines and lead to opportunities for publications. Accordingly, faculty who understand the concept of "action research" have little difficulty in developing journal articles from their applied work. In addition, applied projects can involve opportunities to obtain funding outside the mainstream of funding sources. Kornreich argues, however, that some faculty who respond to such opportunities for funding their research really do not want to change their orientation or may be "unable to distinguish between field-based research and action-oriented scholarship and outreach" (Kornreich n.d., 4). Thus, some universities have instituted training programs to accompany the funding programs. The training helps faculty understand that the goal of applied research is not discovery of basic principles but effecting change in the community. Administrators can also support faculty members making the transition to doing applied community research with mentoring programs and faculty discussion groups.

Faculty also may need assistance in finding and writing for journals that publish applied research. In recent years, the number of such journals has increased substantially. Some are interdisciplinary journals; some are traditional journals with a section or department on applied research; some are journals of professional associations or their subdivisions; some span the academic and practitioner audiences; and some focus on pedagogy. Administrators can enlist experienced faculty and librarians to act as consultants and help researchers identify appropriate outlets for their articles. They can also work with promotion and tenure committees to validate such publications and to be more inclusive, rather than exclusive, in consideration of where articles are published (the "flagship" syndrome). When appropriate, universities can assist in founding new journals or presses that specialize in publishing applied scholarship.

Applied community research can also contribute to a reinvigorization of teaching among tenured faculty. A recent article in the *Chronicle of Higher Education* (Cordes 1998) reports on a chemistry professor at Loyola University of Chicago who began doing research with students on lead contamination and eventually revamped her entire undergraduate chemistry course on instrumental analysis. She reports that both she and her students are learning more chemistry without sacrificing any of the basics covered in a traditional course. Faculty frequently feel overwhelmed by the multiple demands of their positions because they feel they are expected to pursue high levels of teaching, research, and service simultaneously (Baker and Zey-Ferrell 1984). Conducting applied community research with student teams provides faculty with opportunities to integrate those roles of teaching, research, and service, thus increasing their satisfaction and productivity.

The final way in which administrators can help faculty with applied research is through the recruitment process. Before starting faculty searches, it is important to discuss institutional and departmental needs in the light of the service mission and to encourage departments to favor candidates who can contribute to that service. The candidate whose area of interest is compatible with community issues or who can become part of a collaborative team within the department may be a better institutional fit than the candidate from the highest-ranked doctoral program. The ability to provide high–quality teaching, research, and service are still the criteria for selecting new colleagues, but with the large number of Ph.D.s on the market in many disciplines, high quality and an applied orientation are not at all incompatible.

Creating Interdisciplinary Structures

Despite a great deal of discussion in the academic press about the need for interdisciplinary collaboration and the need to prepare students for a complex world (for example, Haaland et al. 1995), universities and colleges are still overwhelmingly organized into departments according to discipline. Not just teaching but also hiring, tenure and promotion, and resource allocation follow, in the main, the boundaries of the disciplines. It is probably unwise for administrators in established institutions to try to fight or dismantle the disciplinary structures. It is possible, however, to encourage cross-disciplinary collaboration by establishing interdisciplinary structures that will compete with and provide alternatives to an exclusively disciplinary-based organization.

At the University of Massachusetts Lowell, senior administrators have established a number of structures that support interdisciplinary work. Centers, institutes, and departments are three forms of such structures.

Research centers are organizations of full-time faculty and students drawn from different departments. They provide faculty the opportunity to form teams to seek funding, conduct research, and do outreach to the community. Faculty from the social sciences at UMass Lowell have created three centers. The Center for Family, Work, and Community is heavily involved in outreach to neighborhood organizations, nonprofits, local government agencies, and disadvantaged populations such as immigrants and lower-income youth. The Center for Industrial Competitiveness collaborates with groups in private industry to study how Massachusetts industries are positioned relative to those in other locations and to assist industry groups in strategic economic planning. The Center for Women and Work provides a focal point for a broad range of faculty, students, and community members to share current scholarship and collaborate on research projects dealing with women's participation in the workforce. Although these centers are coordinated by faculty in the social sciences, they are truly interdisciplinary: Individuals from the Colleges of Management, Health Professions, Education, and even Engineering participate in their projects. More information about the specific activities of the centers can be obtained on the World Wide Web at (main heading): www.uml.edu/Centers/ (subheading) CFWC/, CIC/, and CWW/.

Institutes are larger, more formal research and service organizations that run on "soft money." Their lead researchers, while associated with academic departments, are not typically full-time faculty members, and their staff members are hired as needed, depending on the inventory of grants at any given time. Two examples of interdisciplinary institutes are the Toxics Use Reduction Institute and the Kerr Ergonomics Institute. The former receives funding from the Commonwealth of Massachusetts and other sources to assist firms in identifying and reducing the pollutants used in their production processes. The Kerr Ergonomics Institute does research and education in the creation of healthier workplaces for companies, agencies, and nonprofit organizations in the public and private sectors. While both institute heads are members of the engineering faculty, they have assembled diverse functional teams including natural scientists, social scientists, and health professionals.

The university provides support for the centers and institutes through a coordinating body called the Council of Federated Centers and Institutes. This group receives an annual budget from the university, which it uses to assist centers and institutes through a program of small grants. It also provides an outreach mechanism for the centers and institutes as a group, publicizing their accomplishments to the community and to potential research partners.

The third structure that UMass Lowell has created is the interdisciplinary department. Interdisciplinary departments combine the disciplinary diversity of the centers and institutes with the stability of regular university funding and the authority to grant degrees. As distinct from interdisciplinary programs, which borrow faculty from a number of departments, these interdisciplinary departments are the permanent home for the faculty assigned to them. UMass Lowell currently has two such departments: Work Environment and Regional Economic and Social Development. The Department of Work Environment is a graduate-level department whose province is producing professionals capable of conducting research and consulting on workplace health and safety. The Department of Regional Economic and Social Development trains practitioners from industry, government, and nonprofit organizations to understand and intervene in the process of regional development, both in the United States and abroad. Both departments include faculty members drawn from at least six different disciplines. More information about these departments can be found on the Web at www.uml.edu under "academic departments."

The modest numbers of faculty, staff, and students working in the centers, institutes, and interdisciplinary departments have had a large impact across the institution. These structures provide the majority of the university's applied community research and, more important, provide a focus and a clear statement that such research is of value to the institution. Moreover, they provide necessary opportunities to integrate theory and practice, or, as two faculty members (Forrant and Silka 1999) say, opportunities for "thinking and doing, doing and thinking" about real-world issues.

In addition to the specific issues addressed by the centers and institutes, the creation of these innovative structures has helped to foster a culture of collaboration and a series of cross-disciplinary links that have furthered the spirit of experimentation. The Committee on Industrial Theory and Assessment (CITA) is a case in point. Established in 1993 as a spin-off of the Council for Industrial Development, CITA has two goals: to build a theory of industrial development relevant to the regional economy and to assess the contributions of enterprises and institutions (including the university itself) to the innovative performance of the regional economy. Thus, CITA's role is to spur innovation in the region through its studies. CITA has provided a number of supports to faculty interested in applied community research, including a seed-grant program, a release-time program, a funded symposium series for presentations of faculty research, and an ongoing interdisciplinary discussion forum for faculty and graduate students.

These new structures have unleashed a spirit of collaboration among the faculty. Because of the cross-disciplinary connections that have been created, faculty and students from different parts of the university have embarked on a number of projects that would not have been possible inside their own "silos." One example is the aquaculture project initiated by the Center for Family, Work, and Community. This project involves a partnership among the university, the Cambodian Mutual Assistance Association, the U.S. Department of Agriculture, and the public schools (Cambodian immigrants, numbering over twenty-five thousand, are the largest ethnic minority in Lowell). The goals of the project are to establish a viable fish farm for the sale of fish to Southeast Asians in the local area, to study the nutritional value of the fish as a way of improving their growing and feeding conditions, and to provide educational materials on aquaculture to public school teachers for use in their classes and field trips.

Assisting Faculty with Funding

Research costs money. Sclove and Scammell (1998) report that the private research and development industry spends over $200 billion per year, much of it going to universities. By contrast, only about $10 million annually goes to support applied community-based research. Yet researchers must find the money somewhere, and the guidelines of many funders favor basic research over applied studies.

One way that administrators can help is to provide start-up or seed funds for applied research projects and related community outreach. At the University of Massachusetts Lowell, senior administrators have allocated several substantial pools of money, both directly and within the council structure, to support mission-related projects.

Administrators can also work with their sponsored research offices to identify and disseminate information about funding sources for applied community research. A number of foundations and government agencies have consistently supported such efforts or have recently made community research a priority. Among the foundations, the Kellogg and Annie E.

Casey Foundations fund numerous applied community studies. Among government agencies, the Departments of Education, Housing and Urban Development, and Health and Human Services, the Environmental Protection Agency, and the National Institute of Environmental Health Science (a division of the National Institutes of Health) sponsor such research.

One of these sources, with which we have had experience at UMass Lowell, is the Community Outreach Partnership Center (COPC), sponsored by the Department of Housing and Urban Development. According to Mazey (1997), HUD created the COPC program in 1992 as a demonstration program to encourage universities to act as resources to communities by assisting communities rather than studying them. The funding guidelines provide that Community Outreach Partnership Centers offer joint learning opportunities—for the university staff and students on the one hand and for community residents on the other. The UMass Lowell COPC was established in 1997. Its mission includes using university resources to improve conditions in the local area, partnering with community groups to define problems and potential solutions, and expanding the capacity of community groups to do constructive work. Besides changing the community, however, the center strives to transform the university itself: One goal is to build institutional memory and resources within the university to sustain projects and relationships with community groups.

In addition to collaboration between an institution and the community, funding can be sought for projects that provide collaboration among various educational institutions. One of the most complex partnerships is the Policy Research Action Group, a collaborative of four universities (Loyola, University of Illinois Chicago, DePaul, and Chicago State) and fifteen community organizations in Chicago. The community organizations define the studies to be carried out, and the universities provide teams of student interns to conduct the research. This group has completed more than 130 research projects, has established "think tanks" on gentrification and employment opportunities, and meets regularly with local leaders to advocate for policy initiatives. Funding for this collaborative effort was provided by a major grant from the John D. and Catherine T. MacArthur Foundation in 1988 (Nyden et al. 1997).

Administrators can be of assistance to faculty seeking funding for such innovative projects. The institution must, first of all, provide a match, ranging from release time to office space to staff support, for such ventures. In addition, administrators must be aware of and accept the radically transformative initiatives that such projects engender. The assumptions behind the Community Outreach Partnership Center, for example, turn the traditional institution–client role on its head, proclaiming that we are there to learn from community members as well as to teach them. Hathaway et al. (1995, 13) comment on the implications of this commitment, saying,

"Metropolitan Universities are agents of change . . . but the transformation is not unilateral. Just as the university is a transformer of the society of which it is a part, so it will be transformed by that society."

Providing a Gateway to the University

While it is difficult to get faculty outside the institution's walls, it is also sometimes difficult to get outsiders in. To whom are community organizations, businesses, or local agencies supposed to appeal when they need assistance? A large university contains not only several colleges with dozens of academic departments but also administrative offices, research centers, public information offices, sponsored-research offices, continuing education programs, and university relations offices. Community groups run the risk not only of missing the people who could assist them but also of becoming convinced that the university as an institution is unresponsive to their needs.

Administrators can assist faculty in linking with the outside in two ways. One is to compile a directory of the interests and skills of faculty, staff, and students, defined not just their academic area of expertise but also by the ways in which they can assist nonuniversity groups (Kanniyappan et al. 1998). Examples of the latter might include assistance in writing proposals for funding, assistance with negotiating the maze of local or state government bureaucracy, assistance in starting a small business, assistance in identifying public health or environmental hazards in a neighborhood, housing rehabilitation assistance, assistance in identifying cases of racial or gender discrimination at work, and so on. Such a directory could be published and distributed widely through a network of nonprofit agencies.

Another way for the university to become more "community friendly" is to set up a "gateway center"—an office dedicated to finding within the university the expertise to address problems raised by outside groups (whether in the private, public, or nonprofit sector). Such a center would be the urban analogue of the agricultural extension programs conducted by land-grant institutions. The gateway center might consist of a single individual, but one whose job it would be to keep track of the entire scope of practical knowledge in the institution and to market it to outside groups.

Recommendations for Action

Like other large and complex organizations, universities are freighted with inertia. Even when faculty and administrators agree that community involvement such as applied community research is an important undertaking, they may encounter obstacles caused by the cultural and structural barriers to change discussed earlier. People in positions of power within the university (who may well include senior faculty in addition to administrators) must allow the institution to change in ways that support the goal of increasing community involvement. A spirit of innovation and experimentation is helpful regarding the structures of the university itself; innovations are even more valuable when they occur at the grass roots rather than being set in motion from higher levels. A spirit of humility can be helpful also, since, as Clavel (1998) notes, many real innovations originate outside the university, and we need to learn from them.

Earlier several recommendations for action were presented, based on the experiences of the University of Massachusetts Lowell and other institutions that support community involvement by their faculty and students. For the sake of clarity and emphasis, the most important recommendations are summarized below. Most of them can be accomplished, or at least initiated, by administrators, but changes will be more powerful if they are initiated or strongly supported by members of the tenured faculty.

The recommendations for increasing applied community research are:

Top administrators must clearly articulate the mission of the institution, particularly as to how the research mission relates to the service mission. All internal and external communication channels that reach faculty, staff, students, alumni, community groups, corporations, local agencies, nonprofits, and the public should convey the message that the university is serious about creating partnerships with outside groups to conduct research that will benefit the community.

Administrators and tenured faculty should examine the faculty reward structure, including criteria for hiring, tenure, and promotion to ensure that the institution's emphasis on applied research and community outreach is adequately reflected in faculty appointments and evaluations. If the reward structure does not clearly support outreach and community research, key administrators should initiate a discussion with faculty leaders about how to provide such support without reducing standards for performance. Such discussions should involve representatives from other institutions that have successfully integrated community research into the reward structure.

Administrators and department chairs can specifically assist faculty who are interested in doing applied community research through resource allocation. Assistance can take the form of release time, start-up or seed funds, equipment, travel funds, research assistance, space, and other resources.

Administrators and faculty who have been successful in doing applied community research can establish a mentoring system by which more experienced faculty assist less experienced faculty in identifying appropriate projects, finding faculty in other disciplines whose expertise applies to the projects, working collaboratively with nonacademic groups, integrating the research into their teaching, and finding journals or other outlets in which to publish the results.

Senior administrators can encourage the formation of innovative organizational structures that foster interdisciplinary collaboration, such as research and outreach centers. To be successful, these alternative structures need both grassroots support and a level of resources (funding, space, and personnel) sufficient to establish their credibility and produce tangible results.

An administrative channel can be used to create a gateway into the university for the community. A gateway might be a single office or individual to serve as a contact point for all incoming requests and ideas. This facilitator could then identify a faculty member with expertise or interest in the topic and arrange a meeting. Alternatively, the university might publish and frequently update (on paper or electronically) a directory of the kinds of services and aid that its staff and students can provide.

These suggestions are drawn from a number of institutions that have undertaken to interact better with the community and to make their research both more useful and more productive. Any particular change may or may not work in a given university. The most important element in effecting change is a spirit of innovation and flexibility: the propensity to ask questions, the willingness to change some long-standing practices, and the perseverance to keep monitoring the projects launched.

REFERENCES

Baker, Paul, and Mary Zey-Ferrell. 1984. Local and cosmopolitan orientations of faculty: Implications for teaching. *Teaching Sociology* 12:82–106.

Boyer, Ernest L. 1990. *Scholarship reconsidered*. Princeton: Carnegie Foundation for the Advancement of Teaching.

Brown, Phil. 1997. Social science and environmental activism: A personal account. In *Building community: Social science in action*, ed. P. Nyden, A. Figert, M. Shibley, and D. Burrows. Thousand Oaks, Calif.: Pine Forge Press.

Chaskin, Robert, Prudence Brown, Sudhir Venkatesh, and Avis Vidal. 2001. *Building Community Capacity*. New York: Aldine DeGruyter.

Clavel, Pierre. 1998. Comments at the Conference of the Committee on Industrial Theory and Assessment, October 23, University of Massachusetts Lowell.

Cordes, Colleen. 1998. Community-based projects help scholars build public support. *Chronicle of Higher Education*, September 18, A37–39.

Dewey, John. [1916] 1944. *Education and democracy*. Reprint. New York: The Free Press.

Diamond, Robert. 1995. Aligning the institutional mission with the reward structure. Presentation at the Forum on Faculty Roles and Rewards, American Association for Higher Education, January, Phoenix, Ariz.

Diamond, Robert, and Bronwyn Adam. 1993. *Recognizing faculty work: Reward system for the year 2000.* San Francisco: Jossey-Bass.

Edgerton, Russell. 1994. The engaged campus: Organizing to serve society's needs. *American Association for Higher Education Bulletin* 47 (1): 2–3.

Forrant, Robert and Linda Silka. 1999. "Thinking and doing—doing and thinking": The University of Massachusetts Lowell and the community development process. *American Behavioral Scientist* 42 (5): 814–26.

Gray, Peter, Robert Froh, and Robert Diamond. 1994. Myths and realities: First data from a national study on the balance between research and teaching at research universities. Syracuse: Syracuse University Center for Instructional Development.

Haaland, Gordon, Nell Wylie, and Daniel DiBiasio. 1995. Faculty scholarship: The need for change. In *Metropolitan universities,* ed. D. M. Johnson and D. A. Bell. Denton: University of North Texas Press.

Hathaway, Charles, Paige Mulhollan, and Karen White. 1995. Metropolitan universities: Models for the twenty-first century. In *Metropolitan universities,* ed. D. M. Johnson and D. A. Bell. Denton: University of North Texas Press.

Holland, Barbara. 1997. Analyzing institutional commitment to service: A model of key organizational factors. *Michigan Journal of Community Service Learning* (Fall):30–41.

Johnson, Daniel M., and David A. Bell, eds. 1995. *Metropolitan universities: An emerging model in American higher education.* Denton: University of North Texas Press.

Kanniyappan, Rajeshwari, James Ngeru, Cheryl West, Jane Worthley, and Robert Forrant. 1998. Enhancing interactions between local businesses and university centers: A survey analysis and discussion paper. University of Massachusetts Lowell: Center for Family, Work, and Community.

Kennedy, Donald. 1995. Another century's end, Another revolution for higher education. *Change* (May–June): 8–15.

Kornreich, Toby. n.d. The role of the university in community outreach and empowerment—Lessons learned in the Arizona State experience. Tempe: Arizona State University, Morrison Institute for Public Policy Discussion Paper.

Lynton, Ernest. 1995. What is a metropolitan university? In *Metropolitan universities,* ed. D. M. Johnson and D. A. Bell. Denton: University of North Texas Press.

Lynton, Ernest, and Sandra Elman. 1987. *New priorities for the university.* San Francisco: Jossey-Bass.

May, Marlynn, and Robert E. Koulish. 1998. Joining academy and community in an educational venture: A case study. *Teachers Sociology* 26:140–45.

Mazey, Mary Ellen. 1995. The role of the metropolitan university in facilitating regional cooperation. In *Metropolitan universities,* ed. D. M. Johnson and D. A. Bell. Denton: University of North Texas Press.

———. 1997. HUD's role in building the public service mission of colleges and universities. Presentation at annual meeting of the Urban Affairs Association, April, Toronto.

Nyden, Philip, Anne Figert, Mark Shibley, and Darryl Burrows. 1997. *Building community: Social science in action.* Thousand Oaks, Calif.: Pine Forge Press.

Putnam, Robert. 1995. Bowling alone: America's declining social capital. *Journal of Democracy* 6 (1): 65–78.

Rice, Eugene. 1995. The new American scholar: Scholarship and the purposes of the university. In *Metropolitan universities,* ed. D. M. Johnson and D. A. Bell. Denton: University of North Texas Press.

Sclove, Richard, and Madeleine Scammell. 1998. *Community-based research in the United States.* Amherst: The Loka Institute.

Simmel, Georg. 1955. The web of group affiliations: In *Conflict and the web of group affiliations.* Translated by K. Wolff. New York: Free Press.

Ziegler, Jerome. 1995. Winds of change: The university in search of itself. In *Metropolitan universities,* ed. D. M. Johnson and D. A. Bell. Denton: University of North Texas Press.

Innovation in Engineering Education for Sustainable Regional Social and Economic Development

Krishna Vedula, Michael A. Fiddy, Al Donatelli,
Struan Robertson, Robert Nunn, and Rafael Moure-Eraso

THE need for innovation in engineering education in this country has been clearly established by several issues that have intensified over recent years. Mounting college costs will create a large class for whom college is out of reach unless engineering education is made more affordable. On the other hand, making that education affordable may lead to a decrease in its quality, which is certainly undesirable. The resulting increasing tension between quality and access, hence, is at the center of this predicament. At the same time, changes in the nature of global technologies and the marketplace have resulted in more than two-thirds of the new jobs in the engineering industry requiring some higher education. It is clear that if we are unable to match students and opportunities in the industrial sector, we are going to lose productive capacity, leading to social and economic instabilities in the region. Sustainable regional development can only occur if the education of the workforce in the region is agile and responsive to the changing needs of the region. Since the engineering industry dominates the economy of our region and since 80 percent of our Francis College of Engineering students enter this industry, our education in these disciplines needs to be agile and responsive to the needs of regional industry.

In response to such needs, several universities in the nation are redesigning curricula to make them more relevant to employers, communities, and students. In fact, in many educational institutions, education is behaving more like the business it has become. The faculty and administration of the Francis College of Engineering, in particular, have been exploring several innovative businesslike approaches to their program. These approaches are based on the belief that engineering out in the real world requires engineering colleges to rethink what they are doing, and for this to prove effective, businesses must be involved in redefining what takes place in the classroom. There is a role, in turn, for the engineering college in helping businesses redefine themselves.

406

Innovation, that is, the practical application of creative ideas, in engineering education is a very significant part of regional social and economic development, which is the mission of the Lowell campus. The innovative approaches that are being explored promote collaboration, involve multiple disciplines, address significant questions, push new frontiers, encourage intellectual risk taking, and present sustainable solutions.

Alignment of Engineering Education with the Mission of the Lowell Campus

All the initiatives of the Francis College of Engineering, described below, build on the Lowell model as articulated by the chancellor in his vision for the campus. The mission of the Lowell campus of the University of Massachusetts as the chancellor defined it, is: "To provide an affordable education of high quality and to focus its scholarship and public service on assisting sustainable 'regional' economic and social development."

Five strategic imperatives are identified as critical to the pursuit of this mission. These imperatives respond to the problems faced by the campus in the late 1980s and early 1990s:

Stabilize enrollments in the long term.
Improve the physical plant.
Implement new regional economic assistance plan.
Enhance teaching/learning and technology applications across the campus.
Increase external funding.

Over the past ten years, the chancellor has implemented a plan to pursue these five strategic imperatives and to reposition the campus in order to be able to deliver high-quality programs and services at affordable cost and to expand and maintain the physical plant so that it is a "safe" and "inviting" environment.

The goal of Lowell's proposed Next Phase of Development program (1998–2005) is to mature and demonstrate the effectiveness of a new model for a public university's role in the sustainable economic and social development of regions. The plan to pursue this next phase of development is expected to focus on: sustainability, the physical plant, the UMass Lowell image, and the measurement of success.

Businessizing of Engineering Education

The "businessizing" of engineering education at UMass Lowell, in accord with the chancellor's vision, began with the engineering faculty retreat held in May 1996. There, faculty explored the concept of "customer" and

"product" as relevant to engineering education. Some of these initial delib-
erations were followed through during the 1996–97 academic year and
were further refined at the faculty retreat in May 1997. During these re-
treats, the faculty worked together and developed a shared vision, that
identifies students, industry, and society as the primary "customers" of engineering education. The "product" has been identified as "learning" because this helps to focus on the student who "learns" rather than the
teacher who "teaches." The shared vision of Francis College of Engineering
developed in 1996 is articulated below and is clearly aligned with the campus mission defined above:

> We are committed to providing our students the highest quality of "Learn-
> ing" which is relevant to the needs of industry and society. We will do this
> in an atmosphere that cares about the needs of our customer, the student,
> and in the spirit of cooperation among our faculty, staff, andstudents.
>
>   We will maintain our historical tradition of being relevant to the needs
> of the "regional" economy by partnerships with "regional" high schools,
> colleges, industry, alumni and the community. We will strive to be known
> for providing the highest quality and most affordable engineering education
> in New England.

This vision is not only well aligned with the Lowell mission, but it contains the following important innovative concepts, which are in keeping
with the businessizing of engineering education and its role in regional economic and social development:

1. Emphasizes real learning, real world, and real work
2. Focuses on customers (students and industry)
3. Focuses on product: quality of learning
4. Builds on industrial history and setting
5. Emphasizes regional economic social development.

The continuous quality improvement (CQI) movement, which has
dominated global industry in the past two decades, emphasizes the need to
focus on the improvement of product quality by following three important
elements: customer focus, people involvement, and continuous improvement. The parallel for engineering education would be to focus on continuous improvement of our product, that is, "learning" by involving faculty,
staff, and students and by ensuring that these improvements are consistent
with the needs of the students, industry, and society. This approach makes
it an integral part of regional social and economic development. The
unique industrial history and setting of the Francis College of Engineering
with its roots in the region's textile industry allow us to refine a niche for
our engineering education as being hands-on, practice-oriented, and relevant to the needs of the region. The faculty has used this to define the

key elements of our engineering program as real learning, real world and real work.

The shared vision has allowed the engineering faculty to identify five areas for improvement and designate them a QILTS (quality improvement of learning tasks). These five areas were an outcome of a brainstorming session in which the faculty listed all the things they felt they either currently did or need to do in order to improve the quality of learning by the students. The most important factor contributing to a better quality of learning is informal learning interactions outside the classroom, between faculty and students and among students. This has been supported by many studies on effective teaching and learning in recent years. A second important factor that contributes to better learning in the field of engineering, particularly, is more exposure to the real world of engineering practice and more hands-on laboratory experiences. This has also been substantiated by several recent studies on the need for contextual learning. In addition to these two factors, which are key to the actual delivery of engineering education, the faculty have identified a strong need to motivate and recruit high-quality students into the engineering program to ensure a continuous supply of this resource. Another important component of quality learning is better contacts with alumni and friends from industry in order to provide important advice and opportunities for enhancing resources. These friends can also play a critical role in another very important aspect of quality improvement, that is, outcomes assessment. In fact, the accreditation of engineering education is going through a major transformation with the introduction of the Accreditations Board of Engineering and Technology (ABET) 2000 criteria, which emphasize outcomes assessment.

The Quality Improvement of Learning Tasks (QILTS)

Significant recent activities related to these quality improvement of learning tasks are summarized below. QILTs have been color coded for convenience (a technique used in industry).

Increase Faculty–Student and Student–Student Contacts (BLUE QILT)

This QILT focuses on the "student" as the "customer." In order to keep in close contact with the students, the college has established an Engineering Student Council (E-Council), which calls itself "The Voice of the Students." The E-Council consists of upper-class representatives of the student body from each engineering discipline. Students are selected on the basis of leadership ability as well as academic ability. These representatives are in continuous contact with the dean and provide support for recruitment and retention of engineering students. In addition, they initiate special projects of benefit to other engineering students. A particularly important role played by the students of the E-Council is in the recognition of outstanding teachers during Engineers Week (E-Week).

As a critical part of this QILT, the faculty of the college have increased their focus on retention of students, particularly at the end of the freshmen year. Members of the E-Council play a key role in the modified Introduction to Engineering course for freshmen, so that entering students obtain advice and mentoring during this early phase of their college experience.

Increase Hands-on, Real-World Experiences (GREEN QILT)

This QILT focuses on industry as the "customer." College and departmental industrial advisory boards, consisting of distinguished leaders from industry, have been formed. Alumni of the institutions that have become part of UMass Lowell serve on these boards and provide regular and valuable feedback from industry. Members of each board are encouraged to participate in the Dean's Speaker Series, which is a forum for faculty and students, in which they learn from the real-world experiences of board members. Other industry engineers are invited as guest engineer speakers in classes during E-Week with a similar objective.

An extremely successful program initiated as part of this QILT was the very popular Cooperative (COOP) program for students. This flexible program offers "hands-on" relevant engineering experiences to students via summer internships and part-time or full-time internships during the school year. The Office of Career Services has been particularly helpful in organizing several job fairs, which have allowed students to explore a variety of job and co-op possibilities.

Another unique feature of engineering education at Lowell is its emphasis on relevant design experiences. Students in all disciplines work in teams during their senior year to solve real-world engineering problems identified by faculty via their industrial contacts. Projects associated with the Assistive Technology Program have attracted particular attention because of their human and social impact. As an example of the ATP projects, a group of students in electrical engineering designed and built a voice-activated system to help a quadriplegic man in Lowell perform several key household functions.

This emphasis on real-world experiences as part of the green QILT has increased active partnerships of various types with regional industry. Here are some examples.

M/A-COM, a local electronics company, part of Tyco Electronics, has been a leader in the newly formed COOP program; the college has helped develop worker-retraining programs for M/A-COM, which has also hosted several field trips for high school students.

Analog Devices Inc. has initiated a unique Scholar–Intern Program; the college has developed a project for evaluating worker retraining needs and for offering suitable programs; Analog has also sponsored the Micromouse competition at the Institute for Electrical and Electronics Engineers (IEEE) meeting.

Cisco has provided scholarships and contributed to the preparation of a video targeted at high school students.

Wyle has contributed to the renovation of a new Networking Lab in the college.

NYPRO has collaborated with the college in developing Internet courses in plastics engineering.

Boston Communications Group has set up a training–job site at UMass Lowell.

Starting in 2000, Raytheon is providing UMass with $1 million per year for five years, to be shared among the five campuses for the improvement of engineering education.

Increase Resources via Increased Enrollment (RED QILT)

The focus of this QILT is to increase enrollment and thereby gain more favorable access to university resources. The college has, therefore, emphasized activities marketing the affordable and high-quality engineering education it offers. Public relations materials including a video and several brochures have been developed with the participation of E-Council students and faculty. Closer coordination has also been developed with the University Office of Admissions.

The college has developed strong partnerships with several high schools including Lowell High School, Lawrence High School, Chelsea High School, Nashoba Valley Technical High School, and Greater Lawrence Vocational High School. These partnerships have resulted in many activities, including visits by engineering faculty and students to the high schools and field trips by high school students and teachers to the college and to regional industry. Moreover, active partnerships with community college partners are resulting in articulation agreements between the college and Northern Essex, North Shore, Middlesex, and Mass Bay Community Colleges. These articulations will help students choose alternate career pathways and enable them to transfer more easily among these institutions.

The focus on enrollments via this QILT showed outstanding results. There has been a 10 percent increase in freshmen enrollment each year since 1998.

Increase Resources via Development of Alumni, Friends, and Research (YELLOW QILT)

The dramatic decrease in state funding allocated to the university during the past decade has required an increased emphasis on private fund-raising.

The Francis College of Engineering has taken the initiative in several areas by contacting distinguished alumni and friends. Some of the successful efforts have included increased endowments in alumni scholarships, more successful fund-raising from industry for facilities renovation and equipment, and research funding increased to reach the college's targeted Research II Classification.

The development of alumni relations has been further facilitated by launching several events with departmental alumni (E-Roundtables); by publication of a semiannual newsletter ("E-Vision"); and by the organization of more events to honor distinguished alumni (such as the E-Week Annual Awards Banquet).

Outcomes Assessment (ORANGE QILT)

The major emphasis in this QILT is outcomes assessment, which is the cornerstone of the new Accreditation Board of Engineering and Technology (ABET) 2000 criteria. The emphasis on outcomes assessment shows that engineering education is being required to adopt the same approach that industry is required to take as part of ISO 9000 and similar quality criteria, that is, (1) say what you do, (2) do what you say, and (3) prove it, (4) and improve it.

The following were key elements in our preparation for the engineering accreditation visit of fall 2000:

1. Assessment of students
2. Assessment of program educational objectives
3. Program outcomes and assessment
4. Professional component
5. Faculty
6. Facilities
7. Institutional support and resources
8. Program criteria

A variety of assessment tools were to be used in the outcomes-assessment process, including alumni surveys, capstone design courses, employer surveys, student surveys (including exit interviews), and Fundamentals of Engineering Exams.

This emphasis on outcomes and accountability and the "customer" focus of this approach in the field of education are clearly topics for debate. They run contrary to the traditional academic culture that defines education in a more open and free format. This approach can also lead to some perplexing situations, which may not be easy to resolve. Since students and industry are viewed as customers, conflicting situations could arise, which will need to be resolved within the societal context, since society is the "customer" in a broader sense. The authors expect that the debate generated by such situations will allow us to continue the process of improvement in engineering education.

Statewide Collaborative Leadership

The businessizing approach to engineering education at Lowell has attracted statewide attention in light of a lack of an adequately trained workforce in the region. The college has launched a statewide initiative called Engineering in Mass Collaborative: A Public–Private Collaborative to Plant the Engineering Workforce of Massachusetts, which is described below.

The Objective

The collaborative focuses on the critical shortage in the technical workforce, which is jeopardizing the future of the regional economy in Massachusetts, where the number of degrees awarded in engineering decreased by 37 percent from 1987 to 1996 (Massachusetts Technology Collaborative 1998). Companies in Massachusetts are trying to fill more than two thousand engineering jobs immediately, and mid-sized and start-up companies have many more openings. It is estimated that we have a shortage of about a thousand engineers per year in the Commonwealth.

Obviously engineering careers are not attracting sufficient numbers of talented high school graduates. Several surveys of students and teachers have revealed that the lack of interest in science and engineering careers among middle school and high school students stems from a lack of knowledge about engineering careers among those students and their teachers as well (Patterson 1990). Engineers are perceived as nerds with pocket protectors; they are perceived as not having fun. Obtaining an engineering degree is seen as hard work—and not worth it. Clearly, these misperceptions arise because there is no educational process in place to provide real-world context to the math and science education in grades K–12. Preliminary discussions with representatives from private industry, trade associations, high schools, colleges, and the state government in 1997 have led to agreement that there is need for public-private collaboration to help the colleges of engineering counteract this problem.

Providing real-world context and thereby increasing the supply of scientists and engineers in the kindergarten through college pipeline in Massachusetts is the major objective of our new Engineering in Mass Collaborative. Now this Private–Public Collaborative in Massachusetts has provided a focal point for all the stakeholders to establish suitable partnerships in support of the objectives of the collaborative; that is, to increase the supply of scientists and engineers to meet a critical need of industry, to encourage the study of science and engineering in our school systems, and to increase the rate of enrollment of engineering students.

The Approach

There are two approaches being used by industry in solving the workforce shortage problems: "Hunting" for the workforce, that is, engaging headhunters, which is clearly a short-term approach, and "Planting," that is, helping long-term programs for producing engineers move through the education pipeline.

The focus of the Engineering in Mass Collaborative is on Planting. Obviously, math and science teachers are key to the Planting model, the essence of which is exposing students in middle and high schools to potential careers in science and engineering through their curriculum and nurturing and guiding those showing an aptitude and interest in the direction of such careers. Their teachers need to be familiar with state-of-the-art technological advances so that they can incorporate these concepts into their classes. Contextual teaching and learning are at the core of the Planting model emphasized in this proposal. By incorporating real-world context in math and science curricula, the students are only exposed to real-world careers, but they learn better as well. Teachers need to be the "Gardeners" who plant the seeds and nurture the fledgling engineer and scientist plants. However, the teachers themselves can acquire these Gardening or contextual skills only by direct experience with engineers and scientists in industry and with the help of science and engineering colleges faculty. A large body of literature exists in this field. Some examples are provided in the References (Carter, Park, and Copolo 1998; Clark 1993; Epstein 1993; Kearney et al. 1993; O'Connor and Brie 1994; Patterson 1990).

Although there are some good "Planting" or contextual learning programs underway today in various parts of the country, many of them are not easily sustainable or replicable because of the lack of suitable infrastructure. The collaborative recognizes that adopting contextual learning requires time-consuming and far-reaching changes in practice related to curriculum, instruction, assessment, linkages with workplaces and other contexts, staff development, school organization and communication (Carter, Park, and Copolo 1998). Our collaborative has looked at many existing models; the 3M Science and Engineering Encouragement Programs, for example, have proven successful.

The Programs

Some elements of the Planting model, which have been successfully attempted by key members of the Engineering in Mass Collaborative partnerships in recent years, have included:

Teacher externs in industry. The Teachers in Industry project, which is linked to the Engineering in Mass Collaborative, is an initiative that focuses on helping teachers bridge the gap between classrooms and the workplace, that is, provides context to their teaching. In 1997, a pilot version of the project was managed by the Charles River Consortium, which is the School-to-Work Grant recipient for Brookline, Newton, Waltham, and

Wellesley. Seven teachers (one middle school and six high school) were assigned to six emerging technology companies, generally for a six-week externship. During these assignments, teachers performed productive projects for their respective companies and learned about the work environment and job skills required of new employees. The companies were Aspen Technology Inc., GammaGraphX, Inc., Kronos Inc., MicroE, Inc., VenturCom, Inc., and Zoom Telephonics, Inc.

In 1998, 1999, and 2000 the project expanded to include the Metropolitan School-to-Career Partnership (which services Cambridge, Chelsea, Everett, Revere, Somerville, Watertown, Winthrop, and the Northeast Metropolitan Regional Vocational School) and The Education Cooperative (which serves Dedham, Dover, Holliston, Hopkinton, Medfield, Natick, Needham, Norwood, Sherborn, Walpole, Wayland, Wellesley, and Westwood). More than twenty-five teachers, who participated since 1998, have shared their exciting experiences. Their manual is available that describes how a local version of the Teachers in Industry project can be launched.

The Engineering in Mass Collaborative expects to build on the successes of the current program and expand it to include many more companies and school districts in 2000 and 2001. Careful coordination, orientation, and follow-up strategies are key elements of this program, and the collaborative will help ensure that the program is able to reap long-term benefits, among them the professional development of teachers; an increasing awareness of state-of-the-art technologies among teachers; the development of professional contacts between teachers and professional engineers; collaboration among teachers, engineers, and college faculty for developing new approaches for contextual teaching of relevant topics and their development of relevant kits and videos for classroom use.

Scholar interns in industry. Analog Devices Inc. (ADI) initiated a unique scholarship program in 1998, which provides an important element of contextual learning in the preparation of future engineers in the state. The company has offered four scholarships of $1,500 each to high school seniors graduating from any of the five high schools in the region of their Wilmington manufacturing facility. ADI engineers play an active role in selecting these students from among the pool of applicants to the Electrical and Computer Engineering Department at UMass Lowell. These engineers will then be mentors to the selected scholars, who will work as interns at the ADI facility during all the academic breaks winter and summer and receive up to $7,000 in stipends through the four years of the B. S. program. Each year the internship will be served in a different location within the ADI facility in order to ensure that the students obtain an integration of experiences related to ADI's manufacturing process. Faculty at UMass Lowell will work with the ADI engineers to ensure that the appropriate contextual connections are made and utilized for effective learning throughout the academic work at Lowell.

By 2002, there will be sixteen ADI Scholars, four in each year of the program, in the Electrical and Computer Engineering Department. As a result, in 2001 the collaborative is implementing Analog Devices–type scholarship programs at twenty firms.

This, we believe, is an outstanding example of a partnership program leading to effective contextual teaching and learning. The direct benefits from the scholar–intern program are obvious and include: motivation for high school students to compete for scholarships, and, for college students, contextual learning opportunities, financial support, and professional mentoring by engineers.

The activity and publicity associated with these scholarships will lead to additional long-term benefits such as networking between area high school teachers and company engineers, better publicity for engineering careers among high school students and teachers, and more mentoring of high school students by college students.

Engineering awareness programs. Partnership among M/A-COM, UMass Lowell, and area high schools has provided another effective means by which contextual teaching can be used by high school teachers and their students in math and science. For example, during the past year, engineers from M/A-COM and faculty and upper-class students from UMass Lowell have lectured in classes at Lowell High School about what engineers do in the real world (e.g., in developing new wireless technologies). Students as well as math and science teachers from Lowell High School have been on field trips, which have included hands-on demonstrations in UMass Lowell engineering labs as well as at M/A-COM facilities. Teachers and faculty are utilizing this contextual learning opportunity to link parts of the curriculum at Lowell High School and UMass Lowell in the context of the needs of M/A-COM.

Similar one-day programs have been provided for other high schools in the region. The contextual learning and teaching opportunities that can result from these programs are wide reaching. The collaborative expects to be a resource for videos and kits as learning aids for high school teachers and industry engineers to use as effective tools for introducing real-world experiences to students in these programs. It will also provide coordination for regular tours and visits among high schools, regional companies, and engineering colleges and to act as a clearinghouse for preparation and dissemination of suitable resource materials including kits and videos. These activities will also assist with the long-term objectives of the collaborative to provide real-world experiences for students and teachers, encourage networking among high school teachers, engineers, and college faculty for exploring contextual learning opportunities, and to share resource materials developed for contextual learning. The collaborative hopes to help expand the hands-on, academic-year engineering projects for high and middle school students. In addition, such projects should lead to more participation in science and engineering fairs and competitions.

We emphasize that most of these programs are designed to acknowledge the guiding principles of the Curriculum Frameworks developed by the Massachusetts Department of Education:

Investigating and problem solving are central to science and technology.
Access to the expertise of others is needed in order for teachers to implement the cross-domain and interdisciplinary approach advocated.
Science and technology connect with other disciplines and have a particularly integral relationship with mathematics.
Communication and collaboration are essential to teaching and learning in science and technology.

The Planting model for the future technical workforce in Massachusetts allows us to address the critical need to provide intensive, quality professional development programs for teachers, as documented in several recent reports. The above programs are only some of the outstanding "Best Practices," which are being promoted by the collaborative. The members of the collaborative expect that it will play a critical role in seeking out and promoting "Best Practices" aimed at "Planting" the future technical workforce in the region.

The Collaborative Infrastructure

The Engineering in Mass Collaborative has attracted many private and public agencies in the state and will allow us to tap into the experience and commitment of all of them. The formation of the collaborative and its call for action have been publicized in several ways, thus attracting additional talent and expertise. The collaborative has hosted semiannual meetings since 1998 and attracted over a hundred participants from business, academia, and government. Dean Krishna Vedula has been invited to make presentations about these innovative approaches at many business and educational meetings. Several business and political leaders in the region have expressed strong support for these initiatives, and the collaborative Executive Committee includes many distinguished business, education and community leaders.

Summary

The Francis College of Engineering has embarked on an ambitious program of reinventing engineering education within the framework of the mission of the Lowell campus. This program incorporates a business approach to the improvement of the quality of learning acquired by the students. The Quality Improvement of Learning Tasks integrated with the outcomes assessment emphasized by ABET 2000 criteria provide the framework for a truly innovative approach. The uniqueness and relevance of this approach are attested to by the rapid leadership role that has been thrust upon the college in the statewide Engineering in Mass Collaborative.

We believe that the activities and programs in engineering education outlined meet the goals we set in the beginning. They are truly innovative because they promote collaboration, involve multiple disciplines, address significant questions, push new frontiers, encourage intellectual risk taking, and present sustainable solutions.

SELECTED REFERENCES

Carter, G., John Park, and Cynthia Copolo. 1998. Using industry experiences to promote change in the types of instructional strategies used by secondary science teachers. *Journal of Science Teacher Education* 9 (2): 143–52.

Clark, M. 1993. City science: Defining a role for scientists in elementary education. In *Science education partnerships,* ed. A. Sussman, 86–92. San Francisco: University of California Press.

Epstein, K. 1993. Bay area science and technology education collaboration. In *Science education partnerships,* ed. A. Sussman, 110–16. San Francisco: University of California Press.

Kearney, B., M. Earl, K. Rosman, K. Sloane, Z. Segre, and W. Allsteter, 1993. Industry initiatives for science and math education. In *Science education partnerships,* ed. A. Sussman, 43–54. San Francisco: University of California Press.

Massachusetts Technology Collaborative. 1998. *The innovation economy.* Boston: Massachusetts Technology Collaborative.

O'Connor, J., and R. Brie. 1994. Mathematics and science partnerships: Products, people, performance, and multimedia. *Computing Teacher* 22:27–30.

Patterson, H. 1990. School/industry links: A survey among science teachers in Surrey. *School Science Review* 72 (258): 41–46.

Notes on Contributors

Michael Best is codirector of the Center for Industrial Competitiveness at the University of Massachusetts Lowell and senior research associate of the Judge Institute of Management Studies, Cambridge University. His book *The New Competitive Advantage: What Lies Behind the Renewal of American Industry* was published by Oxford University Press in 2001. He has recently completed a monograph on industrial competitiveness for the Northern Ireland Economic Council, *The Capabilities Perspective: Advancing Industrial Competitiveness in Northern Ireland.* He has worked with the United Nations Industrial Development Office on projects in Cyprus, Jamaica, Honduras, Indonesia, and Malaysia, was on the Advisory Panel for the 1996 Human Development Report of the United Nations, and serves as an adviser to the World Bank on industrial restructuring projects in Eastern Europe and republics of the former Soviet Union.

Meg A. Bond is a professor of psychology and codirector of the Center for Women and Work at the University of Massachusetts Lowell. Her work focuses on the interrelationships among issues of diversity, empowerment, and organizational dynamics. Her current research program is an analysis of organizational approaches to diverse employees with a primary focus on the dynamics of gender and race. Her past publications have addressed sexual harassment dynamics, collaboration among diverse constituencies, and empowerment issues of the underrepresented groups in community and organizational settings. She is a senior editorial board member of the *American Journal of Community Psychology* and is currently co-editing two special issues of this journal on theory and methods at the interface of feminism and community psychology. She is a past president of the Society for Community Research and Action (Division 27 of the American Psychological Association). E-mail: Meg_Bond@uml.edu.

Cathy Crumbley is program director of the Lowell Center for Sustainable Production at the University of Massachusetts Lowell. She has previously managed international environmental education and training programs at the United Nations Environment Program at Tufts University and the Coolidge Center for Environmental Leadership. She has also served as an environmental consultant in the private sector. She holds an M.S. in environmental science from the University of Massachusetts and a B.A. from Pomona College, Claremont, California.

Al Donatelli is currently chair of the Chemical and Nuclear Engineering Department at UMass Lowell. He obtained a Ph.D. in chemical engineering from Lehigh

University and has served on the faculty at UMass Lowell since 1977. He is recognized for his outstanding teaching and his expertise in chemical processes and transport phenomena.

Louis Ferleger is professor of history at Boston University and executive director of the Historical Society. From 1978 to 1999 he was professor of economics at the University of Massachusetts Boston. He is the author of dozens of articles and coauthor, editor, or coeditor of four books, *A New Mandate: Democratic Choices for a Prosperous Economy* (with Jay R. Mandle); *No Gain, No Pain: Taxes, Productivity and Economic Growth* (with Jay R. Mandle); editor, *Agriculture and National Development: Views on the Nineteenth Century;* and *Statistics for Social Change* (with Lucy Horwitz). His coedited book *Slavery, Secession, and Southern History* (with Robert Paquette) was published by the University of Virginia Press in 2000. In addition he is coeditor with Jay Mandle of a special issue of the *Annals of the American Academy of Political and Social Sciences* devoted to globalization that appeared in July 2000.

Michael A. Fiddy graduated from the University of London and was appointed Lecturer in Physics at Queen Elizabeth College in 1979, moving to Kings College, London University in 1983. He has held visiting positions at the Institute of Optics, University of Rochester, in the Mathematics Department at the Catholic University of America, Washington, D.C. In September 1987, he moved to the University of Massachusetts Lowell, where he is professor of electrical engineering and head of the department. He is codirector of the Center for Electromagnetic Materials and Optical Systems. Dr. Fiddy is currently the editor of the journal *Waves in Random Media* (Institute of Physics) and the topical editor for the *Journal of the Optical Society of America* for signal and image processing. He is a Fellow of the Institute of Physics and of the Optical Society of America and a senior member of the Institute of Electrical and Electronic Engineers.

Robert Forrant received his Ph.D. in history from the University of Massachusetts Amherst and is an associate professor in the Department of Regional Economic and Social Development at the University of Massachusetts Lowell. He codirects the university's Community Outreach Partnership Center and cochairs its Committee on Industrial Theory and Assessment. He is an analyst for *Benchmarks*, a joint publication of the University of Massachusetts and the Federal Reserve Bank of Boston. Recent articles of his have appeared in such journals as *Regional Studies, Industrial and Corporate Change, Economic Development Quarterly,* and the *American Behavioral Scientist.* He has been a consultant to the United Nations Industrial Development Organization, the International Labour Organization, and the International Metalworkers Federation. E-mail: rforrant@external.umass.edu.

Ken Geiser is an associate professor in Work Environment at the University of Massachusetts Lowell, where he is also director of the Massachusetts Toxics Use Reduction Institute and codirector of the Lowell Center for Sustainable Production. He holds master's and doctoral degrees (in public policy) from the Massachusetts

Institute of Technology. His research interests focus on the intersection of environmental and industrial policy.

Jeffrey Gerson received his Ph.D. in political science from the City University of New York Graduate School and is an assistant professor in the Political Science Department at the University of Massachusetts Lowell. He teaches courses on urban, ethnic, and state and local politics, as well as the legislative process, leadership development, and political participation. Recent articles of his have appeared in such journals as the *New England Journal of Public Policy* and *Change: The Magazine of Higher Learning,* and he is coeditor (with Carol Hardy-Fanta) of *Latino Representation in Massachusetts* (Garland Publishing, 2000). E-mail address: Jeffrey_Gerson@uml.edu

Tim Greiner is president of Greiner Environmental—a pollution prevention engineering and training firm—where he has conducted over two hundred pollution prevention assessments for manufacturing firms. He is currently working with several corporations to develop plant-wide sustainability indicators. He has work experience in industry as a process engineer for Fairchild Semiconductor, and holds Masters degrees in Business and Urban Planning from the Massachusetts Institute of Technology.

Laurence F. Gross received his Ph.D. in American civilization from Brown University and is now an associate professor in the Department of Regional Economic and Social Development at the University of Massachusetts Lowell. He works and publishes in areas combining the history of labor, technology, and business. His articles in *Industrial Archeology* and *Technology and Culture* received awards from those journals. He reviews manuscripts for numerous journals and publishers and has consulted on labor and technology exhibits for museums in the United States and abroad. His most recent book is *The Course of Industrial Decline: The Boott Cotton Mills of Lowell, Massachusetts, 1835–1955.* He is currently working on a history of Malden Mills in Lawrence as a study in corporate exceptionalism.

Dikshitulu Kalluri received the B.E. degree in electrical engineering from Andhra University, the D.I.I.Sc. from the Indian Institute of Science Bangalore, an M.S.E.E. from the University of Wisconsin Madison, and a Ph.D. from the University of Lawrence, Kansas. He was professor and head of the Electrical Engineering Department at the Birla Institute of Technology, Ranchi, India, also serving as assistant director of the Institute. He is currently professor of electrical and computer engineering at UMass Lowell, codirector of the Center for Electromagnetic Materials and Optical Systems, and coordinator for the ECE Department's doctoral program. He is a fellow of the Institute of Electronic and Telecommunications Engineers.

Nancy Kleniewski is professor of sociology and dean of Fine Arts, Humanities, and Social Sciences at the University of Massachusetts Lowell. Her research focuses on urban policy, especially policies that address social inequalities. Dr. Kleniewski's articles exploring urban renewal, welfare, immigration, economic development, and subsidized housing have been published in such journals as the *International Journal of Urban and Regional Research, Urban Affairs Review, Critical Sociology,* and *Research in Politics and Society.* Her published books are *Community Politics and Policy* (1999); *Cities, Change, and Conflict* (1997); *Urban Problems in Sociological Perspective* (1997); and *Philadelphia: Neighborhoods, Division, and Conflict in a Post-Industrial City* (1991).

David Kriebel is an epidemiologist whose doctoral degree in occupational health and epidemiology is from Harvard University's School of Public Health. Dr. Kriebel is an associate professor in Work Environment at the University of Massachusetts Lowell and codirector of its Lowell Center for Sustainable Production. He formerly worked and taught in Italy on a Fulbright Fellowship.

William Lazonick is university professor in the Department of Regional Economic and Social Development at the University of Massachusetts Lowell, where he is also codirector of the Center for Industrial Competitiveness. He spends part of each year at INSEAD (the European Institute of Business Administration), where he codirects a research project on corporate governance, innovation, and economic performance. Previously, he was assistant and associate professor of economics at Harvard University (1975–1984) and professor of economics at Barnard College of Columbia University (1985–1993), and he has occupied visiting posts at the University of Tokyo, University of Toronto, Harvard Business School, and the Institute for Advanced Study, Princeton. He has been president of the Business History Conference, the main professional association of business historians in the United States. He has published many books and articles on comparative industrial development and international competition, focusing on Western Europe, the United States, and Japan. Professor Lazonick holds a Bachelor of Commerce degree from the University of Toronto (1968), an M.Sc. in economics from the London School of Economics (1969), and a Ph.D. in economics from Harvard University (1975). In 1991, Uppsala University, Sweden, awarded him an honorary doctorate for his work on the theory and practice of economic development.

Charles Levenstein received his Ph.D. in economics from M.I.T. and later earned an M.S. in occupational health at the Harvard School of Public Health. He is a professor in Work Environment at the University of Massachusetts Lowell and has had visiting appointments at Harvard School of Public Health, Tufts University Medical School, DeMontfort University in the United Kingdom, and Central European University in Budapest. At UMass Lowell, he is also special assistant to the provost for regional development, chair of the Council on Industrial Development, codirector of the Public Health Engineering and Policy Initiative, and cochair of the Committee on Industrial Theory and Assessment. He is editor of *New Solutions: A Journal of Environmental and Occupational Health Policy.* He is coeditor (with John Wooding) of *Work, Health and Environment* (1997) and coauthor (with John Wooding) of *The Point of Production* (1999). He has published extensively on occupational and environmental health in Central and Eastern Europe and on occupational health policy in the United States. His work can be found in *New Solutions, American Journal of Industrial Medicine, Journal of Public Health Policy, American Journal of Public Health, International Journal of Health Services,* and the *International Journal of Occupational and Environmental Health.*

John MacDougall holds a Ph.D. in sociology from Harvard University, and is a professor in Regional Economic and Social Development at the University of Massachusetts Lowell. His research covers various social movements, such as ethnic movements in India and local and national peace movements in the United States. His work has appeared in such publications as *Research in Social Movements, Conflicts,* and *Change.* He has also written on service learning and other pedagogical issues and coedited *Teaching the Sociology of Peace and War* (American Sociological Association, 1998). He is codirector of Merrimack Valley 2050, a new University of Massachusetts Lowell–community partnership to promote regional sustainability. E-mail: John_MacDougall@uml.edu.

Rafael Moure-Eraso is an associate professor of the Work Environment faculty at the University of Massachusetts Lowell. He is a certified industrial hygienist. He has served as senior policy adviser on Chemical Exposure Prevention for the Occupational Safety and Health Administration. He also cochairs the World Health Organization Collaborating Center in Occupational Health at Lowell.

Robert Nunn was chair of the Plastics Engineering Department at the University of Massachusetts Lowell. He obtained a Ph.D. in mechanical engineering (polymers) and has been a faculty member at the University of Massachusetts Lowell since 1985. He has experience in a wide variety of industries, including General Motors Corp., HPM, and U.S. Steel Corp. Professor Nunn is internationally recognized for his expertise in injection molding.

Laura Punnett received her Sc.D. in Occupational Health and Epidemiology from the Harvard School of Public Health and postdoctoral training at the Center for Ergonomics, University of Michigan. She is a professor in the Department of Work Environment, University of Massachusetts Lowell, and codirects the Lorin Kerr Ergonomics Institute for Occupational Injury Prevention. Dr. Punnett studies work-related musculoskeletal disorders and other morbidity associated with ergonomic stressors as well as the sociopolitical factors influencing the application of scientific knowledge to improving working conditions. She serves on the editorial boards of *Applied Ergonomics, New Solutions,* and *Salud de los Trabajadores (Workers' Health)*, Venezuela, and on advisory boards for the Massachusetts Department of Public Health, the Center for Video Display Terminals and Health Research (Johns Hopkins University), the Hospital for Joint Diseases (New York University), and the Ergonomics Technology Center (University of Connecticut). She has been a consultant to the National Institute for Occupational Safety and Health, the Occupational Safety and Health Administration, and the World Health Organization. E-mail: Laura_Punnett@uml.edu.

Jean L. Pyle is a professor in the Department of Regional Economic and Social Development at the University of Massachusetts Lowell. She also serves as cochair of the Committee on Industrial Theory and Assessment (CITA) and codirector of the Center for Women and Work. She specializes in the overlapping areas of economic development, labor markets, and policy, with particular attention to gender and diversity issues. Her recently published research includes analyses of the effects of globalization and various forms of restructuring on women; women's roles in the employment networks of multinational corporations; factors that facilitate or constrain the effective use of diverse peoples in U.S. workplaces; and the impact of state policies (including employment, housing, and reproductive rights and family policies) on women's economic roles in Singapore. She is the author of *The State and Women in the Economy: Lessons from Sex Discrimination in the Republic of Ireland* (1990) and is currently coauthoring *Gender in the World Economy*. She has consulted for the United Nations Industrial Development Organization (UNIDO) on gender and development issues and for organizations seeking to facilitate workforce diversity. She is a member of the Board of Directors of the Eastern Economic Association. E-mail: Jean_Pyle@uml.edu.

Margaret M. Quinn is a certified industrial hygienist who holds a doctoral degree in Work Environment from the University of Massachusetts Lowell and is now an assistant professor on the faculty there. She also holds two master's degrees from Harvard University and she held a year-long fellowship in environmental cancer research from the Italian Association for Research on Cancer. She has been very involved in the work of the Lowell Center for Sustainable Production and was in charge of its efforts in occupational health and safety at the well-known burnt-out textiles firm, Malden Mills.

Struan Robertson is chair of the Mechanical Engineering Department at the University of Massachusetts Lowell. He obtained a Ph.D. in mechanical engineering from Rensselaer Polytechnic Institute and has been a faculty member at the University of Massachusetts Lowell since 1983. He has experience in a wide variety of research laboratories, including GTE Research Laboratory, Teledyne Materials Laboratory, and U.S. Army Research laboratories. He is well known for his expertise in mathematical modeling for engineering applications.

Julian D. Sanchez received his B.S. in electronics and communications engineering from Instituto Technologico de Estudios Superiores de Monterrey (ITESM) Mexico, and his M.S.E.E. and D.Eng. degrees from the Electrical and Computer Engineering Departments at the University of Massachusetts Lowell, in 1996 and 1998, respectively. He is currently a faculty member at Instituto Nacional de Optica y Electronica (INAOE) in Puebla, Mexico.

Linda Silka, Ph.D., directs the Center for Family, Work, and Community at the University of Massachusetts Lowell and is a professor in the interdisciplinary Department of Regional Economic and Social Development. From 1978 to 1997, she was a faculty member in the university's Department of Psychology. A social and community psychologist by training, Dr. Silka develops programs that create community–university partnerships. Recent partnerships include the Southeast Asian Environmental Justice Partnership, started with funding from the National Institute of Environmental Health Sciences; a Community Outreach Partnership Center begun through funding from the U.S. Housing and Urban Development's Office of University Partnerships; and the Center for Immigrant and Refugee Community Leadership and Empowerment. Dr. Silka involves community residents, students, and faculty in using new technologies (such as community mapping with geographic information systems) to address long-standing community challenges. Dr. Silka also teaches graduate courses in community mapping, grant writing, and program evaluation and consults widely on capacity-building strategies in program evaluation.

Krishna Vedula is dean of the Francis College of Engineering at the University of Massachusetts Lowell. Dean Vedula obtained his B.S. in metallurgical engineering from Indian Institute of Technology, India, his M.S. in materials engineering from Drexel University, and his Ph.D. in materials engineering from Michigan Technological University. He was assistant professor, associate professor, and professor in Materials Science and Engineering at Case Western Reserve University. Prior to coming to Lowell, Dean Vedula was chair of Materials Science and Engineering at Iowa State University. He also spent a year as program director at the U.S. Office of Naval Research. During the past ten years Dean Vedula has been actively involved with improving the quality of engineering education by developing partnerships with the engineering industry. More recently, he founded the Engineering in Mass. Collaborative, which is a statewide partnership of businesses, K-12 schools, and higher education in Massachusetts aimed at increasing the number of students entering careers in science and engineering. The mission of this collaborative is to "Plant" the future technical workforce in the region by seeking out, recognizing and promoting "Best Practices" aimed at increasing interest in science and engineering careers among middle and high school students.

Vesela Veleva received her M.Sc. in pollution and environmental control from the University of Manchester, United Kingdom, and her B.S. in electrical engineering from the Technical University of Varna, Bulgaria. She is presently a doctoral candidate in Cleaner Production and Pollution Prevention, Department of Work Environment, University of Massachusetts Lowell. Her dissertation work involves development of sustainable production indicators. She has published in *New Solutions*. Earlier, she had worked for the Ministry of Environment in Bulgaria and been a teaching assistant at the Central European University, Budapest, Hungary, and at the Technical University of Varna, Bulgaria.

John Wooding received his Ph.D. in political science from Brandeis University. He is professor of political science and chair of the Department of Regional Economic and Social Development at the University of Massachusetts Lowell. His research interests include the politics of environmental health, the political economy of the work environment, comparative political systems, and the international political economy. He has coauthored two books, *Work, Health and the Environment* (1997), and *The Point of Production* (1999), and numerous articles in the field of occupational health and safety regulation and is currently studying the politics of child labor.

Index